W9-CBZ-337

De Cristoforo's
HOUSEBUILDING ILLUSTRATED

De Cristoforo's
HOUSEBUILDING

A POPULAR SCIENCE BOOK

ILLUSTRATED

by R.J. De Cristoforo

Drawings by
Richard J. Meyer,
W. David Houser,
and the author

Meredith ® Press, New York

Copyright © 1977 by R.J. De Cristoforo
ISBN: 0-696-11101-2

Brief quotations may be used in critical articles and reviews.
For any other reproduction of the book, however, including
electronic, mechanical, photocopying, recording or other means,
written permission must be obtained from the publisher.

Library of Congress Catalog Card Number: 77-6559

Published by

Meredith ® Press
150 East 52nd Street
New York, New York 10022

Distributed by Meredith Corporation, Des Moines, Iowa.

Manufactured in the United States of America

CONTENTS

De Cristoforo's
HOUSEBUILDING ILLUSTRATED

PREFACE

Building a house is probably the most ambitious project an amateur can undertake. It's a formidable assignment, like the challenge of climbing a mountain with peaks concealed by clouds. Yet amateurs have climbed mountains, and they have built houses. Like the sequence of steps in the climb, putting a house together is basically a matter of concentrating on the next nail. Maybe the overall task is intimidating, but it is just a long series of bits and pieces you would not hesitate to tackle if each were an individual project.

The prime factor is not whether you *can* get there, but whether you *wish* to. Determination and enthusiasm are priority tools.

However, don't think you can construct a house simply if you can saw a board and drive a nail. Even though this simplification may apply to the carpentry involved, it naively overlooks areas that require more than a hammer and a saw. You don't, for example, mix concrete with woodworking tools; nor do skills with those tools make you a plumber, an electrician, or an architect.

I state the above in a positive vein and for a number of reasons. For one, seeing the project in depth prepares you psychologically for the entire job, not just phases of it. For another, visualizing the complete assembly as a contribution of various crafts lets you organize a successful program without committing you to accomplish it all by yourself. This concept eases the mind of anyone standing on naked ground with only a dream. The builder who wants to do it all himself can. But builders who want to involve professionals or even other knowledgeable amateurs can do so while still making a contribution that leads to satisfaction and big financial savings.

Planning and mental preparation are an important part of housebuilding. You can make an excellent start by setting up an idea file—a collection of separate folders to contain (1) personal thoughts, (2) notes on existing structures and building details you have liked, (3) sketches you've made of floor plans, (4) tearsheets from publications showing artist's sketches of homes for which you can buy ready-to-use plans, (5) tearsheets of advertisements showing manufacturer's products in use, and (6) above all, catalogs by manufacturers that supply building materials and by associations that represent suppliers.

Since it's the purpose of catalogs and associations to give exposure to products and prompt their use, much of the available information contains enticing ideas you can use directly or modify. In the Appendix of this book, you will find a list of organizations that will be happy to provide literature to bulk out your idea file. The list may call out the name of a particular folder or catalog you might find especially useful. Although the list is usually not absolutely complete, you can supplement it merely by scanning advertisements in popular home, garden science, and mechanics magazines, and then mailing away coupons or postcards.

One value of such a file is that it can acquaint you with the many types of house materials that are available. You may know in advance that you want wood siding. But what kind? Lumber, plywood, shingling? In each category there are choices and variables in relation to basic costs, appearance, and installation time. Some materials require finishing after installation; others stand as installed. Your reaction to initial cost should be influenced by follow-up requirements. A few more cents per foot for a prefinished product might be a more economical way to go, long range, than to pay less for an alternate material that requires painting now and painting again in later years.

With a good idea file and a book like this on hand you will be able to make critical decisions at leisure, not just in relation to materials but also concerning the scope of your physical involvement. I have seen and I've been involved in many successful amateur-constructed houses. Some, especially those designed as vacation or "second" homes, were erected from the first shovelful of soil to the last nail in the roof by the owner or his whole family. Others were examples of various group efforts; still others illustrated that the amateur can be both contractor and part builder. You can determine what parts of the job you wish to do and hire professionals to fill in.

The popularity of do-it-yourself activities has bred some companies that offer "finish-it-yourself" houses. The point where you take over is determined in advance. It might be after the frame is up or after the house has been closed in.

There are many ways to go. You don't have to give up the dream of a custom home if the thought of doing it from scratch turns you off. You can take a middle road by providing all of the design but only part of the work.

In any case, if you are to be involved with house construction to *any* degree, or if you are just in the market for a finished home, or if you are happy where you are but wish to make changes, you should know what building construction is all about.

I have been guided in some areas by the fact that it is not the intention of the amateur to take up house construction as a career. Some suggested work procedures are not those used by pros, but they get you there with quality results and with minimum time devoted to preliminary study. My simplified explanation of the carpenter's square is a good example. The square is an ingenious instrument and has some highly complicated functions — but you need to know only the basic ones, so that's all I attempt to teach you. Besides, when you buy a good square, you get with it an impressive little booklet that tells all there is to know. So why repeat it here?

R.J. De Cristoforo

1

BEFORE YOU BUILD

A prehistoric man who stumbled onto a sizeable opening in a hillside or rock wall and adopted it as shelter did not have to contend with the codes and regulations of today's housebuilders. But neither did he have the options that make the difference between just a shelter and a custom home. We no longer settle for any kind of dwelling that will keep us alive, but demand comfort, appearance, and the intangible factors that make us feel "right" in particular surroundings.

Few people live in areas so remote that consideration of present or future neighbors is not a factor in construction and selection of the site. For good or bad, we're past the time when it was possible, or necessary, to let house location and design be determined by whim and the nature of on-site materials.

Existing codes often provoke frustration, if not anger. It's true that going through necessary channels takes time and might prove annoying, yet their existence can contribute much to the feasibility of amateur housebuilding. Many codes are established from past experiences, and they exist to help and protect you, as a builder, as well as others. A structure that fails is the basis for new code requirements that prevent a recurrence. A code may result from experimental buildings erected by federal or private agencies to test construction practices or new ideas. The programs may be for general construction or they may be for a particular area.

Soil conditions and climate are two factors that affect codes in a particular area. Minimum permissible trenching for a foundation in a mild-climate area will not apply in severe-climate areas. A roof frame that is adequate in one area may prove inadequate where it must contend with heavy snow loads. Strong winds and freezing temperatures call for special considerations that do not apply where breezes are balmy and temperatures moderate.

Of course we could all go back to school and become professional structural designers. But we could have the house built in that time. It's best to regard building codes as a store of available knowledge. Don't regard them as restrictions on visual design or floor plans. Think of them as safeguards to guarantee structural quality in your house. Some local codes do restrict architectural innovations so neighborhoods can maintain a degree of harmony and a standard of value, but rarely is this so extreme that you must build a box to match the box next door. Most ticky-tacky lookalikes result from house farms established by developers on a production-line basis.

To discover how your own ideas and needs might be met in a neighborhood, explore the area by car and on foot. Talk to residents and pay a preliminary call on the building inspector. Discover *now* whether he is guided strictly by the book or has imagination and enough knowledge to show interest in what you have in mind, even though your idea may not conform with what already exists.

Chances are there will be literature available to spell out some of the basics, such as how close to property lines you can build, how high the house can be, how high perimeter fences can be, facts about utilities, and so on. There will be more of such information available in incorporated areas than in outlying districts. But here too, it makes sense to discover what others have done in the same area. Learning from the experience of others is one way to minimize if not eliminate bad moves. Very few projects are accomplished without some redesigning done in retrospect. There is much camaraderie among owner-builders, whether the houses are close or separated by several acres. The sharing of skills, experience, and knowledge does much to speed the task and make it more enjoyable.

THE DREAM AND THE FACTS

The castle you want to build should be viewed realistically in terms of cost and labor. Your physical contribution will decrease out-of-pocket costs, but material expenses, regardless of design, will be in direct ratio to the square footage of the plan. It simply requires more material to build a 6-cubic-foot box than a 3-cubic-foot box, assuming, of course, a similarity in structural design and materials.

The first home (which we assume to be a cave) was the absolute no-frills residence. There was no thought of a site which would be occupied by generations of the same family. The selection was for immediate needs, and esthetics had little to do with the decision. This has been the case through much of man's history, actually; emphasis on comfort, scale, and appearance really doesn't go back so far. And these days, because of inflation and many other practical considerations, we're again trying to do the most with the least.

The word "big" is giving way to "minimal," a term which in a housebuilding sense translates to maximum utilization, minimum maintenance, and minimum demand on diminishing resources. Ideally, it would be nice if we could build a structure that could be made bigger or smaller to meet current needs. Some modular concept might still emerge whereby we might be able to build for today and later have a truck deliver a ready-to-attach add-on and then take it away after it serves its purpose. Mobile homes come quite close to such modular construction. But the family that wants a custom home is not apt to be satisfied with production-line modules.

A house built to accommodate current needs but designed for expandability is a more realistic approach. The idea calls for much logic in the initial construction if the end result is to appear unified and is to be achieved with minimum disruption. Pay special attention to the minimum setback required by the building codes, since you must determine in advance if you have sufficient space for the additions. Another way is to design for the future but finish for the present. For example, a couple might frame and close in a three-bedroom house but finish only those rooms that serve immediate needs. However, financing might be awkward here, since lenders may hesitate to provide money for a house that may not be finished for several years.

ROOF LINES AND SUCH

Consider the design of the house in relation to the ease or complexity of construction. If the house makes a turn — for example, L-shaped or U-shaped — rather than running in a straight line, you introduce construction complications. The gable roof, which is simply two slopes meeting at a high point called the ridge, is economical and comparatively easy to build until you choose to turn a corner. The junction forms a valley and the affected rafters require compound angles which take more time to cut accurately than the simple angles needed on the common rafters.

The low-slope shed roof is another option for the amateur who wishes to minimize construction intricacies. This is the way my family chose to go when we erected an addition for use as a studio (**Figure 1-1**). Building it against an exist-

FIG. 1-1

ing garage saved material, since we had one wall up to begin with. The idea should be considered by anyone wishing to build small now and add later. Shed and gable roofs like this one are often combined and are visually and architecturally acceptable.

The point is, while you do have choices, decisions should be made realistically after considering all options. The dream may become a compromise but it need not become a nightmare.

BUILDING UP, OUT, AND DOWN

Assuming similar square footages of living area, building "up" is cheaper than building "out." For example, a two-level square house with a given amount of floor space requires half as much foundation and roof as it would if built on a single level. Consider the single-level house in relation to three levels or two levels plus a finishable attic. Of course, multilevel houses do break up the total square footage into separate blocks, which can reduce the feeling of spaciousness you might otherwise have. However, you can compensate by avoiding the tendency to make rooms mere closed-in boxes. Where possible, screen dividers can define areas and direct traffic without closing off space. A popular example is the area which serves as living room and dining room and, sometimes, kitchen. Here an open, spacious feeling can be achieved even though sections are defined by such means as rails, planter boxes, and counters.

Building "down" means a basement. These days, the only justification for a basement is its value as living area. Here, the term "living area" is broad in scope and covers storm cellars, recreation rooms, shops, and so on. If the basement is to become, as many do, a deposit area for junk you can't bear to part with, then forget it. Doing without a basement will make housebuilding easier and less expensive.

THE ROOM AND THE SPACE

Somewhere near the beginning must come the decision on how much space will be roofed over. There are two ways to go. One, decide on the number and types of rooms you want and the size of each. Two, arrive at a dollar-per-square-foot figure and decide how much you can or wish to spend. The second system seems wiser generally, even if you choose to work with an architect. One of the first questions he would ask anyway is how much you can afford. The cost factor prevails. Maximum utilization of space can make up for space limitations.

Many houses—like the closets, cabinets, and drawers they contain—are wasteful of space and materials. A walk-in closet is luxurious, but it does require traffic as well as storage space. A utility room may be nice to have, but if you think of it as so many dollars per foot to build and maintain, you may consider including a utility area in the garage instead. Also, there are modern laundry units that stack, and hot-water heaters that may be located under a counter. It's wise to use space already there instead of adding space.

A good way to economize on materials and labor is to view space in terms of actual requirements. Decisions can be based on real needs, not fancies. Part of the fun of building for yourself is the challenge of design. Determine in detail what a room will be used for before you decide its shape and size. Quite often you can scale down in one area so you can expand in another. Your life-style must be considered. Do you really need a family room *and* a living room? In most such situations that I know of, the living room has regressed to the status of parlor — dusted, lighted, and used only for "company." You should view fad designs with a degree of skepticism. The oversize kitchen may prove a burden to those who use and clean it. The "conversation pit" is not guaranteed to add sparkle and wit to any discourse. If you realize that an elaborate foyer will cost $50 per square foot, you may hesitate to be generous in that area.

LOWERING MAINTENANCE COSTS

Anyone who designs today should give maximum consideration to heating and cooling requirements. For example, regardless of the type of heating system you plan, the more you invest in insulation, the lower your long-range costs will be and the less demand you will be making on diminishing resources.

Actually, efficient insulation is not just so many batts of fibrous material but a whole category of materials and design factors — weatherproofing, ventilation, type and area and location of window glass, house orientation, and so on. Glass is a major source of heat loss in any home, especially glass on a north wall. You do need windows, but there must be a happy compromise between floor-to-ceiling view windows on all sides of a house and the porthole concept. The placement of windows should be influenced by how the house is located on a site — which means, of course, that the house orientation should be determined in the first place with north, south, east, and west exposures in mind. The rules for glass, generally, are as follows.

● Keep window area on the north wall to a minimum. The south wall can have more glass, since here you can benefit from the sun's heat in the winter. For protection from too much sun in the summer, you can work with greenery, window coverings, or overhangs. This applies to some degree to east and west walls also. The cost of double-pane and storm windows is justified even though they are an aid, not a solution. Try not to distinguish too strictly between view, light, and ventilation windows. The functions can often be combined to result in a respectable saving in construction and materials cost.

● No one should live in a windowless box, but there is a difference between "happily adequate" and "excessive."

● Thicker walls—in some places—make sense. The conventional wall structure calls for 2×4-inch framing material. This gives you only 3½ inches for insulation (2×4-inch lumber actually measures only 1½×3½ inches). But no law says you can't do the north wall with 2×6-inch studs so you can increase insulation thickness by 2 inches. And for more insulation space all around, many new houses are being framed entirely with 2×6s.

THE HOUSE PLANS: DOING OR GETTING

There is a difference between knowing what you want and communicating the thought to the building inspector, the contractor, or the helper. Most people get around this by employing an architect, either for the entire job or for part of it. Having a stranger decide what your house should be has negative aspects. Having someone work from your basic plans, doing the detailing and specifying structurally suitable sizes of materials, is something else. If you visit an architect cold, without having done considerable homework, the chances are good that his point of view will predominate. To avoid this, you had better establish a strong concept you can be stubborn about.

If you decide to work with an architect, it's best to visit several to discover which one will be most sympathetic with your scheme. Ideally, from the build-it-yourselfer's point of view, it should be possible to sketch the needs, and wants, and shapes, and pay the professional to do the final plans so they "read" and conform to the building codes of the area. The pro, of course, will spot flaws and you can correct them before being told to do so by the building inspector.

What should interest the amateur is the variety of ready-to-use building plans which are available for as little as $40 or $50. You may think this is the same as giving someone else control over the design, but on the other hand, it's possible you may find something that suits you or that can be modified, and in any case, the more you consider professional designs, the more sophisticated your own design will become. This chapter shows a few examples, but the best bet is to send away to the companies listed at the back of the book for catalogs that show all that is available. Most such literature will tell briefly what the house is and show an artist's rendition plus a floor plan. Some organizations will even supply, at slight extra cost, itemized lists of materials, plumbing and wiring diagrams, and so on.

The rambling one-story house in **Figure 1-2** affords abundant space and considerable privacy. Note that entry to the garage is from the rear so the line of the house is uninterrupted by garage doors.

FIG. 1-2   RAMBLING ONE-STORY, PLAN 3301
(Hudson Home Guides)

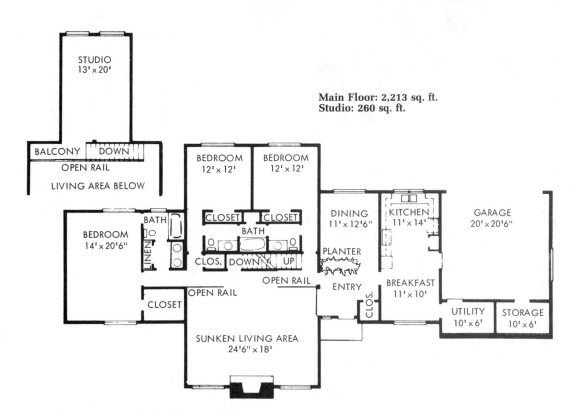

Main Floor: 2,213 sq. ft.
Studio: 260 sq. ft.

STUDIO
13' x 20'

BALCONY   DOWN
OPEN RAIL
LIVING AREA BELOW

BEDROOM
12' x 12'

BEDROOM
12' x 12'

DINING
11' x 12'6"

KITCHEN
11' x 14'

GARAGE
20' x 20'6"

BATH

CLOSET   CLOSET
BATH

BEDROOM
14' x 20'6"

LINEN

CLOS.   DOWN   UP

PLANTER

CLOSET

OPEN RAIL

OPEN RAIL

ENTRY

CLOS.

BREAKFAST
11' x 10'

SUNKEN LIVING AREA
24'6" x 18'

UTILITY
10' x 6'

STORAGE
10' x 6'

Available designs include multilevel homes like the one in **Figure 1-3**, which is designed to make the most of a steep, difficult site. Entry to the house is from the third, or street, level.

FIG. 1-3   TRI-LEVEL, PLAN 5004,
(Hudson Home Guides)

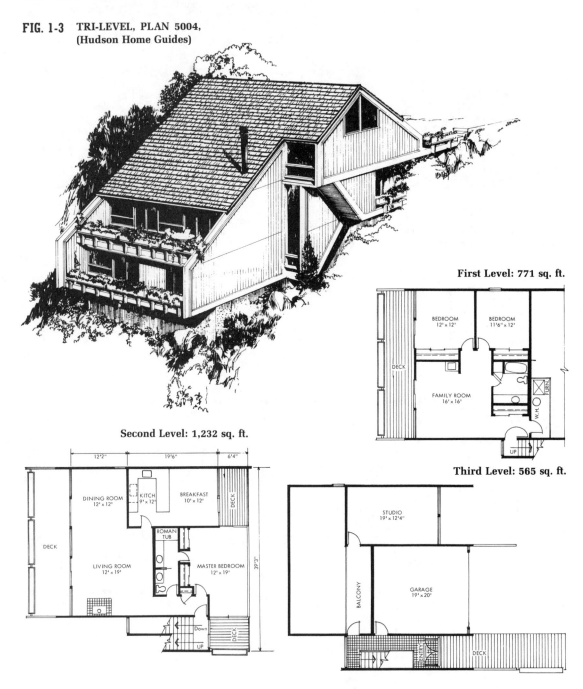

Traditional designs like the Cape Cod in **Figure 1-4** may be purchased also. The upper floor of this house can easily be viewed as a "finish later" project.

FIG. 1-4 **CAPE COD, PLAN 2238**
(Hudson Home Guides)

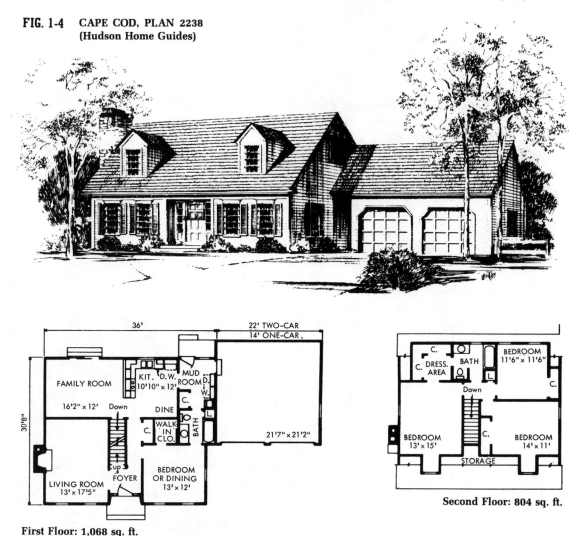

First Floor: 1,068 sq. ft.

Second Floor: 804 sq. ft.

Many designs for vacation or "second" homes are available. Some, like the one in **Figure 1-5**, start as a small basic unit to which you can add rooms at will until you end up with the completed plan. This plan, or something like it, makes much sense for the person whose dream is to move permanently, someday, into the house he now escapes to on weekends and vacations. It should also appeal to a couple who want or can afford only the minimum now but have expansion in mind for the future.

FIG. 1-5   BASIC UNIT PLUS OPTIONAL ADDITIONS (Western Wood Products)

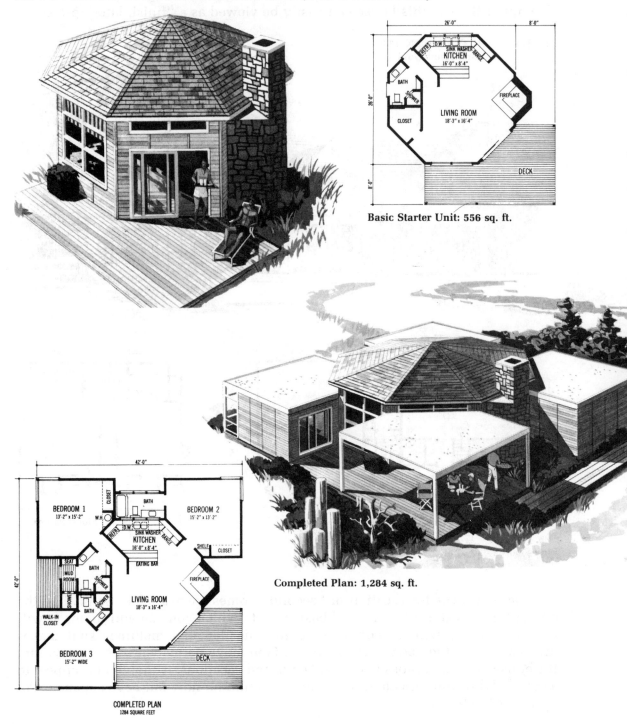

Basic Starter Unit: 556 sq. ft.

Completed Plan: 1,284 sq. ft.

No one should react negatively to the thought of working with ready-drawn plans. For one thing, checking catalogs doesn't commit you. For another, you can use one you like as the basis for doing your own thing. It's a lot cheaper to modify an existing plan than it is to have an architect draw one from scratch—and you will be more welcome in an inspector's office if what is under your arm resembles professional blueprints instead of doodles on a scratch pad.

THE SPECIALS

Before you decide on your house and how you will build it, you should check out some of the new, and not so new, ideas that are on the scene. Systems such as these: (1) Mod 24, in which studs are placed 24-inch on centers instead of 16-inch on centers; (2) pole construction, in which there is no conventional foundation and the structure is supported by pressure-treated telephone poles sunk a substantial distance into the earth; (3) post-and-beam, in which there are no bearing walls in the conventional sense; and (4) the all-weather wood foundation, which utilizes pressure-treated plywood and lumber instead of masonry and against which you can backfill with soil, directly.

We'll discuss these in more detail in Chapter 28, and we'll tell you where you can get in-depth information concerning each. But it's good to be aware they exist so you can examine the possibility of working with them. One of the problems you will encounter is getting acceptance from the public and from inspectors. There are such things as national codes, but these are interpreted, accepted, and rejected on a local level. It's possible for a design that is acceptable to a federal lending institution to be substandard to a local examiner. Often this is just due to the length of time it takes for changes to filter down into the local code books. At any rate, if any of the specials appeal to you, do some checking before you rent a power auger to form holes for poles.

DO A DOLL HOUSE FIRST

A good way to see what the end result will be is to do a mock-up with stiff cardboard. All you need are cardboard, tape, a sharp knife, a ruler, a pencil, and ideas. Here you can experiment with different floor plans, house shapes, and even sizes, and all you waste is cardboard.

Adopt a scale—say, ¼ inch equals 1 foot. If, for example, you are thinking of a 20×80-foot house, then your cardboard floor would be 5×20 inches. Play with the floor plan before you set up walls. If you like, you can carry this to the point of indicating furniture placement. Make items you can move around by measuring your own furniture and cutting pieces of cardboard accordingly. Color the "furniture" with crayon so it will stand out against the cardboard floor. It may pay to place the furniture before you consider partitions. Make the roof of the model removable.

The project may also be used to decide orientation on the site. Draw the shape of the land on a sheet of paper; indicate existing trees, slopes, north, south, east and west. Here, too, you can sketch in driveways, future gardens, utility areas, and so on. By working with dried moss, twigs, small branches, and the like, you can get a pretty good idea of what the final picture will be.

WORKING SAFELY

Proper work clothes are important. Have a special uniform that fits you snugly and includes heavy, high, steel-toed shoes and a hardhat. There are areas of housebuilding where it makes plain sense to have someone help you lift things. Be sure any helper knows what's being done and how you plan to accomplish it.

FIG. 1-6

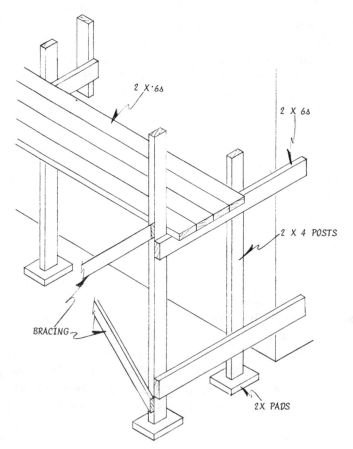

Watch yourself when you get to upper areas where there may be a need to walk across open joists. Take the time to set down some heavy boards as a temporary platform. Be aware that you can slide off a roof, trip over a board, walk into a beam. Stack materials neatly and have a special place where you can deposit cut-offs.

Don't be careless when you find it necessary to construct a scaffold. It *is* a temporary thing but your safety demands a structure that will not collapse. The design in **Figure 1-6** is typical of the types erected by pros. Work with duplex (double-headed) nails so the breakdown will be easier. Use enough boards for the platform so you won't be walking a tightrope. Note that the scaffold is tied into the house frame. You can, if you wish, rent scaffolds that are an assortment of pipes you assemble on the site.

Sawhorses are important tools and may be used as bases for platforms and as benches for cutting wood. Ready-made brackets are good to consider since they can be used with various size legs to suit particular applications. The same brackets can be used to make a permanent sawhorse like that in **Figure 1-7**. Adding a shelf increases rigidity and gives you a place to put tools or boxes of nails.

Safety is the individual's responsibility. No set of rules will guarantee it. Being aware, always, that you can get hurt is probably the best precaution. Preview each phase of the job before you tackle it. Knowing what is ahead will prepare you for doing the job correctly and safely.

FIG. 1-7

2

THE HOUSE
AND THE SITE

One of the surest ways to waste money, materials, and labor in house construction is to regard the site too casually, or to be so impressed with one feature that you become blind to practical considerations. Stories of people who became enamored of a storybook setting and were subsequently disenchanted by such prosaic things as the cost of putting in a road are not so rare. This can be discouraging; hidden costs can upset an otherwise sound financial program.

The question is not so much what can be done, but what you want to do and are prepared to pay for. Land that pleases you and is suitable for building without hidden costs is ideal. A landsite should be approached with caution if not suspicion; if possible you should talk to a local builder and nearby residents.

Sometime during the thinking or building process, questions about access, water, power, zoning, and so on will have to be answered. The best time is before you buy. Seek out and calculate all costs carefully. Building on land with serious natural defects may be expensive enough to cause you to explore elsewhere.

SOME TYPICAL SITES

Five basic sites are shown in **Figure 2-1**. The site that is the least expensive to build on is usually flat topographically, and often flat esthetically as well. A flat site needs little grading and excavation and does not require constructions like retaining walls or terraces. On the other hand, a flat site calls for careful investigation of drainage conditions, especially if it has hilly surroundings or is lower than adjacent properties.

A slope that provides an adequate view and good drainage can be an excellent house site. Too much slope, though, and you might encounter problems with excavation, and situations requiring fill and retaining walls. Some hillsides can be very challenging from a design point of view. And when built on with minimum disturbance of natural grade, they often become prisons for the resi-

14

FIG. 2-1

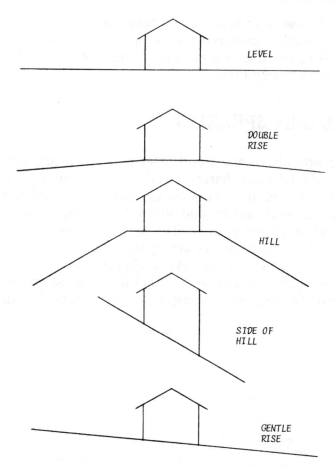

LEVEL

DOUBLE
RISE

HILL

SIDE OF
HILL

GENTLE
RISE

dents, since areas around the house are not level enough to walk on. Compensations take the form cf expensive grading, which often destroys the naturalness of the site, or a system of decks for outdoor living.

It's okay to choose land that requires imagination to make it suitable as a homesite, but do think ahead. Be aware that extensive excavation, cutting of access roads, and possibly importing of fill are not easy chores for the average builder. On the other hand, there is pole construction, a building technique becoming more popular because it solves steep-slope construction problems with minimum disturbance of natural surroundings.

Sites that are high above an existing street require steep driveways and front walks or steps. There are safety hazards here as well as additional construction expenses. Even in mild climates, rain can make such a driveway slippery. In snow country, the problem can be more extreme. Solutions take the form of lower-level garages and parking areas or switchback driveways, features that can add interest but inconvenience as well.

Sites that are below an existing street have problems too. There are still the driveway and entry-walk considerations, except now the pitch is down instead of up. Below-grade sites are also potential reservoirs, especially if existing gutters are not able to handle heavy rains.

TOPOGRAPHICALLY SPEAKING

The homebuilder won't have much enthusiasm unless he can feel that the site is an asset—a piece of land with character and some natural advantages. The ideal would not be difficult to achieve if there were no concern for finances, reasonable proximity to schools, work, stores, and so on. Compromises should be accepted with grace and undiminished zeal, or else the project won't be worth doing.

My own present building site is a far cry from the "view" lot we sought in the foothills of the Santa Cruz Mountains. We could get to such a lot, but only on foot. So we lowered our sights, literally, and settled for a three-acre chunk of property, as shown in **Figure 2-2**, that was almost a castoff because a county road had re-

FIG. 2-2

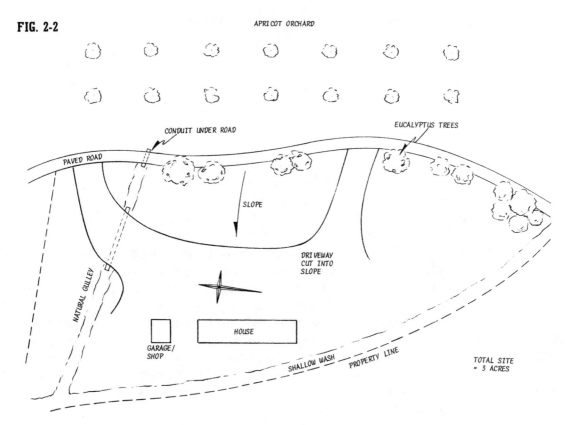

APRICOT ORCHARD

CONDUIT UNDER ROAD

EUCALYPTUS TREES

PAVED ROAD

SLOPE

DRIVEWAY CUT INTO SLOPE

NATURAL GULLEY

HOUSE

GARAGE/ SHOP

SHALLOW WASH   PROPERTY LINE

TOTAL SITE = 3 ACRES

moved it from a large apricot orchard. The slope was more than we would have accepted on paper, and a quick look gave the impression that it was a natural drainage basin for surrounding terrain.

Upon closer examination, with open minds, we determined that we could correct obvious defects in a fairly straightforward manner. We wanted the house at the property's widest point, which also happened to be the area with sharpest slope. But since a circular driveway was a part of our plan, we figured that required grading could do two things—create the driveway and provide fill for a level area behind and along the sides of the house. Grading was planned carefully so there would be good drainage, a cut in the bank that would not require a re-taining wall, and undisturbed ground for the house foundation, as shown in **Figure 2-3**.

FIG. 2-3

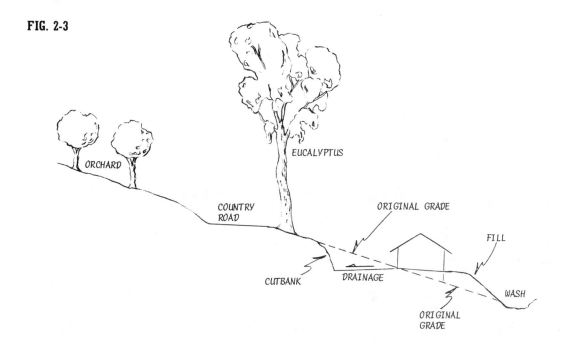

Ultimately, the layout gave us adequate drainage away from all areas and enough soil from the grading to fill the gully and so create the entry driveway.

The plan worked fine. During heavy rains we did have a small stream flowing at the base of the cut in the bank, but, even though it was not a serious problem, we eliminated it by digging a trench and filling it with large gravel and a line of drain tiles.

The foundation area was still a slope, so we had to decide between full concrete perimeter walls or a combination of concrete and wood. We made the

FIG. 2-4

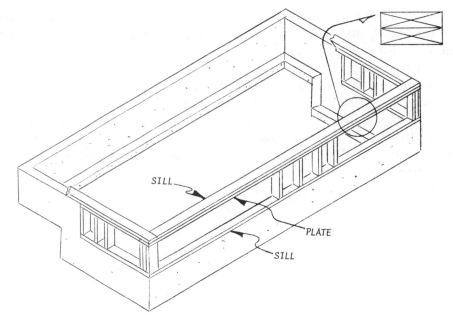

SILL

PLATE

SILL

latter choice, shown in **Figure 2-4**, mostly because I am more a woodworker than a mason, but either system would work. A split-level design of one kind or another would have been fine on our lot, of course, but that is not what we wanted.

EXCAVATION: ECONOMICS AND ESTHETICS

Make a close study of the characteristics of the site. Drainage, utility entrances, and grading all should be designed to preserve the natural contours and the natural growth as much as possible. Undergrowth and trees in the way of the house must be removed, but careless filling around, and injury to, well-located trees destroys valuable assets.

Grading is always simplified, and least expensive, when necessary cuts provide required fill. Expenses mount when excavated material must be hauled away from the site and when fill must be purchased and trucked in.

House sites do differ, yet there is enough similarity, generally, so a few examples can point up good excavation procedures. The two-level house in **Figure 2-5** is designed so the ground floor is at new-grade level. Material from the excavation is used to level front areas and as a fill at the rear for drainage.

On level ground you can use excavated material to raise the general grade, as shown in **Figure 2-6**. This actually reduces the amount of excavating you must do and provides drainage you might not have if you accepted the original grade.

FIG. 2-5

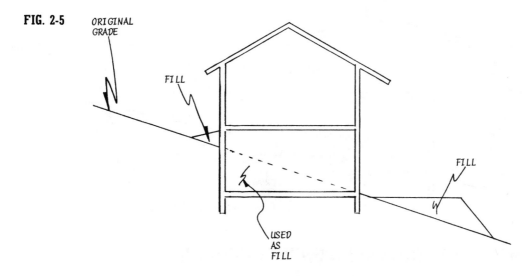

ORIGINAL GRADE

FILL

FILL

USED AS FILL

FIG. 2-6

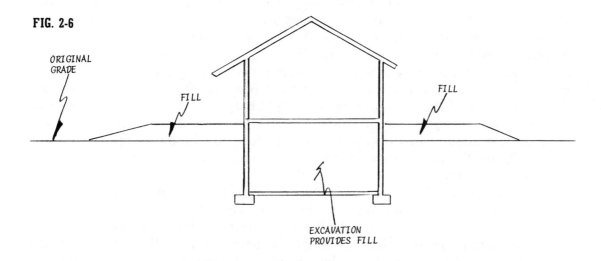

ORIGINAL GRADE

FILL

FILL

EXCAVATION PROVIDES FILL

A good example of minimum excavation is provided by the three-level design in **Figure 2-7**. Note how the cut in the original grade is made to provide a drainage slope away from the house and how the removed material is used as a fill at the front. Another interesting feature is the use of both slab floor and crawl space.

In all situations where any excavating must be done, scrape off and save the topsoil. Push it into a mound so you can spread it around the property after the building chores are done.

FIG. 2-7

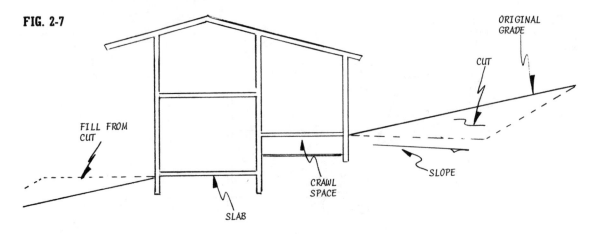

SUMMARY OF DRAINAGE CONSIDERATIONS

A site that slopes up from an existing street usually provides good drainage toward runoff facilities. A site that slopes away from the street may have to be drained toward a rear yard or by a special underground pipe system to a storm sewer. A basement may not be advisable.

Check the topography of surrounding areas and study the directions of natural surface drainage. It's also wise to do some research on the possibility of *subsurface* drainage. City or county records might help here, but the surest way is to bore a few test holes so you can study the soil texture and look for accumulations of water. Negative results do not necessarily rule out the building site but they may point up the importance of storm-sewer locations, or the need for special construction techniques.

Always check established grades, such as streets, that you can't alter. These, plus any storm sewers that have been or will be installed, should be starting points when you investigate the drainage situation of your own site.

Level sites generally call for raising the existing grade, at the perimeter of the house at least, so surface water can be directed to a runoff point. The soil removed for a basement can be used for that purpose.

Remember that drainage is, or should be, a community project. Your neighbors won't appreciate being inundated any more than you will.

HOUSE TYPE AND PLACEMENT

There are many variables here—personal preferences, available finances, codes, and so on—so it's impossible to provide all answers. It is true, though, that no factor contributes as much to success as the right house for the lot. The ideal can result from either of two approaches: Knowing exactly the kind of house you want and then searching for the ideal site; or falling in love with a site and designing the house for it.

Compromises should not reduce incentive. Often a design modification leads to a practical and satisfactory solution. Avoiding pitfalls may be more important than an artist's rendition of the ideal and picturesque homestead.

The small house will look even smaller if you set it like a box on a mound surrounded by steep embankments, as shown in **Figure 2-8**. But you don't have to increase square footage to make the house look larger, if you design the garage as an extension. Construction costs will actually decrease, since the driveway considerations will be similar but there will be no need for expensive, and maybe dangerous, retaining walls.

FIG. 2-8

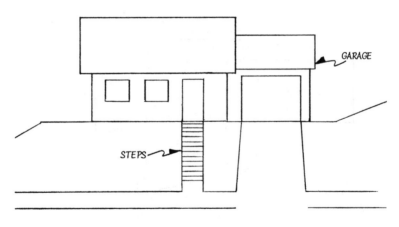

BETTER THIS

GARAGE

STEPS

THAN THIS

WALLS

WALLS

STEPS

GARAGE

Try to plan for specific separations of outdoor areas while still making maximum use of the land. The more you reduce the length of a driveway, the less the project will cost. A garage or carport that utilizes an existing wall costs less to build. A detached garage, especially if it is far from the house, requires a long driveway, a special walk from garage to house, and even extra costs for electrical work, and security is reduced.

You might want to spread things out if the lot is large and especially if the site is elevated. A separate garage level can be made to avoid a steep driveway and ramplike entry steps to the house. But to avoid long flights of entry steps and a steep driveway cutting across contours that may require retaining walls, plan for something along the lines shown in **Figure 2-9**. Here you have an entry to the house from the driveway and close to the garage. Here you break up total elevation into two levels with minimum pitch for each. Also, when house entry walk and driveway are combined, you leave an undisturbed area in front of the house.

FIG. 2-9   THIS IS GOOD

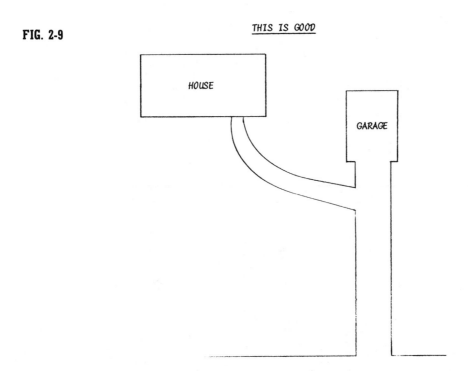

In all situations, it's imperative to have an overall plan so you can decide on orientation of house and site to make maximum use of the most desirable features of each. **Figure 2-10** shows such a plan for a small lot.

FIG. 2-10   *THE OVERALL PLAN AND WHAT TO CONSIDER*

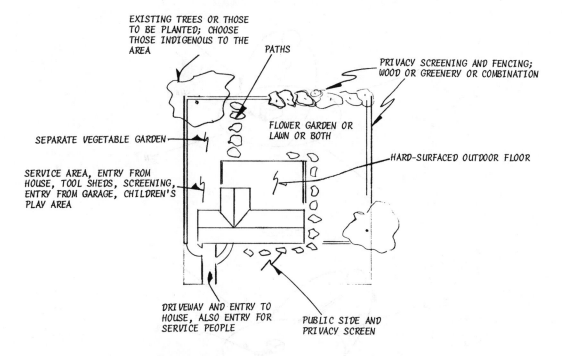

EXISTING TREES OR THOSE
TO BE PLANTED; CHOOSE
THOSE INDIGENOUS TO THE
AREA

PATHS

PRIVACY SCREENING AND FENCING;
WOOD OR GREENERY OR COMBINATION

SEPARATE VEGETABLE GARDEN

FLOWER GARDEN OR
LAWN OR BOTH

HARD-SURFACED OUTDOOR FLOOR

SERVICE AREA, ENTRY FROM
HOUSE, TOOL SHEDS, SCREENING,
ENTRY FROM GARAGE, CHILDREN'S
PLAY AREA

DRIVEWAY AND ENTRY TO
HOUSE, ALSO ENTRY FOR
SERVICE PEOPLE

PUBLIC SIDE AND
PRIVACY SCREEN

NATURE WILL COOPERATE, IF YOU LET IT

Study the features of the land before you draw the site plan. Any building will affect the environment and the landscape. The idea is to respect what nature has already established. Accept the terrain as a partner, and plan knowing you are making a lasting impact that should provide maximum comfort and economy for you with minimum harm to the area. If you can combine your needs with a good degree of reverence for nature, you have it made, and will profit. The effort you invest in planning the building and locating it will pay dividends in the form of enjoyable living and reduced energy demands. Privacy, views, wind directions, traffic convenience, and sun travel in winter and summer are the primary considerations.

● Know how the sun travels over the building site (**Figure 2-11**). Bear in mind that hills, tall trees, and adjacent structures can affect sun and shade. This can be an important heating and cooling factor, especially if you add solar collectors at some time.

● What about the view you desire to look at and scenes you may wish to screen out?

FIG. 2-11

SUMMER SUN

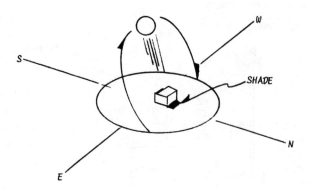

WINTER SUN

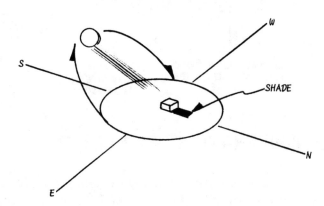

● Are there storms and cold prevailing winds in the winter? Day or night breezes in the summer?

A true study of orientation is a detailed science, one that would require a daily inspection of conditions of each site over a year's period. Yet there are basic considerations, as shown in **Figure 2-12**, and general rules to apply even though they are most applicable in the belt between 35 and 45 degrees north latitude (**Figure 2-13**). This belt includes the majority of the country's most populated places. When the house site is below this belt, taking advantage of summer breezes and seeking protection from intense sun heat are major considerations. When building above the belt, designing for protection from cold winter winds and getting the most from winter sun are prime objectives.

FIG. 2-12

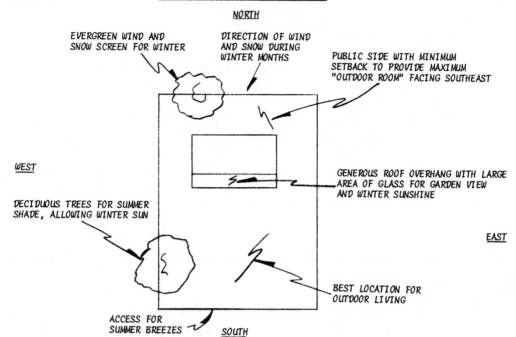

SOME BASICS OF HOUSE ORIENTATION

NORTH

EVERGREEN WIND AND
SNOW SCREEN FOR WINTER

DIRECTION OF WIND
AND SNOW DURING
WINTER MONTHS

PUBLIC SIDE WITH MINIMUM
SETBACK TO PROVIDE MAXIMUM
"OUTDOOR ROOM" FACING SOUTHEAST

WEST

DECIDUOUS TREES FOR SUMMER
SHADE, ALLOWING WINTER SUN

GENEROUS ROOF OVERHANG WITH LARGE
AREA OF GLASS FOR GARDEN VIEW
AND WINTER SUNSHINE

EAST

BEST LOCATION FOR
OUTDOOR LIVING

ACCESS FOR
SUMMER BREEZES   SOUTH

FIG. 2-13

45°
40°
35°

WHAT TO CONSIDER

Summer sun is high overhead at midday and has an extended arc. In winter there is a reduced arc and the midday sun is lower in the sky at the same time. The object is to protect from the overhead summer sun, but to capture as much of the winter sun as possible (**Figure 2-14**).

FIG. 2-14

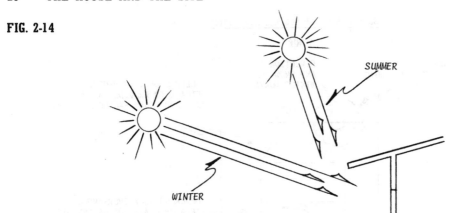

Because of the sun factor, a southeast slope usually makes an ideal house site, but a south slope is also desirable. An east slope is better than a west slope. And a north slope should be avoided.

Roof overhangs make most sense on the south side of a house. They may be used elsewhere for other purposes, but sun protection is achieved only when the exposure is generally south.

West sides can be protected from hot afternoon summer sun through the use of tall plantings or an attached garage.

A garage or tall evergreen trees, or both, can do much to screen the northwest direction and so provide protection against cold winter winds. Here you can also put topography to use; high ground to the north can serve as a screen too.

Open house designs are suitable in the south, since they encourage circulation of summer breezes. Here, outdoor living can be enjoyed for a large part of the year, but overhead constructions should be incorporated to protect open and enclosed house areas from the summer sun.

You should design for a minimum of window area on the north wall. Large glass doors and windows are most suitable on the south side. Windows on the west side should get special protection against hot summer sun.

Always remember that warm air will rise because it is lighter than cold air. Therefore, ventilation considerations generally should provide for cooler air to enter the house close to grade while warm air should be permitted to escape at a high point—at the ceiling or through the attic.

When you do the site plan, think of the lot as having three outside areas: (1) one for your own outdoor living, (2) one as a service area, and (3) one as the public area. Defining these clearly and locating them in terms of convenience, livability, and privacy will contribute much to a satisfactory site development.

PRIVACY: ANOTHER GOAL

Privacy would not be difficult to achieve if you ignored all other factors and just put up solid walls and spent all your home time indoors. But today's life-styles call for outdoor as well as indoor areas to relax, to eat, and to enjoy friends and recreation. No house is so large that spreading living area to the outdoors won't be an asset, psychologically and in relation to house value. Often, providing for privacy also affords a good measure of protection from noise — noise from the outside or noise *you* make that might bother neighbors.

Success doesn't come by accident. It's achieved through the planning you do for the house and its surroundings. For effectiveness and attractiveness, you should have an overall plan that considers the shape of the house, its location on the site, subdivision of the yard, wood or masonry fences, greenery, and trellises. If, for example, you use a lot of glass in a wall because the rooms there will benefit from it, plan to shield the area so your privacy will be assured regardless of which side of the glass you are on. Don't forget there are such things as fast-growing vines and creeping plants you can grow on trellises. And there are fast-growing trees too. First check around to determine which plantings will do well in your area.

Your privacy goals can be combined with the private outlook — taking advantage of what is nice to look at and screening out what is bad. The view does not have to be a snow-covered mountain, a lake, or a river. It can be a low or high flower-bed wall, an attractive trellis, a garden waterfall, groups of trees or shrubs, and so on. Remember that your screening can also conceal eyesores on your own property such as areas designated for outdoor jobs, trash cans, or tool storage.

SUMMARY THOUGHTS

It is wise to know the capacity and the location of local facilities such as sewage and electric lines before you plan the house. In areas without public sewer and water systems, you still must consider electric and telephone lines, your own water supply, and your own sewage disposal. Building codes follow you most anywhere. You may be required to have a septic tank of a certain size and a drain field of a certain length, and there may be restrictions on the location of the tank.

Local codes may demand that utility lines passing through landfill be specially supported and be of special materials. This can increase expenses, and you may still have problems from fill settling.

The length of underground lines can be controlled to a great extent by the house plan and its location on the site. It helps to use minimum setback from a street where community services run. Locating plumbing fixtures on that side will also reduce costs and labor. Blocks of plumbing can serve several house facilities — for example, two bathrooms, a kitchen and bathroom, a kitchen and utility room.

A single trench for both sewer and water lines can save money if local codes permit such an installation and if anti-contamination precautions are taken. The trench can be wider at the top area to supply a ledge for the water lines (**Figure 2-15**). Or you can cut into a minimum-width trench and provide a water-pipe seat as shown in **Figure 2-16**. The ledge height is variable for easiest connection to the supply source, as long as it is below the frost line.

FIG. 2-15

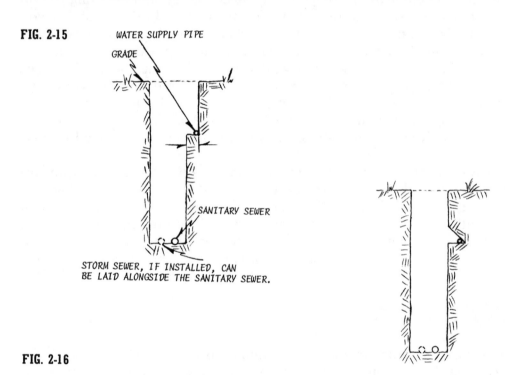

WATER SUPPLY PIPE

GRADE

SANITARY SEWER

STORM SEWER, IF INSTALLED, CAN BE LAID ALONGSIDE THE SANITARY SEWER.

FIG. 2-16

Expensive fill and foundation walls may be required if drainage lines make it necessary to raise the house high above the natural grade. Changing the drainage situation or modifying the house to suit existing conditions may be a better solution.

Successful landscaping depends more on the thought given to planting, screening, and grading than it does on how much money is spent. Save as many existing trees and large plants as possible. Trees are assets that will increase in value along with the house. Any time you must cut into the land for the house site or the driveway or a walk or whatever, scrape off the topsoil first and store it for use when construction work is complete.

Informal landscaping which makes use of indigenous greenery will be cheaper to do and easier to maintain than more formal schemes. Lawns do well on flat or rolling land, but they do require constant care and much water. When you add plants, be sure to research ultimate shapes and sizes and rate of growth. Groups of trees or shrubs usually have more visual appeal than overall coverage, although dense in-line arrangements are more effective as view and sound barriers. Check out local successful ground covers when you plant on embankments.

Obeying the rules that blend house and location while keeping costs low and livability high will be easier on some sites, but a good balance is achievable just about anywhere. Each house and site will have problems. You may find it impossible to achieve absolutely everything you want—but as long as you study all the features and defects of the site in advance, and incorporate them as best you can in your plan, you can find the optimum solution.

3

LAYING OUT
THE BUILDING SITE

We can assume that the lot has been professionally surveyed and that stakes marking property lines and corners have been correctly and legally established. There are true stories about sales with haphazard perimeters that resulted in constructions that were later challenged and, in some cases, had to be moved. Removal may mean just a fence, or a greenhouse, or some new trees. But such problems can be avoided by having a professional survey done, whether it is sponsored by you or the seller.

If you have the job done, it will not cost much more to have the surveyor establish the setback lines, if any, and place stakes to establish grade elevations. If you have already chosen house plans and decided house placement, he can set two stakes to show one side of the structure. From this point you can just about work with a carpenter's level, a line level, a square, and a length of line to do other layout requirements. Yet, it pays to know something about professional leveling instruments that you can rent, since they can be very helpful when laying out angles, establishing foundation heights, checking vertical corners, and so on, accurately and with minimum fuss.

Both the *level* and the *level-transit* are used to do surveying and land layouts, and to check ongoing construction work. Both tools are basically telescopes with built-in levels, read-out scales, and adjustment screws for leveling the instrument after it has been mounted on its tripod support. The telescope may be turned on its base so you can lay out or measure any angle on a horizontal plane. Since the line of sight through the scope is always a straight line, you can use the tool together with a *leveling rod* (**Figure 3-1**) to determine the difference in elevation between the point where the level is set up and the point on which the rod is placed. Leveling rods may also be rented, but many workers improvise by using one of the make-shift methods shown in **Figure 3-2**. Hold a common level against one edge of the rod to be sure it is vertical. Needless to say, readings will be useless unless you take the time to set up the tripod correctly to begin with. Make sure the leg points seat firmly on or into the ground. Work with the adjustable legs until the

FIG. 3-1

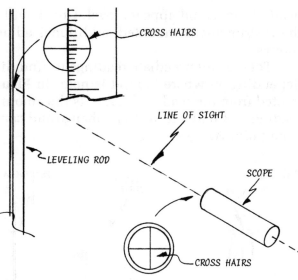

CROSS HAIRS

LINE OF SIGHT

LEVELING ROD

SCOPE

CROSS HAIRS

FIG. 3-2

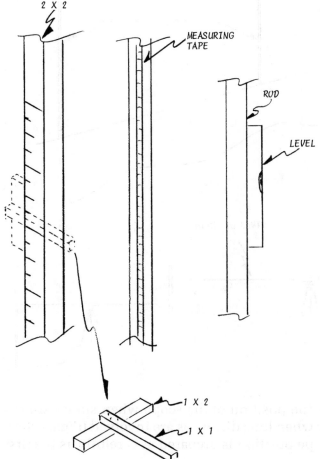

2 X 2

MEASURING TAPE

ROD

LEVEL

1 X 2

1 X 1

head of the tripod appears level. Check with a common level if you wish. Once the instrument is mounted, additional adjustments are made with the built-in screws.

Total or intermediate readings to find differences in elevations can be taken depending on where the rod is held. In **Figure 3-3**, the height of the scope, subtracted from the reading at A, tells the amount of change in grade at that point. A reading at B will tell the total change and can be used to determine the difference from point A.

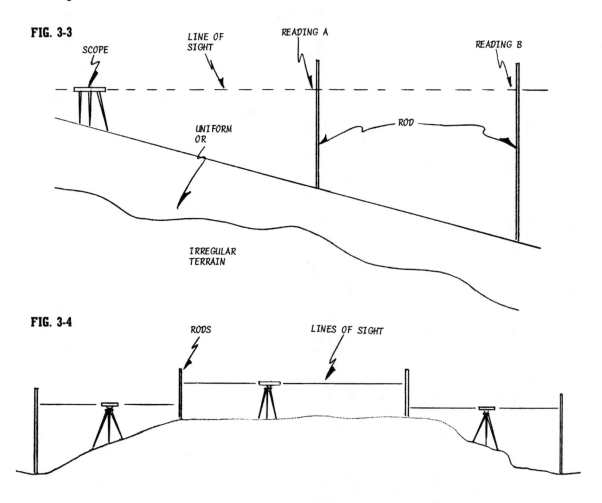

FIG. 3-3

SCOPE

LINE OF SIGHT

READING A

READING B

UNIFORM OR

ROD

IRREGULAR TERRAIN

FIG. 3-4

RODS

LINES OF SIGHT

It may be necessary to change the position of the scope when slopes are extreme, or when terrain is erratic, or when long distances are involved (**Figure 3-4**). In such situations, be sure each scope position is prepared as carefully as the first one.

To lay out angles or to establish corners of constructions, attach a plumb bob to the screw or hook that is on the underside of the scope. Set the tripod so the plumb bob is positioned at the exact centerpoint of a stake that has been driven as a starting point. The centerpoint is usually a nail driven into the center of a 2×2 or 2×3 wooden stake. With the instrument leveled and the plumb bob placed correctly, sight through the scope toward a rod at point B. Measure the correct distance with a tape, and, by working with scope and rod, establish a stake and centerpoint (**Figure 3-5**). Then you can swing the scope 90 degrees, or whatever angle is needed, and take a second sighting to establish point C. Reposition the instrument at point C, and sight back to point A (**Figure 3-6**). Then follow the same basic procedure to get point D. The job is complete if you are doing a rectangle or a square, but you may wish to position the scope at point D so you can sight to points C and B as a check. After corners are established, batter boards and lines are set up (see Chapter 4) to define excavation areas. From this point, it's usually

FIG. 3-5

FIG. 3-6

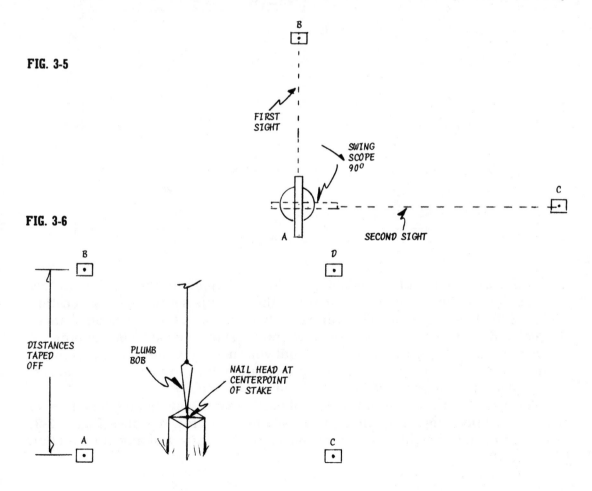

fairly simple to work with a common level and a square to set stakes for small projections and any irregular shapes.

The procedure for laying out an L shape is much the same. As shown in **Figure 3-7** the scope, set up at A, takes sightings to establish points B and C. Then, sightings are taken from C to D, from D to E, and finally from E to F.

FIG. 3-7

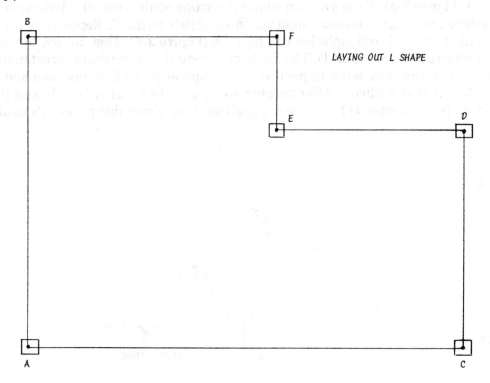

LAYING OUT L SHAPE

With corner stakes in, establishing grade stakes (for a footing) or the locating of batter boards can be done more easily if the scope is set in a central location (**Figure 3-8**). This way distances will be fairly equal, and thus focusing changes will be minimized. Once the correct elevation is established at a control point, it can be transferred to all other points without your having to move the instrument. Grade stakes are usually driven to an approximate point and then adjusted to correct height while being checked with the level and rod.

The level and the transit-level work about the same way, but the transit-level has a vertical pivoting feature that can be used to align a row of stakes (**Figure 3-9**) regardless of their height or to check the vertical accuracy of any construction (**Figure 3-10**).

FIG. 3-8

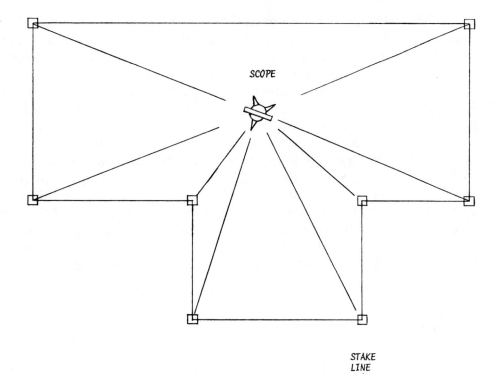

SCOPE

FIG. 3-9

STAKE
LINE

LINES
OF
SIGHT

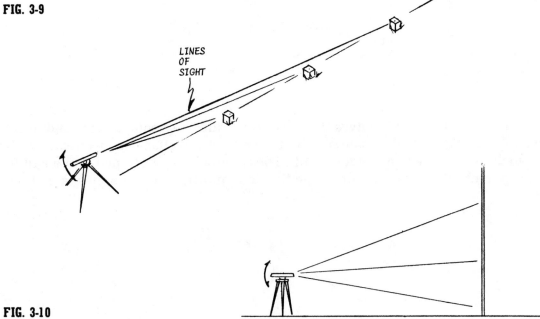

FIG. 3-10

When used for such purposes, the instrument should be leveled in the usual manner and then unlocked from its horizontal position. Pivot the scope both vertically and horizontally and establish a reference point along the vertical plane (if you are checking for plumb) or along a line (if you are establishing stake alignment). Then lock the horizontal position. Now, as you pivot the scope, all sightings will be on the same vertical plane. Vertical lines should be checked from two positions, the second position being 90 degrees to the left or right of the first one.

Much checking can be done by working with a common level, a line level, and a measuring tape. The example in **Figure 3-11** is elementary, but it does serve to illustrate the procedure. Here a stake is set to establish the height of a footing. A line is stretched to a second stake, which has been measured off an appropriate distance. The level will tell whether to raise or lower the line at the second stake. The differences between A, B, and C will tell the amount of change in elevation and will be guides that show heights of walls, or amounts of fill that may be required, or depths of excavations needed, depending on the design of the construction.

FIG. 3-11

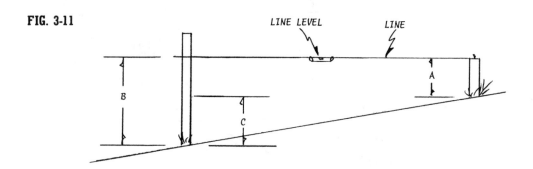

Much layout work involves the use of lines and the frequent tying and untying of knots. You can eliminate some of the knots if you attach an electrical terminal to the end of your line (**Figure 3-12**). The terminal can be slipped over a nail in a stake. From there you can draw the line taut and tie it at the opposite end.

FIG. 3-12

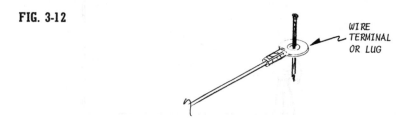

4

BUILDING THE FOUNDATION

BATTER BOARDS AND BUILDING LINES

After corner stakes have been established, batter boards, as shown in **Figure 4-1**, are set up 3 to 4 feet outside the marks. Batter boards are not always L-shaped. Straight ones are used in some circumstances and may be placed inside or outside the building site depending on the type of excavation needed; they are shown in-

FIG. 4-1

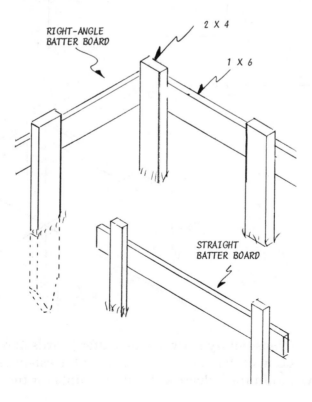

RIGHT-ANGLE BATTER BOARD

2 X 4

1 X 6

STRAIGHT BATTER BOARD

side the site in **Figure 4-2**. The units do not have to be fancy but they should be firmly set, since you will be using them throughout the foundation work. Use sharpened 2×4s for the stakes and 1×6s or wider stock for the ledgers (horizontal pieces). Work with a level when you set up the first set of ledgers and then with a line level to establish the height of others. Keeping all of them on the same horizontal plane will be an aid when you check depths of excavations, heights of footings, and so on.

FIG. 4-2

BATTER BOARDS
MAY BE PLACED
OUTSIDE OR INSIDE

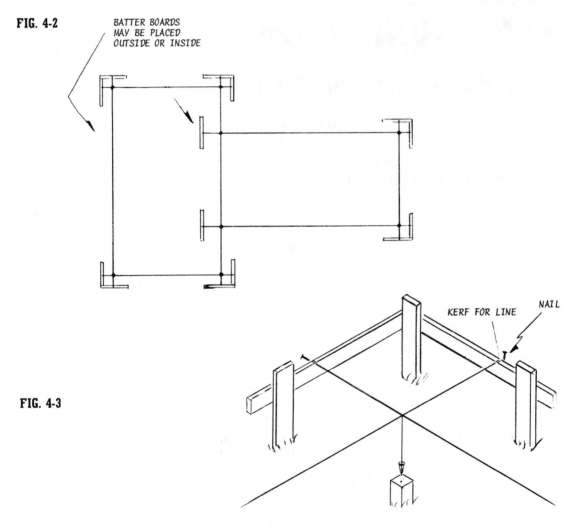

FIG. 4-3

KERF FOR LINE   NAIL

Next, stretch the lines tightly between the batter boards and use a plumb bob to be sure the intersection of the lines is exactly over the established stake (**Figure 4-3**). Saw a shallow kerf in the ledgers as location points for the lines, and drive a nail so the strings can be secured. This way, when the lines have been removed for excavation purposes, they can be replaced accurately for positioning foundation forms, as shown in **Figure 4-4**.

FIG. 4-4

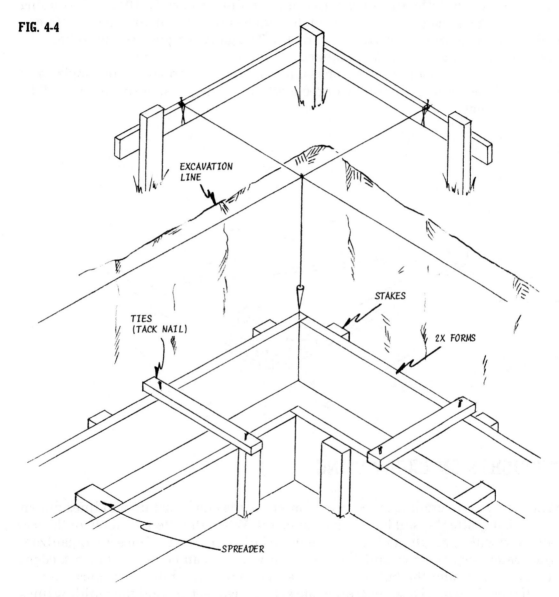

EXCAVATION LINE

STAKES

TIES (TACK NAIL)

2X FORMS

SPREADER

Foundations that are out of square can cause many headaches throughout construction procedures, so you want to be sure when setting up lines that corners do form 90-degree angles. This can be done by using the 6-to-8-to-10 system. The hypotenuse of a right triangle will measure 10 feet if one leg is 6 feet and the other is 8 feet. Therefore you can determine if a corner is square by marking points 6 feet from the corner on one side and 8 feet from the corner on the other side, then seeing if the distance between the points is exactly 10 feet. Just doing this with tape is risky, however; accurate corners are easier if you take the time to make a check gauge as shown in **Figure 4-5**. The gauge can be used for other purposes also, such as to check plumbness of vertical constructions.

As a further check for squareness, measure the diagonals of rectangles and squares. If the corners of a square or rectangle are 90 degrees, the diagonals will be the same length.

FIG. 4-5

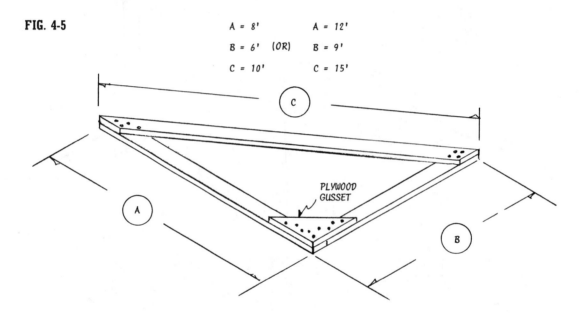

THOUGHTS ON EXCAVATING

When little or no grading is required, batter boards can be set up and lines drawn immediately after topsoil has been scraped off. When the site is a slope, or the terrain is irregular, rough grading can be done before perimeter lines are organized. After batter boards are established, temporary stakes can be driven to mark edges of excavations, and the building lines can be removed while work goes on.

If there is to be a basement, excavate at least two feet beyond the building lines so there will be plenty of room for the construction of forms for concrete, or for the laying of block. This extra digging isn't necessary if the house is to be on a slab or over a crawl space. Local codes will tell how deep to dig for foundations or footings so that bottom areas will be below the frost line. This is important to avoid the damage that can occur should water in the soil *under* a foundation freeze. Frost pressures can push walls upward and cause various types of damage that may be impossible to repair.

Excavation depths, which, of course, control foundation heights, are usually established by using the highest point of the building-site perimeter as a check point; all depth adjustments are made from there. All foundations, whether for basements, slabs, or crawl spaces, should extend above the final grade enough so that wood members of the house are some distance from the soil. This distance, too, may be regulated by local codes.

Always control excavation depths so that footings and foundations can rest on undisturbed soil. Having to fill or to bring in gravel to compensate for uneven digging can waste time and money, and may result in uneven settlement.

REINFORCEMENT FOR CONCRETE-MASONRY

The most common materials used to strengthen concrete and masonry constructions are steel bars, shown in a block wall in **Figure 4-6**, and welded wire fabric, shown in **Figure 4-7**. Both of the materials should be clean, free of scale and rust and any foreign matter. The steel bars come in diameters up to ¾ inch, but ⅜-

FIG. 4-6

FIG. 4-7

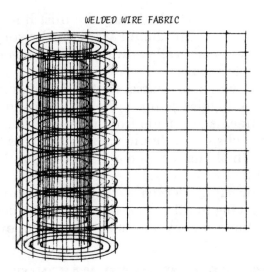

WELDED WIRE FABRIC

inch or ½-inch sizes are used for most home-construction projects. Use one-piece bars whenever possible; when the length or the design of the project doesn't permit it, overlap the bars and bind them with wire as shown in **Figure 4-8**. The diameter of the bar has a bearing on the length of the overlap. Bars ¼ inch in diameter should be overlapped 1 foot, bars ⅜ inches in diameter should be overlapped 18 inches, and so on, increasing overlap by 6 inches for every ⅛-inch increase in diameter. It pays to use more than two ties for any overlap more than 12 inches. Try to plan placement of bars for continuous runs around corners. If you can't, lap the bars at one or both ends as shown in **Figure 4-9**. Be generous with both the lap and the ties. Always use ties wherever bars cross (**Figure 4-10**).

FIG. 4-8

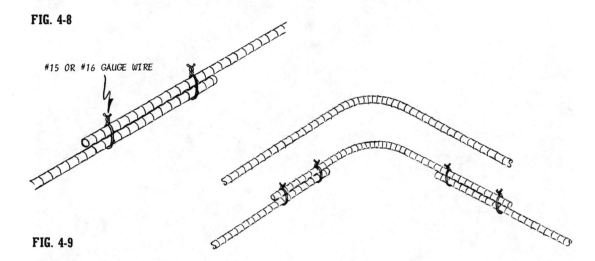

#15 OR #16 GAUGE WIRE

FIG. 4-9

FIG. 4-10

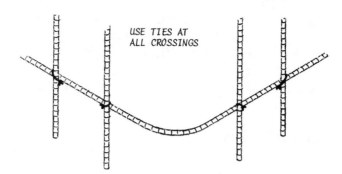

USE TIES AT
ALL CROSSINGS

The bars can be bent rather easily even though they are rigid and strong. Use a hacksaw to cut them to length by making a cut about halfway through the bar and then bending sharply at the cut. Place bars so they will be midway in the pour. For example, if they are used in a slab, elevate them on blobs of concrete or pieces of stone so the concrete can flow under them. Small pieces of wood are often used for the purpose but such materials should be removed as the pour progresses.

The welded wire fabric (or mesh) is used most often in slab work required for basement floors, full slab floors, patios, driveways, and the like. Overlap joints from 6 to 8 inches, or at least the width of one of the openings, when you can't do the job with one whole piece as shown in **Figure 4-11**. The material tends to curl because it comes in rolls, but you can get rid of the curl by spreading it over a flat area and then walking on it. Use wire snips for cutting, but be careful; cut strands are apt to snap back against your body or face.

FIG. 4-11

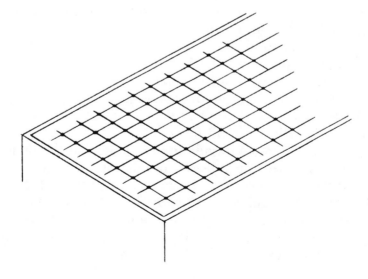

Like the steel bars, wire fabric should be placed so it will be midway in a pour. This can be accomplished by using stones or blobs of concrete as elevators, but many professionals place the fabric directly on the ground and then raise it with a rake or a wire hook as the concrete is being poured. Be sure that you don't bring the fabric closer than one inch to the surface of the slab.

PLASTIC UNDERLAYMENT

Plastic sheeting that comes in 4-mil or 6-mil thicknesses and in various widths and lengths forms an effective moisture barrier when placed over a subbase before concrete is poured. It has other uses as well, as we will discuss later, but a common application is under slabs used as basement or house floors.

The material is easy to cut and place and cheap enough so neglecting to use it is unjustified when you consider how it helps with moisture control and insulating. In most situations you should be careful during and after application not to pierce the plastic. Be very generous with overlaps and at perimeters. On a slab, for example, curl the sheet up and over forms. The excess is easy to cut away after the pour is complete.

THICKNESS OF FOUNDATIONS

Foundations do more than support the building. They protect against frost, guard against termites, may be designed to provide a basement, and can be constructed to resist shocks from earthquakes. Don't skimp on them just because they usually can't be seen once the building is complete and they stand between you and the fun part of the job. Local building codes often specify minimum standards, and even if they don't, if you're haphazard about this important phase of construction, which can be the case even when the job can pass a visual inspection, just be aware that the rewards can be a house that tilts, floors that sag, foundations that crack, doors that won't open, and windows that jam. The good foundation is not difficult to plan or to build, and you can get expert advice about specifications merely by going along with the building codes.

When all parts of a concrete foundation are placed at the same time—like a one-piece casting—it is called an integral foundation. This is desirable in some areas because it is resistant to earthquake damage. **Figure 4-12** shows an integral foundation. However, even though the design is a one-piece unit, various parts should be viewed in relation to the jobs they must do. The main foundation, the part on which the house itself actually sits, for example, will require thicker and probably deeper walls and special footings. Other walls, like retaining walls

FIG. 4-12

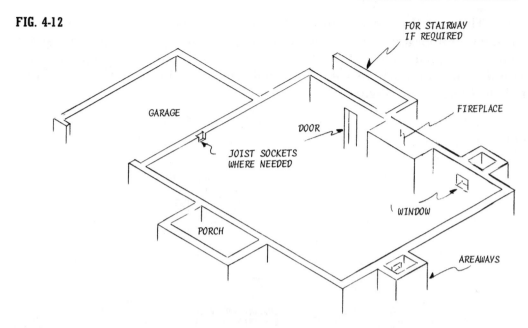

FOR STAIRWAY
IF REQUIRED

GARAGE

FIREPLACE

DOOR

JOIST SOCKETS
WHERE NEEDED

WINDOW

PORCH

AREAWAYS

needed for an outside stairway such as shown in **Figure 4-13** and walls for a garage can be thinner and may not even require a footing. Walls so designed are called *trench walls* and may be okay in your area to support light loads.

Areaways, such as the one shown in cross-section in **Figure 4-14**, are often designed into a house foundation to provide light and ventilation for basement

FIG. 4-13   **FIG. 4-14**

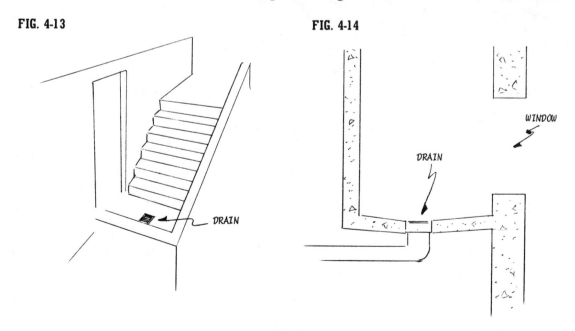

DRAIN

WINDOW

DRAIN

windows. They are thinner in cross-section than load-bearing walls. Note that a sloped concrete floor and a drain are included in the design; these do not have to be part of the original pour. Actually, such features are often omitted and the light and ventilation problems are solved through good grading and the use of corrugated steel backstops (**Figure 4-15**).

FIG. 4-15

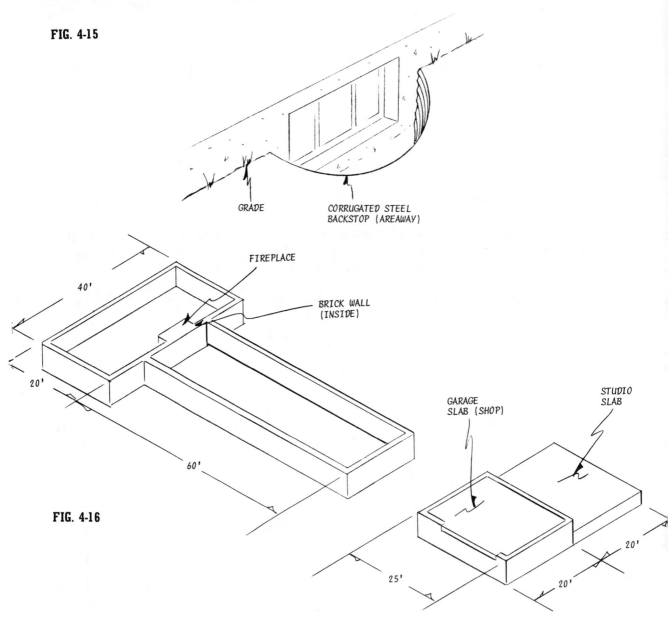

GRADE

CORRUGATED STEEL
BACKSTOP (AREAWAY)

FIREPLACE

40'

BRICK WALL
(INSIDE)

20'

GARAGE
SLAB (SHOP)

STUDIO
SLAB

60'

FIG. 4-16

25'

20'

20'

Not all foundations can be so neatly integrated. In my own situation, shown in **Figure 4-16**, the garage and work areas are detached from the house so that we could have a breezeway leading to a side entry and to a back patio. But wall height and thickness and footing considerations apply regardless.

On some slope situations, and even fairly level grades if the idea conforms to a split-level design, the foundation itself can be stepped, as shown in **Figure 4-17**, and then filled in with wood materials.

FIG. 4-17

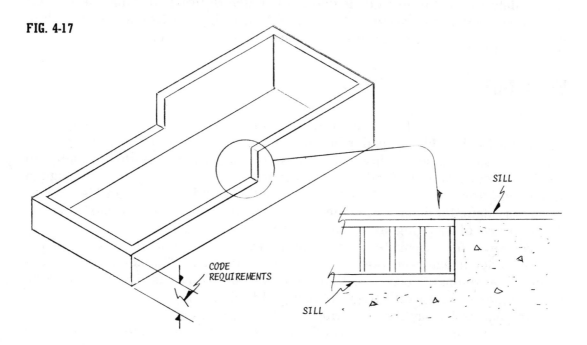

Overall, good procedure calls for a detailed look at all of the foundation requirements. Even if the design doesn't permit a *one-piece casting*, form work should be done to permit a *one-time pour*.

If you are building in or near an area with established building codes, you will have no problems getting correct specifications for a good foundation. If not, here are some guide rules. But do remember these are guides that may require some further research, such as checking successful structures in the same area, and that all foundations should provide for a depth that is below the frost line.

Foundations for a wood-frame house that does not include a basement should be 6 to 8 inches thick. With a basement, the minimum is 8 inches but should be increased to 10 inches if the walls are longer than 20 feet. If house walls are masonry, minimum thickness of foundations should match the thickness of what they support.

A two-story wood-frame house with a basement calls for 10-inch-thick foundations that should be increased to 12 inches thick if the walls are more than 7 feet below grade. Minimum foundation thickness for two-story masonry walls should match the wall thickness, as long as the foundation depth is not more than 7 feet below grade. If more, increase foundation wall thickness to 12 inches.

Foundations only 5 inches thick will usually do for a porch. Walls for areaways and exterior below-grade stairs usually must act as retaining walls, so some safety factor should be considered when determining wall thickness. A bit too much is better than too little; 6 inches will probably work for areaways, while 8 inches will be better for stair walls as long as they are not more than 10 feet long.

FOOTINGS

Footings carry the weight of the building and are always wider than the foundation so the load will be spread over a broader area. Soil conditions affect footing size and design, so codes should be checked before work is started. Usually, in residential construction, loads will be carried safely if the footing is designed along these lines. Notice that the thickness of the foundation wall in **Figure 4-18** is the base factor for determining the thickness and width of the footing. However, the thickness of a footing should never be less than 8 inches.

The key shape is made by pressing a beveled 2×4 into the surface of the concrete after it has been poured and is being leveled. This key is not always included in commercial constructions, but its value is obvious. The safety factor a key provides is well worth the time required to make it.

National codes say that a footing should be placed at least 12 inches below the frost line. This can vary anywhere from on grade to as much as 5 feet below grade depending on the locality, so again, a check of local codes is in order.

Reinforcement rods are required in earthquake areas and wherever footings are placed on poor load-bearing soil, and they are recommended no matter where you build. Two 5/8-inch bars are normal practice in footings up to 12×24 inches, but this can vary; 1/2-inch bars may be acceptable, or three bars instead of two may be called for. The bars should be placed so they will be covered by at least 3 inches of concrete at all points. This is easiest to do if you use "foundation chairs," shown in **Figure 4-19**, which are made specifically for the purpose. The chairs may be spaced any distance as long as the bars are reasonably level.

At this point it is wise to consider the footings you may require for interior posts and bearing and nonbearing walls so that forming and concrete-pouring for them can be done along with the perimeter footing. It would seem reasonable to assume that the specifications of the perimeter footings can be guides for other

FIG. 4-18

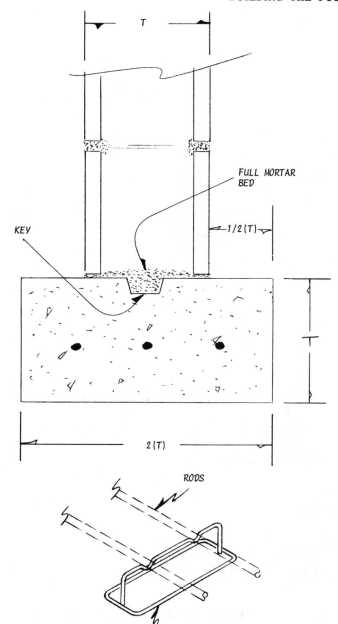

T

FULL MORTAR
BED

KEY

1/2 (T)

2 (T)

FIG. 4-19

RODS

CHAIR

types, but the soil conditions and the structural design of the house must be considered. Bearing posts, since they carry concentrated loads, and bearing walls will
require huskier footings than non-load-bearing walls.

FIG. 4-20

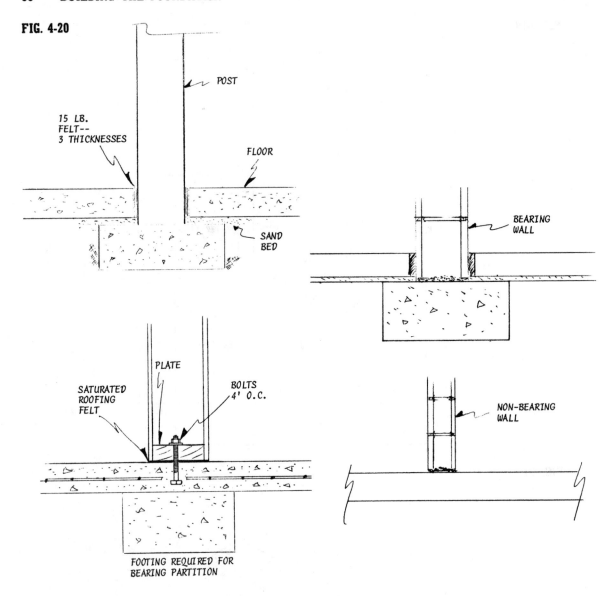

15 LB. FELT-- 3 THICKNESSES

POST

FLOOR

SAND BED

BEARING WALL

PLATE

SATURATED ROOFING FELT

BOLTS 4' O.C.

NON-BEARING WALL

FOOTING REQUIRED FOR BEARING PARTITION

The designs shown in **Figure 4-20** illustrate typical construction procedures, but the thickness and depth of the footings should be determined after consulting with the building inspector.

Such considerations apply to houses built on slabs and those with basements. In slab work, special footings are often provided for as shown in **Figure 4-21**. A square excavation in the ground is the form; reinforcement bars are included to supply strength to the single-pour casting.

FIG. 4-21

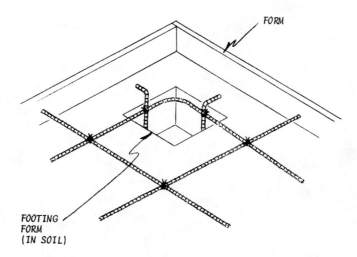

FORM

FOOTING
FORM
(IN SOIL)

Perimeter footings may also be designed as integral parts of the slab. This is the design I used when I poured a foundation for a studio addition on the back of the garage (**Figure 4-22**).

FIG. 4-22

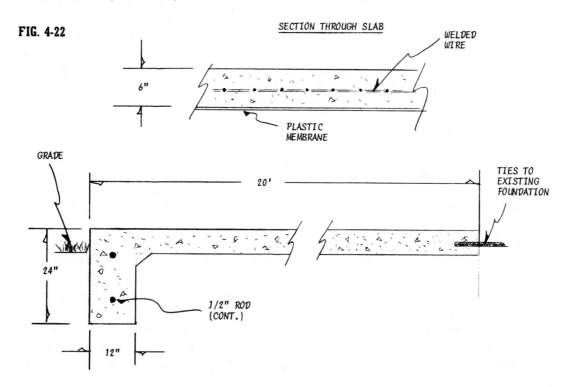

SECTION THROUGH SLAB

WELDED
WIRE

6"

PLASTIC
MEMBRANE

GRADE

20'

TIES TO
EXISTING
FOUNDATION

24"

1/2" ROD
(CONT.)

12"

FIG. 4-23

Houses built over a crawl space require supports between perimeter walls, but this is usually accomplished with precast piers that are set on footings a bit larger than the base area of the piers and from 6 to 8 inches thick (**Figure 4-23**). The supports are spaced in relation to the size of the beams that will support the floor joists. More about them later.

FOOTING FORMS

Establish corners by dropping a plumb bob from the intersection of the lines stretched between batter boards (**Figure 4-24**). Drive one corner stake to establish the height of the footing and then work with a builder's level or a line level to set the height of others. Stretch lines from nails in the stakes so you will know where the outside of the forms must be.

Either 1-inch or 2-inch boards can be used for the forms, the difference being that the thinner material will require stakes about 2 feet apart while the heavier material can be staked at 4-foot intervals. Use 2×2 or 2×4 pieces of wood, sharpened at one end, as stakes. Attach form boards by nailing from the outside of the stakes, using regular common or box nails as long as you don't drive them completely, or double-headed (duplex) nails. The latter, shown in **Figure 4-25**, has one head that you drive to, and a second head that remains exposed so nails are

FIG. 4-24

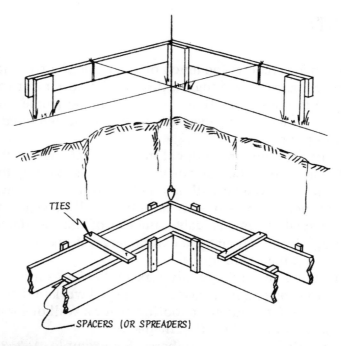

TIES

SPACERS (OR SPREADERS)

FIG. 4-25

DUPLEX HEAD

easy to pull when forms are removed. In all situations, remember that forms do have to be removed, and nails accordingly.

A good procedure is to set up outside formboards first, and then work with a carpenter's level to set the height of inside boards. Use precut spreaders between boards to maintain width and so you won't have to measure constantly; use ties across the top for strength during the pour. The spreaders are removed as the concrete is placed.

THE CONCRETE

House-construction jobs require too much of the material for you to even consider mixing it in a wheelbarrow or trough unless you're so far out there is no other recourse. A step in the right direction is to rent a mixer (some can be run with a gasoline engine), but the wisest procedure is to have it delivered to the site in a big truck, ready to pour.

Many amateurs fear this professional method; they picture the mass of fluid stone as a monster and think the truck driver just wants to dump the load and run. Actually the driver's knowledge will be a help and he will be aware that placement takes some time. Some ready-mix suppliers will make a surcharge if truck and driver are delayed an unreasonable amount of time in relation to the load they are delivering. But just figure the extra charge, if necessary, is worth it if it eliminates a hand-mixing operation.

Another advantage of buying ready-mix is that the supplier will know, as well as anyone, the correct mixes for various types of work in that area. You might just get by with asking for a mix for a footing or for a wall or for a patio or whatever. And there is always the building inspector.

Mixes are specified by three numbers—for example, 1–2–3, which means one part of Portland cement to two parts of sand to three parts of gravel. A common mix for footings, foundations, and walls is 1–3–4. A mix of 1–2–2¼ is better for projects that will be exposed to extremes in wear and weather, while a 1–2¼–3 mix is generally okay for walks, patios, and floors.

Your pour will be successful if you are sure you are ready for the material before you ask for delivery. Check all formwork and plan truck positions so the concrete can be poured directly into forms or, at least, with minimum transportation required. Have help and extra wheelbarrows on hand if they are needed. Be sure shovels and rakes are handy so the concrete can be spread and tamped to fill forms solidly. Lengths of 2×4s or 2×2s can also be used to tamp but, regardless of the tool used, don't overdo; you may move the gravel enough to destroy equal distribution. Tapping the outside of the forms with a hammer is another technique that helps settle the concrete and produce a smoother surface. When the forms are full, place a 2×4 across the top edges and move it to and fro to even the concrete. Keep a trowel handy so you can fill depressions and remove any excess.

Concrete must be allowed to cure before it is subjected to any stress even though you can remove forms after about two days. Curing means no more than keeping the concrete moist or taking steps to slow up evaporation. A frequent wetting with a very fine spray from a garden hose will do the job, and the applications can be minimized if you cover the project with a moisture-absorbing material such as burlap. Newspapers are often used in similar fashion, and if you cover the concrete with special kraft curing papers or even plastic sheeting, you may not have to use water at all. There are also special commercial curing compounds that are sprayed on the concrete surface immediately after finishing to form an antievaporation shield.

Concrete work should not be done in cold weather without special precautions. It should never be dumped on frozen ground, and if temperatures fall below 50 degrees you'll have to work with mixes heated to between 50 and 70 degrees, since that temperature range is where concrete will harden correctly. Also, the temperature must be maintained for a period of time. You can see that it is wise to plan concrete work during temperate weather.

If the thought of doing concrete work bothers you, just remember it is one phase of house construction you might wish to pass on to professionals. You can still do the formwork, planning it and executing it exactly the way you want. A lot of more detailed information concerning concrete and masonry work is available from special associations. These will be listed later on.

FOUNDATION FORMWORK

Forms for walls must be strong and braced enough to resist the tremendous pressures created when the concrete is poured. The pressures increase along with the height of the wall, so the higher the wall, the stronger the forms must be. Most forms for residential foundations can be done by using 1-inch boards or $5/8$-inch or $3/4$-inch plywood as sheathing, and 2×4 materials as studs. Since a considerable amount of money can go into the materials used to build forms, it's wise to think beyond the primary use and plan to use the material, after forms are broken down, as house sheathing, subflooring, joist bridging, and so on.

You don't have to be a master cabinetmaker to construct good forms, but they must be tight and smooth and correctly aligned. Joints between sheathing pieces must be tight enough to prevent cement paste from leaking through. Interior surfaces must not have projections that might lock pieces of the form to the concrete.

Forms for low walls, up to about 3 feet high, can be done with boards or plywood, braced with 2×4 studs about every 2 feet (**Figure 4-26**). You can use this same design to go a bit higher, but more strength should be provided by using a closer stud spacing. Forms that are over 4 feet high should be reinforced further with wales, which are staked 2×4s or 4×4s placed longitudinally along the base of the studs (**Figure 4-27**). Note that this particular form was constructed in a subgrade trench. In such situations, minimum trench width must provide room for work to go on.

Foundation walls that are above grade are often cast integrally with a footing that is contained by a trench formed in the earth, or forms are built over a footing

FIG. 4-26

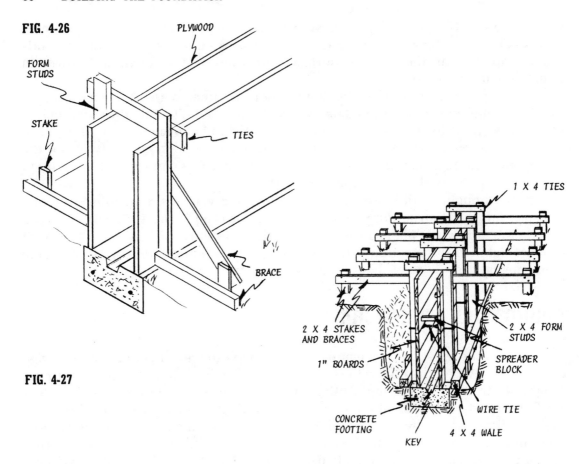

FIG. 4-27

that has been so cast (**Figure 4-28**). Note that the bottom of the trench has been undercut to provide an adequate base which simulates the form of a separate footing. This kind of thing is not advisable everywhere, so check local procedures before doing it. If the design is okay, be sure you cut trench walls straight and true and that the bottom of the trench is undisturbed soil.

In all situations, wire ties and spreader blocks as shown in **Figure 4-29** are used in addition to the ties across the top of the forms. The wire keeps the two sides of the form from spreading apart; the spreaders are gauges that maintain a uniform wall thickness. Pass the wire through holes that you drill on each side of the studs and then twist with a screwdriver or something similar until tight. Best bet is to use a 1×2 spreader near each tie so you won't pull the form sides out of alignment because of excessive tightening.

The wire ties stay put until the forms are removed. Then, they are cut flush with the surface of the concrete. The spreaders are removed as the concrete is poured.

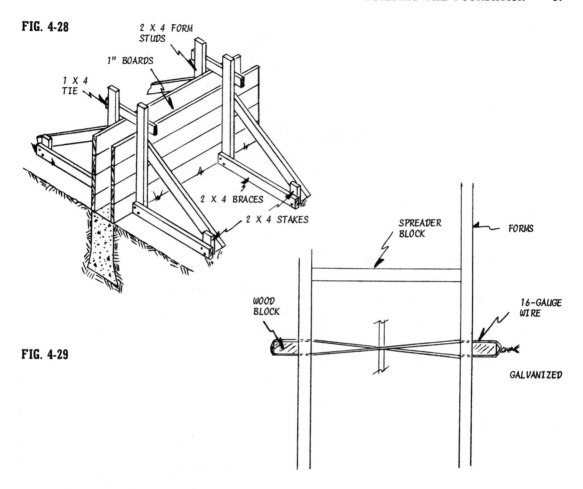

FIG. 4-28

2 X 4 FORM STUDS

1" BOARDS

1 X 4 TIE

2 X 4 BRACES

2 X 4 STAKES

FIG. 4-29

SPREADER BLOCK

FORMS

WOOD BLOCK

16-GAUGE WIRE

GALVANIZED

READY-MADE FORMS

Professionals work with reusable forms that are flexible enough for various types of constructions. Such products are usually available from concrete and masonry supply yards or other establishments on a rental basis. It pays to check out availability, since the use of the forms can save time and may result in a higher-quality foundation. There have been amateur housebuilders who have purchased ready-to-assemble forms, or the hardware required for them, and then sold them after personal use. I know of one who retained the equipment but rented it out to other builders and soon got more than his investment back.

The units that will be available in your area will probably conform to established foundation requirements. Forms needed for houses over crawl spaces will be one design while those for basements or high foundations will be another.

FIG. 4-30

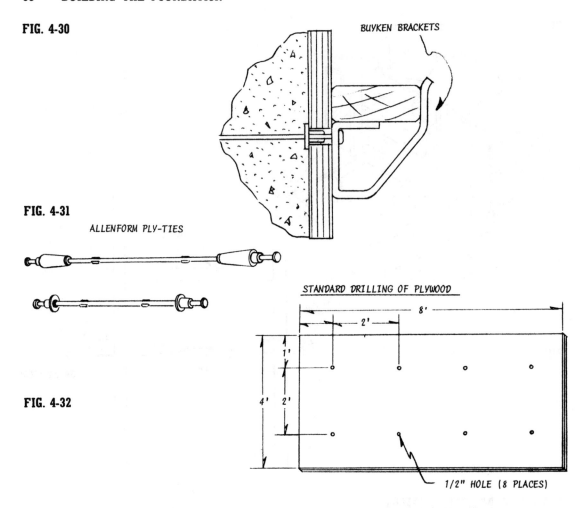

BUYKEN BRACKETS

FIG. 4-31

ALLENFORM PLY-TIES

STANDARD DRILLING OF PLYWOOD

FIG. 4-32

1/2" HOLE (8 PLACES)

Some will be reinforced panels that you bolt together, others may be pieces of hardware that you use with sheets of plywood and 2×4s. Typical of the latter is the Conform system, which utilizes Buyken brackets, shown in **Figure 4-30**, and Allenform Ply-ties, shown in **Figure 4-31**, together with standard sheets of plywood. The plywood is drilled as shown in **Figure 4-32**. You can drill several sheets at a time if you stack them. The holes are ½ inch in diameter, drilled on 2-foot centers but with edge distance of 1 foot. The standard drilling method produces eight holes in each sheet of plywood. Shown in **Figure 4-33** are typical arrangements for a low and a high wall. Corners are turned as shown in **Figure 4-34**. Special considerations such as a ledge for brick veneer can be included by using Allenform Ply-ties designed for the purpose (**Figure 4-35**).

FIG 4-33

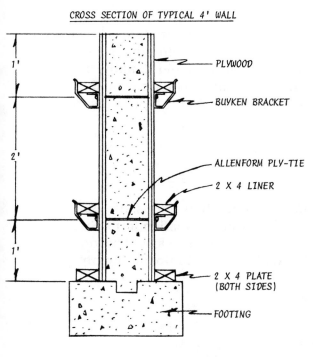

CROSS SECTION OF TYPICAL 4' WALL

1'
2'
1'

PLYWOOD

BUYKEN BRACKET

ALLENFORM PLY-TIE

2 X 4 LINER

2 X 4 PLATE
(BOTH SIDES)

FOOTING

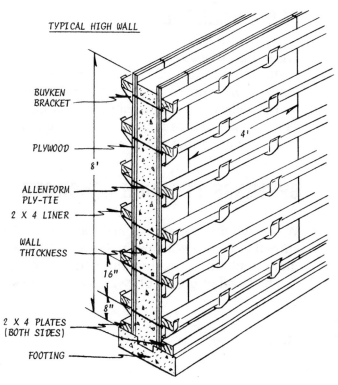

TYPICAL HIGH WALL

BUYKEN
BRACKET

PLYWOOD

ALLENFORM
PLY-TIE

2 X 4 LINER

WALL
THICKNESS

2 X 4 PLATES
(BOTH SIDES)

FOOTING

8'

4'

16"

8"

FIG. 4-34

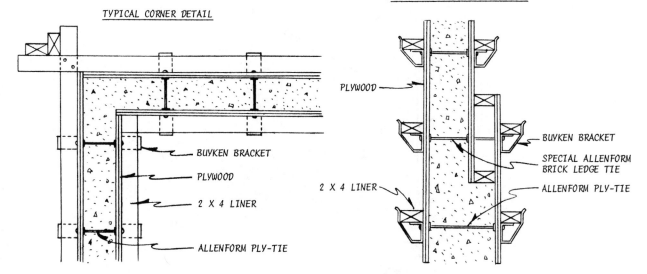

TYPICAL CORNER DETAIL

BUYKEN BRACKET

PLYWOOD

2 X 4 LINER

ALLENFORM PLY-TIE

FIG. 4-35

TYPICAL BRICK LEDGE DETAIL

PLYWOOD

2 X 4 LINER

BUYKEN BRACKET

SPECIAL ALLENFORM
BRICK LEDGE TIE

ALLENFORM PLY-TIE

FORM RELEASE

Wood forms can be separated from set concrete more easily and with less chance of damage to the concrete if they are treated with a "release" before pouring is done. A thin coat of engine oil applied with a brush or rags is an old standby, but more modern materials that are less messy and more convenient are available. These form releases may be purchased in bulk and applied with the type of insecticide sprayers that are used in gardens.

OPENINGS THROUGH CONCRETE

When window openings are required through the foundation, forms like that shown in **Figure 4-36** are installed between the walls of the foundation form. Size them to fit snugly against the main forms to prevent cement-paste loss. The keys can be beveled pieces of 2×2s, 2×3s, or 2×4s that will be locked into the concrete

FIG. 4-36

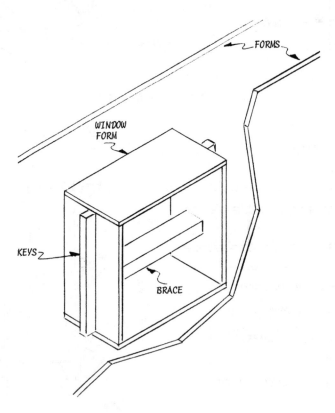

FIG. 4-37

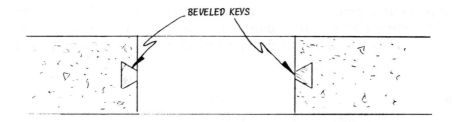

BEVELED KEYS

and be used as nailing strips when window frames are installed (**Figure 4-37**). Door openings are done the same way except that form pieces are permanently attached to each other and actually used as jambs later on (**Figure 4-38**). Use good material and be very accurate when you make the frame. Work with a level and a square when you install it in the form.

FIG. 4-38

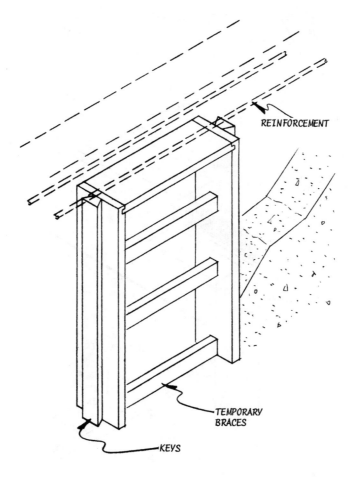

REINFORCEMENT

TEMPORARY
BRACES

KEYS

Keyed nailing strips may be utilized in other ways. The cast window sill shown in **Figure 4-39** includes one so that wood members may be added later. Note forms and braces used for the special sill pour, and the nail anchors which help hold the key securely.

FIG. 4-39

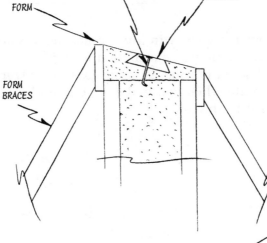

FIG. 4-40

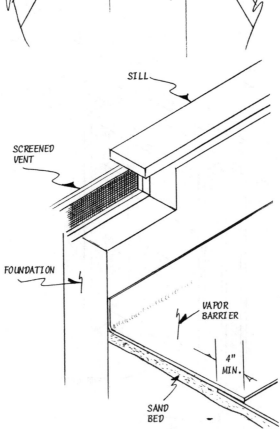

Ventilation openings through crawl-space walls are also done with pre-assembled forms. The form is U-shaped and sized to fit the screened vent that will be used (**Figure 4-40**).

Pockets for beams that must span across foundation walls can be formed as shown in **Figure 4-41,** or simply with blocks of Styrofoam (**Figure 4-42**).

FIG. 4-41

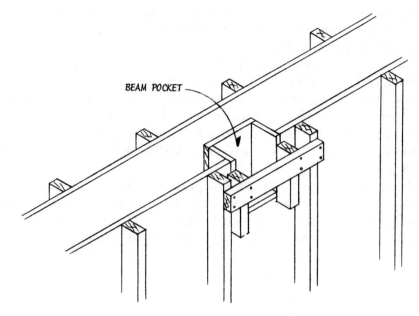

BEAM POCKET

FIG. 4-42

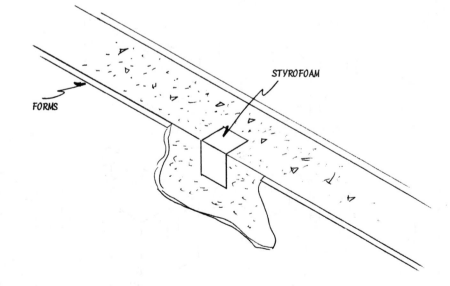

STYROFOAM

FORMS

WALLS OF CONCRETE BLOCK

Concrete blocks are available in various shapes and sizes that are especially suitable for foundation work and general house construction (**Figure 4-43**). They seem more appealing than poured concrete to the amateur builder. Two good reasons are that they eliminate form construction and permit a stop-and-go procedure. You can't quit in the middle of a foundation when pouring concrete unless you have designed the formwork accordingly. With block you can quit anytime and take up again tomorrow.

All blocks are cast masonry units, but they may contain different types of aggregate. Sand and gravel or crushed stone are common ones, and "concrete block", at one time, designated those materials specifically. Today, the term includes all units whether they are made with slag, cinders, shale, or whatever. Much of the weight of a block is in the aggregate. An 8×8×16 gravel block can weigh 50 pounds, but a similar-size unit made with cinders or pumice will weigh only 25 or 35 pounds.

Both lightweight and heavyweight units are acceptable for many types of masonry construction, so your selection might be influenced by, in addition to weight, such things as texture, insulation factors, availability of types or shapes in your particular area, and—as always—how you must work to abide with local codes.

FIG. 4-43

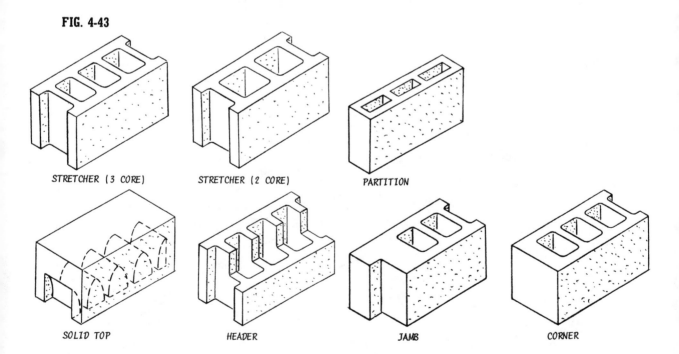

STRETCHER (3 CORE)   STRETCHER (2 CORE)   PARTITION

SOLID TOP   HEADER   JAMB   CORNER

Generally, lightweight units do a better insulating job, but the hollow cores of all units may be filled with a granular insulating material. Both types do a better job of absorbing sound than a smooth, dense wall of concrete, but in this area too, the lightweights are more efficient than the heavyweights.

Check out all special units and shapes that are available or that can be delivered to your site from outside the area. You may be able to incorporate some to facilitate a procedure or to end a job in a professional way. Specials include such units as "headers" that form a ledge as shown in **Figure 4-44**, "caps" that finish off the top of a foundation or any wall, and "corner" blocks with rounded edges for neat endings as shown in **Figure 4-45**.

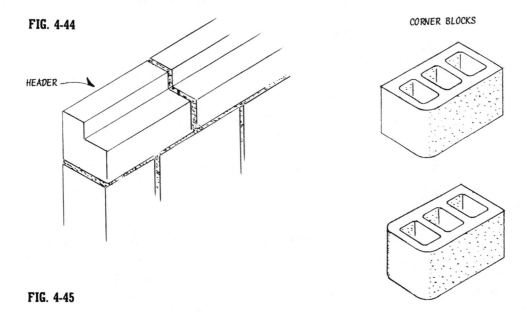

FIG. 4-44

HEADER

CORNER BLOCKS

FIG. 4-45

TYPICAL SETUP

The organization of a typical block wall is shown in **Figure 4-46**. The installation of concrete drain tiles is often omitted in areas with very dry climates and where adequate subsoil drainage is guaranteed, but this seems unwise, especially when the wall is for a basement. When tiles are used they should lead to a suitable drainage point or outlet. Cover the joints between them with pieces of building felt and then add at least 12 inches of coarse gravel or crushed stone.

A parge coat is always applied to the earth side of basement walls. This is a covering done with a 1–2½ mix of mortar applied in two coats to build up a total thickness of about ½ inch. To apply the parge coat, first moisten the wall with a

FIG. 4-46

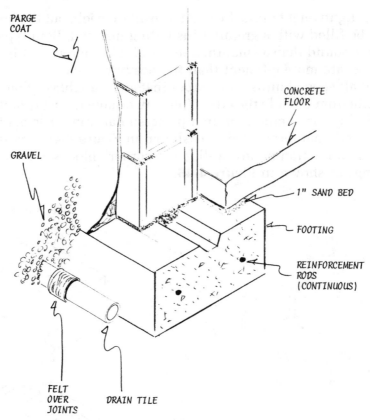

PARGE COAT

CONCRETE FLOOR

GRAVEL

1" SAND BED

FOOTING

REINFORCEMENT RODS (CONTINUOUS)

FELT OVER JOINTS

DRAIN TILE

fine spray of water and trowel on a first, ¼-inch-thick application of plaster. Allow this to partially harden and then rough up the surface a bit so you will have a good bond for the second coat. Professionals do the roughing up with a special "scratch" tool, but you can accomplish something similar by making random sweeps across the surface with a stiff-bristled broom. Keep the first coating damp but allow it to set up for about 24 hours before repeating the plaster application.

Extend the parging to at least 6 inches above grade. Form a generous cove at the base of the wall so water can't collect near the wall-to-footing joint. Keep the second coating moist for at least 24 hours after application. The parge coat can be a single ½-inch-thick application of plaster, but in that case, keep the application moist for at least 48 hours.

MORTAR FOR BLOCK WORK

Good, correctly mixed mortar is necessary if masonry units are to bond as a strong, well-knit wall. Proportions of materials in mortars vary depending on the job to be done. Severe stresses and frost actions call for stronger and more durable mortars than you need for walls exposed to ordinary service. The table in **Figure 4-47** makes recommendations but should be checked out for your area before being used.

FIG. 4-47.   MORTAR MIXES FOR CONCRETE BLOCK (BY VOLUME)

WORK	CEMENT	HYDRATED LIME	MORTAR SAND
Average projects	1 part masonry cement	none	2-3 parts
	1 part Portland cement	1 to 1¼ parts	4-6 parts
Heavy duty work— severe frosts, earthquake area, strong winds	1 part masonry cement plus 1 part Portland cement	none	4-6 parts
	1 part Portland cement	0-¼ part	2-3 parts

When mixing, blend all materials together in dry state before adding water. Use enough *drinkable* water, gradually, to bring the mix to a plastic, workable state. A good mix will hold together but will spread easily. Stroke across its surface with a trowel to see if you get a smooth finish, which is one sign of a correct mix.

Don't mix more than you can use in about two hours. If the mortar stiffens because of evaporation you can restore its workability by thoroughly remixing and adding a minimum amount of water. Mortar that becomes stiff through hydration—the setting action—should be thrown out. It's not easy to tell the difference between hydration and evaporation, so your best bet is to set a reasonable time limit in relation to how fast you work. You can always mix new batches of mortar as they are needed.

All the precautions suggested for cold-weather work with concrete apply to block work as well. For safety's sake, it's best to do the job during temperate weather.

MORTAR JOINTS

The joints between blocks should be weathertight and neat. There is no need for anything but a flush joint, as shown in **Figure 4-48**, when the wall will be backfilled. This is done by cutting off excess mortar with the edge of a trowel and then running the point of the trowel along the line. The idea is to compact the mortar and force it tightly against the masonry.

Joints like the V, concave, raked and extruded, shown in **Figure 4-49**, are often used in above-grade walls but for appearance only. The V and the concave

FIG. 4-48

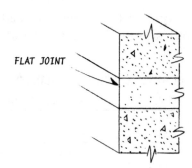

FLAT JOINT

FIG. 4-49

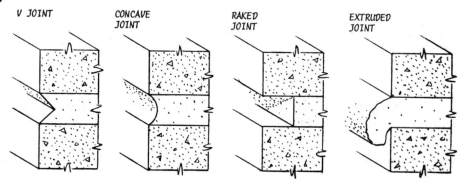

are the most popular since they look neat and do not form a ledge that might accumulate water. The V joint can be done by working with a length of $\frac{1}{2}$-inch-square bar stock, and a concave joint by working with a piece of $\frac{1}{2}$-inch or $\frac{5}{8}$-inch dowel stock or tubing.

Tool the joints when the mortar has become hard enough to retain a thumbprint. Do not attempt to move a block after the mortar has stiffened even partially. This will break the mortar bond and may create a water-entry point.

LAYING BLOCK

Blocks are heavy, so stack them conveniently about the site to avoid excessive carrying. Handle them carefully to avoid damage to them and to yourself. Keep them covered, and use them dry and dust-free. If you have worked with brick you know the units must be wet when placed. This *does not* apply to block.

The thicker face shell of the block should always face up so it will provide more bedding area for the mortar. A full mortar bed will cover all the web areas while a partial one (face-shell mortaring) will not (**Figure 4-50**). For most work, a full coating is used on the first course and a partial coating thereafter, but local

FIG. 4-50

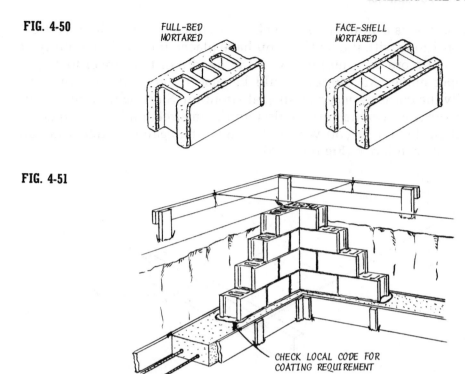

FULL-BED MORTARED

FACE-SHELL MORTARED

FIG. 4-51

CHECK LOCAL CODE FOR COATING REQUIREMENT

codes may say differently, so check. The procedures described here apply whether you are constructing on a footing, as in **Figure 4-51**, or on a slab.

Do a dry run first by placing the first course without mortar. Check for alignment and spacing so you won't have to make corrections after mortaring is started. Some adjustments can be made, design permitting, if they eliminate having to cut blocks to fill out a line. Now you can snap a chalk line to mark the footing or trace around the blocks with a crayon—even mark the position of each block so it will be easier to place them accurately when you do the mortaring.

Start at a corner with a full bed of mortar for two or three blocks. Furrow the mix with a trowel to be sure of getting plenty of mortar along the bottom edges of the face shells. Place the corner block first and then add one or two others, after you have applied mortar to the ends of the face shells (**Figure 4-52**). Use a level or

FIG. 4-52

a straight piece of 2×4 as a straight edge to check alignment. Check both vertical and horizontal positions with the level. If you have placed the blocks carefully, it won't require much more than tapping with the handle of the trowel to nudge blocks into proper grade. All blocks should be placed carefully, but the first course, especially the corners, deserves special attention. Starting right makes all that follows much easier. Do the corners first and then stretch a line as an additional alignment guide and so you will have something against which you can check the height of each block (**Figure 4-53**).

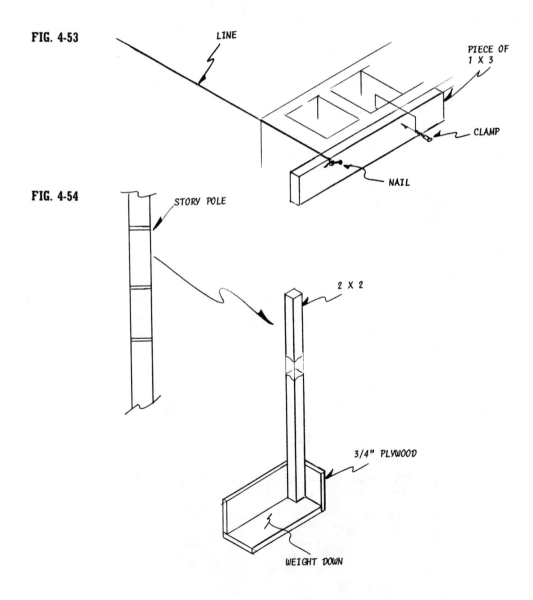

FIG. 4-53

LINE

PIECE OF 1 X 3

CLAMP

NAIL

FIG. 4-54

STORY POLE

2 X 2

3/4" PLYWOOD

WEIGHT DOWN

Build up the corners after the first course is complete. A story pole—often called a course pole—can be an accuracy aid for establishing masonry heights in each course. Such items can be rented, but one you make yourself as shown in **Figure 4-54** will do as well. Mark the vertical piece to designate masonry units and joints. This can be done with a pencil, or you can use different-width adhesive tapes, even tapes of different colors to indicate special places like door or window openings.

Build up the corners three or four or even five courses high, but work constantly with a level, checking vertically, horizontally, and diagonally on each block and on the assembly as it progresses (**Figure 4-55**).

FIG. 4-55

With corners up, you can fill in the wall between, working each course to a stretched line that you can hold in place as shown in **Figure 4-56**. If the wall is long and the line might sag, set up an intermediate block as shown in **Figure 4-57**. Hold down the flashing with another block.

Place each unit carefully to minimize jiggling that must be done to achieve alignment. If you work from the back side of the wall and tip the block a bit toward you, you'll be able to see the edge of the course below, and this should

FIG. 4-56

FIG. 4-57

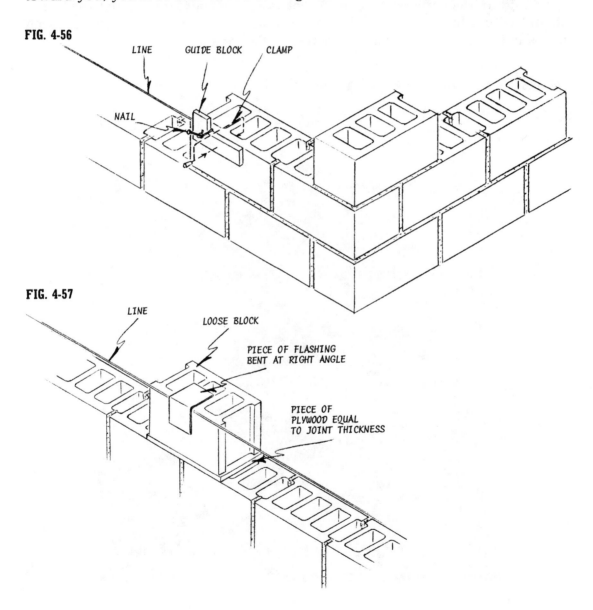

help to do pretty accurate placement right off. The action should combine a slight roll to vertical position with a shove that puts the block solidly against the adjacent one. If you do this right, very little tapping with the trowel handle will be required to finish the placement (**Figure 4-58**). Be sure that you make any block adjustment while the mortar is soft and plastic. If you move things after the mortar has stiffened, you'll break the bond, and that will cause problems later.

FIG. 4-58

FIG. 4-59

The closure block, the last unit in a course, gets special attention. Apply generous amounts of mortar to all four vertical edges of the block and to all edges of the opening (**Figure 4-59**). Lower the block carefully into position to be sure no mortar falls out. If you do lose mortar, which may cause an open joint, take the block out and repeat the operation after applying fresh mortar.

FIG. 4-60

Tall block walls that are speedily erected and quickly backfilled may require bracing for temporary support (**Figure 4-60**). The idea is to be sure the structure is knit solidly before stresses are applied.

The tops of foundation walls should be capped with a course of solid masonry units which will act as a termite barrier and also help to distribute loads. When solid blocks are not used, the cores in the top course can be filled solidly with mortar or concrete. To do this, you must place a strip of metal lath in the mortar joint under the top course. The strip, which is just wide enough to cover the block cavities, forms a base for the fill you use in the top course.

REINFORCEMENT

Long walls, or walls subjected to above-average stresses, can be strengthened with pilasters. You can see in **Figure 4-61** how a typical design forms an interlock of units. In practice, the design and size of the structure plus earth conditions must be considered when determining the need for pilasters and their size and frequency. Check codes; if such reinforcement is required, all units in the pilaster should have full mortar bedding.

Steel is often used but, again, the need for it and the way it is used depend on local conditions. Usually, no matter what, it pays to place a ½-inch bar in each corner and to fill those cores solidly with mortar or concrete (**Figure 4-62**). If you anticipate this need, the bar can be embedded in the footing pour as a tie from

FIG. 4-61

FIG. 4-62

FIG. 4-63

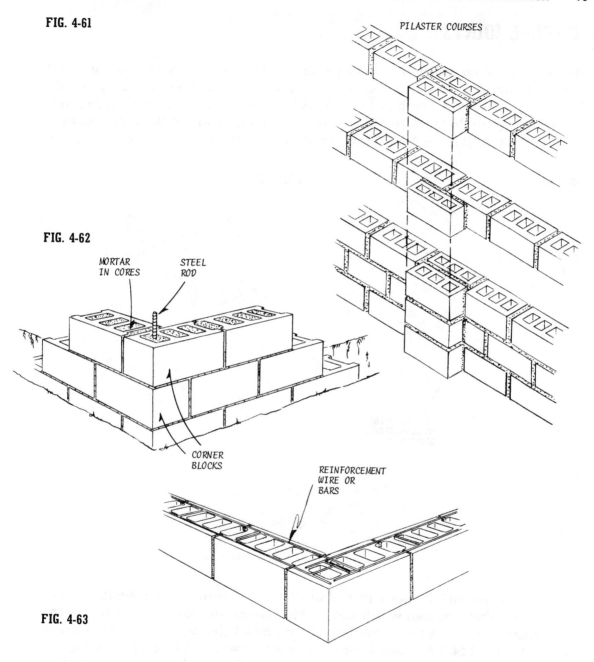

PILASTER COURSES

MORTAR IN CORES   STEEL ROD

CORNER BLOCKS

REINFORCEMENT WIRE OR BARS

wall to base. Often, especially on high walls and those that must withstand considerable pressures from slopes or backfilling, special reinforcement wire is placed in the joint of alternate courses (**Figure 4-63**).

CONTROL JOINTS

The purpose of control joints is to permit stress movements to occur without damaging the structure. Whether or not they are needed at all will probably depend mostly on the length of the wall. A basic stress joint is simply a continuous, vertical joint that is done by combining half-length and full-length blocks at the control point (**Figure 4-64**). Another one can be done with common stretcher

FIG. 4-64   **FIG. 4-65**

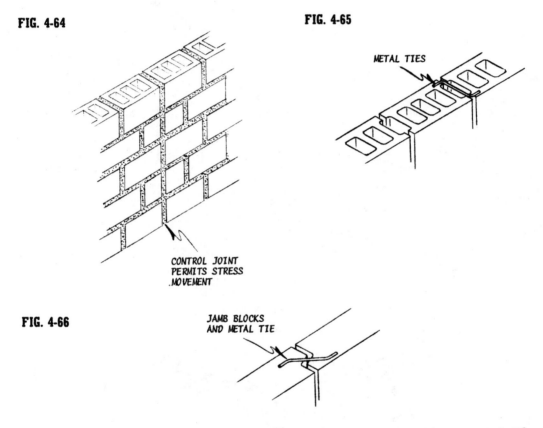

METAL TIES

CONTROL JOINT
PERMITS STRESS
MOVEMENT

FIG. 4-66   JAMB BLOCKS
AND METAL TIE

block if Z-shaped metal ties are placed in alternate courses (**Figure 4-65**). The ties are narrower than the wall width and provide lateral support on each side of the joint. Offset jamb blocks, also reinforced with metal ties, may be used in similar fashion (**Figure 4-66**). A common procedure is to insert building paper or roofing felt into the channel of the block so that mortar can't bond the units together (**Figure 4-67**). Cut the paper so it will be long enough for the full length of the joint, and wide enough to span across it. Fill the channel with mortar to contribute lateral support.

FIG. 4-67

FIG. 4-68

PAPER OR FELT

TONGUE & GROOVE BLOCKS

There are also tongue-and-groove blocks that are available in half and full units (**Figure 4-68**). These are made for the purpose and provide much lateral support because of the way they are shaped. Such special shapes may not be available in all areas. Usually, control joints should be incorporated every 25 to 30 feet.

WALL INTERSECTIONS

Bearing walls should not be tied together with a masonry bond unless they occur at a corner. Let one wall terminate against the face of the other and include a control joint at that point by doing this. Tie bars, which supply lateral support, should not be spaced more than 4 feet apart vertically (**Figure 4-69**).

FIG. 4-69

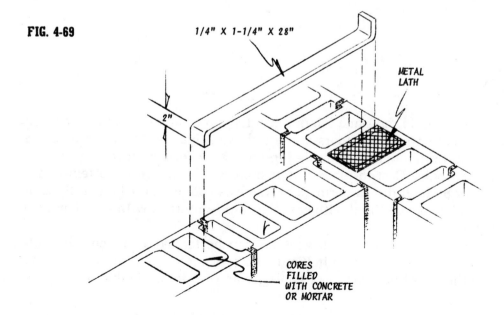

1/4" X 1-1/4" X 28"

METAL LATH

2"

CORES
FILLED
WITH CONCRETE
OR MORTAR

Nonbearing walls can be tied to other walls by using strips of lath or mesh across the common joint in alternate courses (**Figure 4-70**). If the nonbearing wall is to be constructed later, incorporate the ties in the first wall so they will be available when needed.

FIG. 4-70

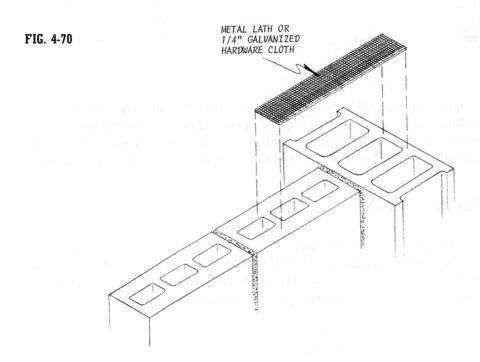

METAL LATH OR
1/4" GALVANIZED
HARDWARE CLOTH

OPENINGS

Precast lintels, which do the same job that headers do in a wood frame, are often used to span the opening over doors and windows (**Figure 4-71**). Note how the use of half and full units in this example produces a uniform appearance without the need to cut blocks. Cast-in-place lintels are done by providing a temporary support frame or by including a permanent frame as you erect the blocks. Special lintel blocks, which can be filled with concrete and reinforced with steel bars, are available (**Figure 4-72**).

Frames installed as block work goes on can be locked in place with galvanized nails used as ties as shown in **Figure 4-73**. In such situations, be sure you use a level on the wood members as frequently as you do on the blocks.

FIG. 4-71

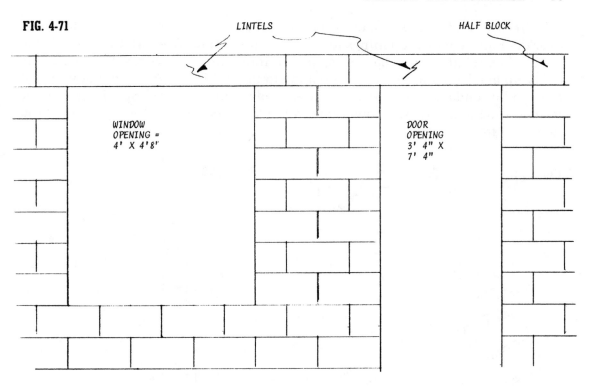

LINTELS

HALF BLOCK

WINDOW
OPENING =
4' X 4'8"

DOOR
OPENING
3' 4" X
7' 4"

FIG. 4-72 **FIG. 4-73**

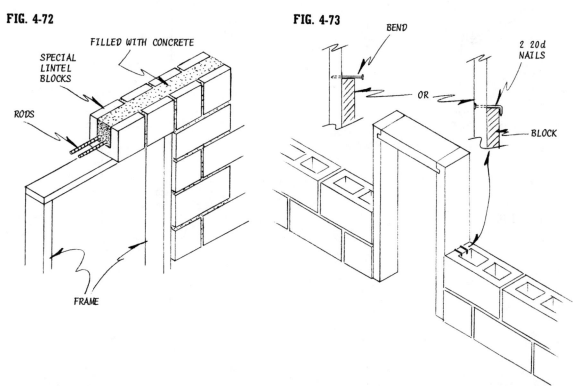

SPECIAL
LINTEL
BLOCKS

FILLED WITH CONCRETE

RODS

FRAME

BEND

OR

2 20d
NAILS

BLOCK

Precast lintels, shaped to correspond with special jamb blocks, are often used over door and window openings for modular units (**Figure 4-74**). Channel blocks are used for windows that have steel or aluminum frames (**Figure 4-75**). Best bet is to lay up block on one side of the opening to the full height of the window. Place the window and align it by using wooden wedges, or whatever. Special wedges or fasteners to secure the window frame in the channel are usually supplied by the manufacturer.

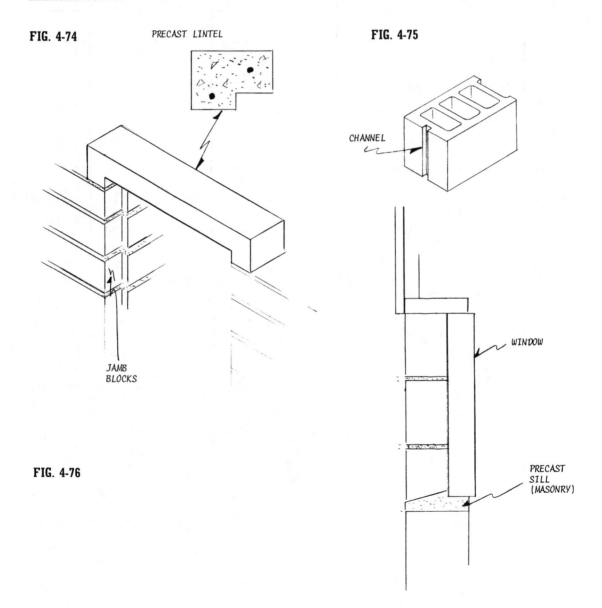

FIG. 4-74

PRECAST LINTEL

JAMB BLOCKS

FIG. 4-75

CHANNEL

WINDOW

PRECAST SILL (MASONRY)

FIG. 4-76

Sills may be cast in place, or they can be precast units as shown in **Figure 4-76** that you bond to the masonry. In such cases, be sure you seal end joints completely by packing them tight with mortar or caulking (**Figure 4-77**).

FIG. 4-77   FIG. 4-78

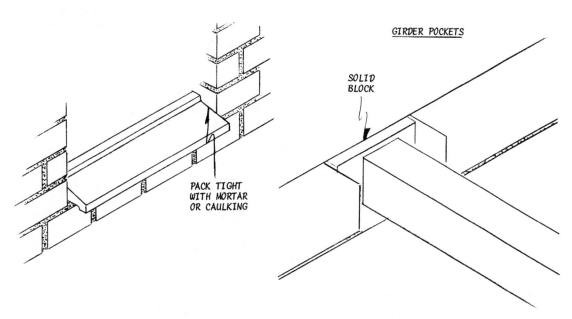

Beam or girder pockets can be done by omitting a block from the top course — or a half block, if beam size permits — and then filling the opening with a solid block (**Figure 4-78**). Much depends on the size of the member to be supported. Sometimes it's necessary to do some block cutting to shape out a pocket.

Two ways you can provide openings through block walls for crawl-space ventilation are shown in **Figure 4-79**.

FIG. 4-79

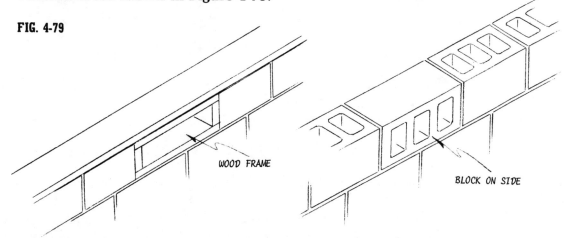

ANCHOR BOLTS

Anchor bolts are special mechanical fasteners, hooked on one end and threaded on the other as shown in **Figure 4-80**, that are installed in the top of block or concrete foundations so that the first wood member of the structure (the sill) can be secured. Spacing should follow **Figure 4-81** pretty closely, but moving some an inch or two one way or another to clear a spot where a stud will be placed is okay.

FIG. 4-80

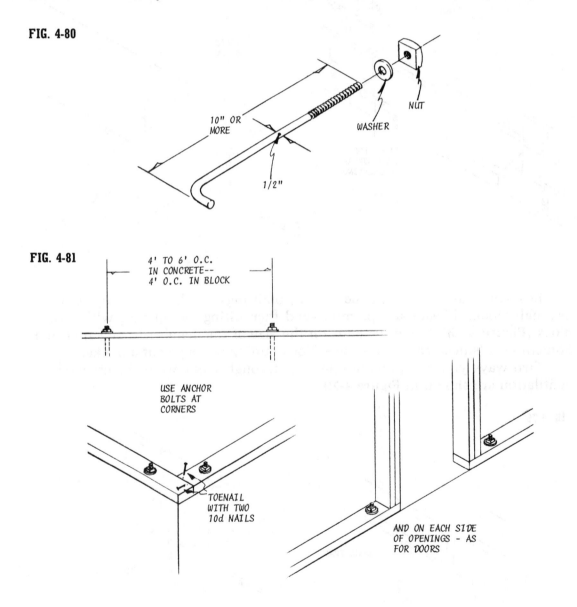

FIG. 4-81

The amount of thread that extends above the sill is not critical as long as the nut can go on all the way with at least ½ inch to spare. Some workers tend to be careless with placement of the bolts so that inaccuracies occur in spacing, centering, and vertical alignment. You will regret carelessness later when you get to placing sills and joists and studs. This is especially true in slab work with integral footings, since studs are attached directly to plates which are secured by the anchor bolts (**Figure 4-82**). Doing the job right in a concrete foundation calls for making temporary holders that will keep the bolts centered, assure correct projection above the pour, and suspend them vertically. The holders can be made from scrap pieces of 1-inch wood—or 2-inch material to simulate the sill—which are drilled to accommodate the bolt and tack-nailed to the top edges of the forms (**Figure 4-83**).

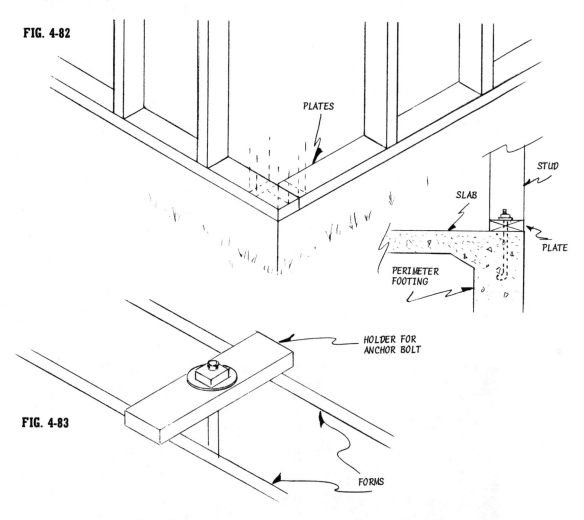

FIG. 4-82

PLATES

STUD

SLAB

PLATE

PERIMETER
FOOTING

HOLDER FOR
ANCHOR BOLT

FIG. 4-83

FORMS

Some professionals place sills (or plates) immediately after the concrete has been poured and leveled. This is called "floating" and is often recommended because it creates a bond between wood and concrete. When this is done, all wood pieces must be cut to exact length and drilled accurately, and held ready to go as soon as concrete work is done.

Anchor bolts in block walls are organized as shown in **Figure 4-84**. Positions are anticipated and a piece of metal lath is placed in the joint under the cores that will receive the bolt. Later, the cores are filled with concrete or mortar and the anchor bolt is embedded (**Figure 4-85**). When all the top cores of the blocks are to be packed, the metal lath should be installed the full length of the joint.

In all situations, be sure you do not tighten the nut until the concrete or mortar has hardened.

FIG. 4-84

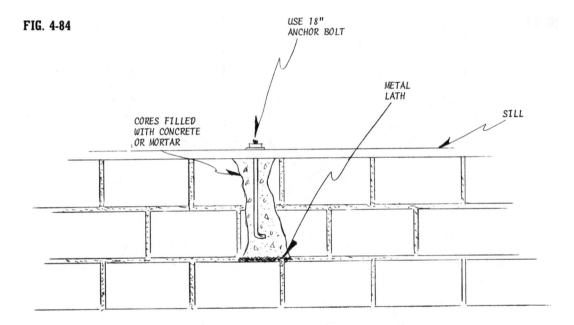

USE 18"
ANCHOR BOLT

METAL
LATH

SILL

CORES FILLED
WITH CONCRETE
OR MORTAR

FIG. 4-85

CONCRETE FLOORS

Slab-on-ground floors are considered desirable by many professionals and by amateurs, regardless of whether or not a basement is involved. Footing and foundation considerations do not change; the prime factor for a basementless structure is that the depth be enough to meet solid soil below the frost line. Much of the designing, which includes reinforcement requirements and drainage considerations, must be in accordance with local codes. For example, many Eastern areas require that welded wire that meets certain specifications be placed in the concrete about 1½ inches below the surface, and ½-inch steel bars, placed parallel and running the full length and width of the slab about 4 feet apart on centers, may be required in areas where earthquakes are likely.

Insulation and moisture controls are critical; subgrades, whether soil or other materials, must be thoroughly compacted. If the slab is placed directly on soil, be sure the area is free of tree roots and debris. If a granular fill is used it may be gravel or crushed stone or slag. Often the choice is based on what is available in the area, but whatever the material, piece sizes should range from ½ inch to 1 inch so that there will be air spaces in the fill; these spaces have insulating qualities and reduce the capillary movement of moisture that might be present, or appear later, in the subsoil.

All mechanical installations such as heating ducts, pipes for radiant heating and water, and other utility entries should be completed before the slab is poured. Water lines, when placed under a concrete floor, must be established in trenches that are deep enough to prevent them from freezing.

Drain tiles around a footing are a minimum precaution to prevent water accumulation under a slab. Soil that drains poorly usually calls for a granular fill and a special drain line leading to a positive outlet to carry away any water that would otherwise accumulate under the slab. If floor drains are used, the slab should be sloped accordingly, but with care to assure a uniform slab thickness.

When there is considerable groundwater that can't be directed away from the site, and the possibility of excessive water pressure under the slab, steel reinforcement to withstand the uplift pressure must be introduced. Often, such a slab is poured in two layers with a built-up bituminous membrane between them as a water block. The membrane consists of hot bituminous material and at least two layers of roofing felt. The membrane is continuous and carried up the foundation wall to the top surface of the slab.

Perimeter insulation is important to reduce heat loss from the slab to the outside. Rigid insulation, stable enough to resist wet concrete, is recommended. Thicknesses can vary and are often specified in relation to local temperatures and the type of heating to be installed.

Methods that provide for movements of the slab vary, but a common one is shown in **Figure 4-86**. The wood pieces are removed after the concrete has set and the groove is filled with a caulking compound. Perimeter nailers, if needed for work that comes later, can be installed as shown in **Figure 4-87**.

The drawings in **Figure 4-88** are examples only and should not be used as working drawings until they have been checked against local codes.

FIG. 4-86

FIG. 4-87

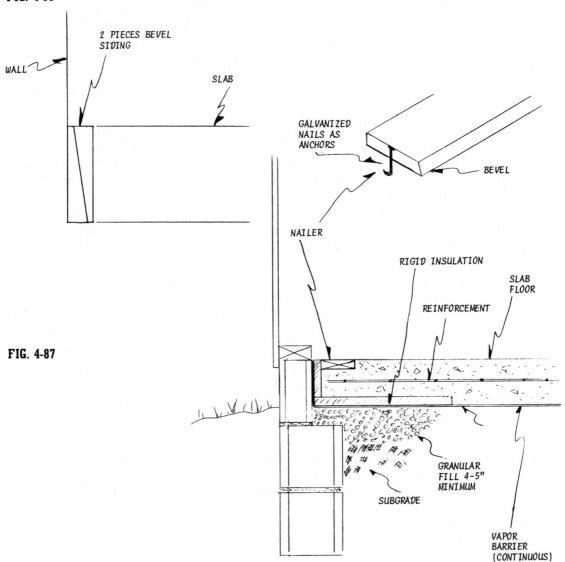

FIG. 4-88

FOR VERY WET SOILS

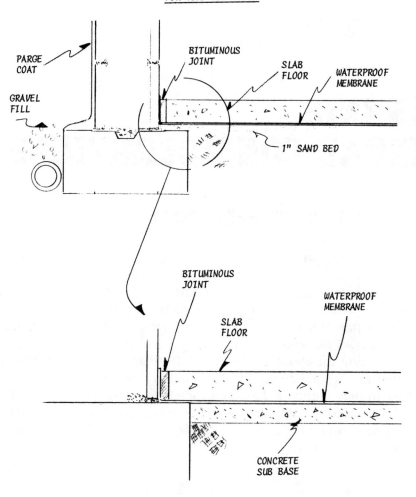

PARGE COAT

GRAVEL FILL

BITUMINOUS JOINT

SLAB FLOOR

WATERPROOF MEMBRANE

1" SAND BED

BITUMINOUS JOINT

SLAB FLOOR

WATERPROOF MEMBRANE

CONCRETE SUB BASE

5

FRAMING: GENERAL CONSIDERATIONS

TERMITES — YOU DON'T NEED THEM

There are few areas in the country where termites can't be a problem, so some steps to protect the house are in order. If you feel, as some do, that the dangers of termite damage have been greatly exaggerated, then you might want to do some research in your area before taking any precautions at all.

A common precaution is a continuous strip of 26-gauge metal placed as shown in **Figure 5-1** between the foundation and the sill. Let the shield extend from the foundation for about 2 inches at a 45-degree angle. Actually, this step is only a supplement to good construction. Shields alone are not adequate. Since the termites live underground, a logical procedure is to treat the soil under and around the house with a chemical toxic to the insects. You can have this done by professionals or you can do it yourself as long as you follow all the safety rules printed on container labels.

FIG. 5-1

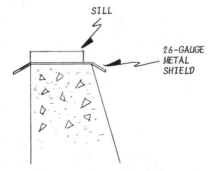

SILL

26-GAUGE
METAL
SHIELD

Use wood which has some natural resistance to termites, like construction-grade *heart* redwood, for sills. This, and other species, may be treated with chemicals before being installed. Remember, though, that the wood is well protected only when the preservative penetrates deeply. A single quick brushed-on coat won't do. You can also purchase wood which has been pressure-treated with suitable chemicals.

General anti-termite procedures call for keeping the work area clean of debris —no wood scraps that might be buried and attract insects. The sill and other wood members should be at least 6 to 8 inches above grade. A sign of termite activity is the earthtubes they build over non-wood materials to get from soil to food. You might check foundations frequently and take necessary steps if you see such tubes.

Avoid making direct connections between in-ground wood (fence posts and the like) and the house. Wood that makes direct soil contact should be pressure-treated.

INSTALLATION OF SILLS

The sills (often called "mud sills") are placed on the top of the foundation continuously around the structure and are secured with the anchor-bolt nuts and washers. The usual material is 2×6 lumber, long enough, if possible, to run the full length of walls, although 2×8s are not so rare when the foundation is concrete block. Butt the sill pieces against the anchor bolts and use a square to mark the locations of the holes you must drill. Make the holes about ¼ inch larger than the diameter of the bolts so there will be some play to make alignment of the sills easier.

Place the sill pieces over the anchor bolts as they are drilled, but do not tighten the nuts until all pieces are placed and correctly aligned. The sills, in addition to being straight, should be level. Carpenters often use wooden shims to fill depressions in masonry, but this is bound to leave openings between the foundation and the sill, so it's better to do the job with mortar. Another common practice is to place a full bed of mortar on the foundation wall. This provides a good seal and the sill can be leveled without introducing other materials.

Special waterproof resilient sill sealers are available. Many are 1-inch-thick strips of fiberglass that will compress considerably when anchor-bolt nuts are tightened. Such material should be used as a seal against insects, dirt, and air, not as a means of leveling the sill.

Use large washers under all nuts. Walk around the foundation several times, tightening each nut an equal amount as you go. Adjoining pieces of sill, whether end-to-end or at corners as shown in **Figure 5-2**, should be secured by toenailing with 10d nails.

Codes in some areas of the country may call for a 4-inch-thick sill, or you may choose to do this on your own if you are building where heavy winds prevail.

FIG. 5-2

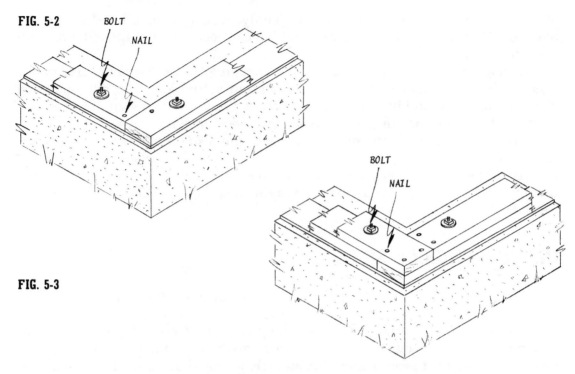

FIG. 5-3

Such sills can be solid material, in which case the anchor bolts provide sufficient security, or they can be built up by doubling 2× material as shown in **Figure 5-3**. Use 10d nails to attach the upper piece, driving two nails at each end of each piece and spacing those between about 24 inches in a staggered pattern.

TYPES OF FRAMING

Before getting to the actual assembly of the house frame, it may pay to look at types of frames so you can preview the steps ahead and become acquainted with some of the nomenclature. *Platform* framing and *balloon* framing are the most common designs, with the former currently the most popular. In both cases, the structural members have a nominal thickness of 2 inches, which actually is 1½ inches. This reduction in thickness also applies to widths. A 2×4 actually measures 1½×3½ inches; a 2×12 will be 1½×11¼ inches. The width of 2× lumber up to 6 inches is reduced by ½ inch. From there to 12 inches, widths are actually ¾ inch less than the nominal specifications. The same applies to 1-inch stock, except that the thickness is reduced by ¼ inch. The heavier materials you are likely to use are reduced in both width and thickness by ½ inch. Thus, a 4×4 measures 3½×3½ inches; an 8×8, 7½×7½ inches.

PLATFORM FRAME

The floor frame consists mainly of joists which span the foundation walls and have intermediate supports where they are required. We'll talk about bridging, openings through floors, and other factors later. Once the joist assembly is complete, it is covered with subflooring, as shown in **Figure 5-4**.

FIG. 5-4

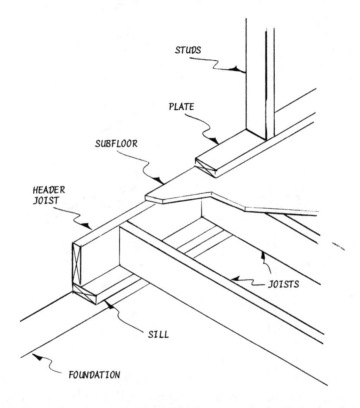

At this point you have a working platform and a solid structure on which to erect walls. It's a popular system with carpenters because the platform becomes a large workbench on which wall-frame sections can be preassembled and then tilted into place.

The same floor-framing procedure is followed if the house has a second story, except that instead of the joists resting on sill and foundation, they are supported by the studs and plates that compose the perimeter walls, as shown in **Figure 5-5**. Note that if you were to remove the first-floor studs and the wall plates from Figure 5-5 it would look exactly like the joint assembly done directly on the foundation. Thus you are creating a second platform on which to erect a second set of walls.

FIG. 5-5

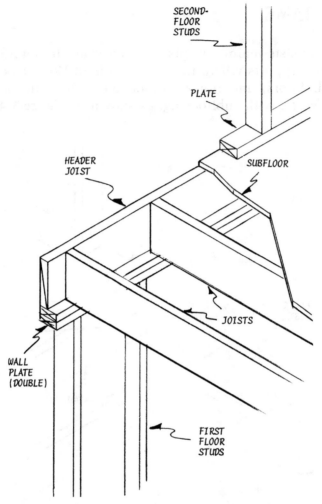

SECOND-
FLOOR
STUDS

PLATE

HEADER
JOIST

SUBFLOOR

JOISTS

WALL
PLATE
(DOUBLE)

FIRST
FLOOR
STUDS

The basic design of interior walls does not differ radically from the stud-and-plate arrangement of perimeter walls. Bearing partitions (generally, those that run at right angles to joists) are designed as shown in **Figure 5-6**. Note that the fire blocks (or firestops), which do much to retard horizontal spread of fire, also act as solid bridging to hold the joists parallel and plumb. Such pieces should be cut carefully so they will be uniform in length and have square ends.

BALLOON FRAME

The studs in a balloon frame run continuously from the sill to the plates which support the roof rafters. Standard framing at the sill has the joists surface-nailed to the sides of the studs and includes fire blocks placed as shown in **Figure 5-7**.

FIG. 5-6

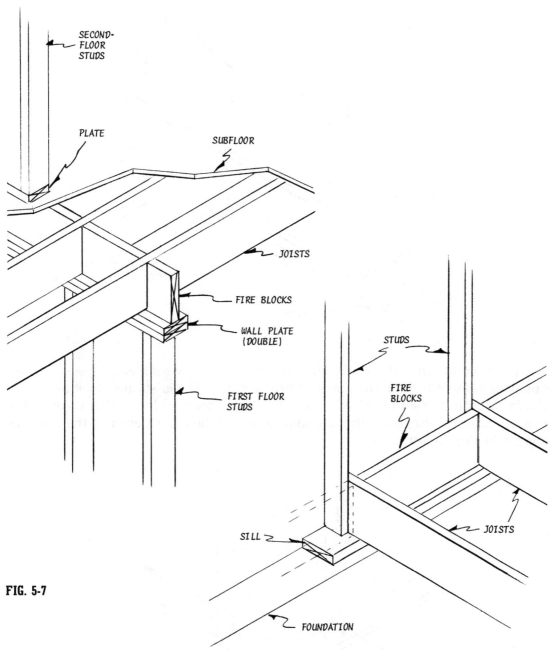

SECOND-
FLOOR
STUDS

PLATE

SUBFLOOR

JOISTS

FIRE BLOCKS

WALL PLATE
(DOUBLE)

FIRST FLOOR
STUDS

STUDS

FIRE
BLOCKS

JOISTS

SILL

FOUNDATION

FIG. 5-7

The fire blocks are easy to install after joists are established, since they can be surface-nailed into studs at one end and secured at the other end by nailing through
the joist.

FIG. 5-8

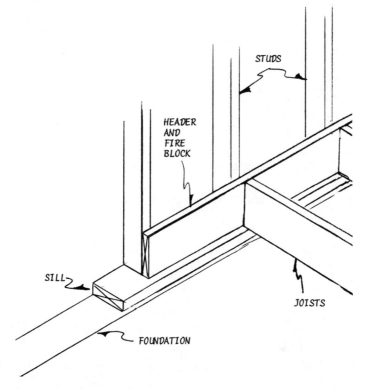

Another design is the T-sill, shown in **Figure 5-8**, which employs a continuous strip to serve as both a header and fire blocking. The joists are offset their own thickness from the stud positions so they may be secured by nailing through the header. In both situations, the subflooring is notched to fit around the studs as shown in **Figure 5-9.**

FIG. 5-9

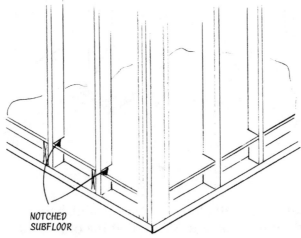

FIG. 5-10

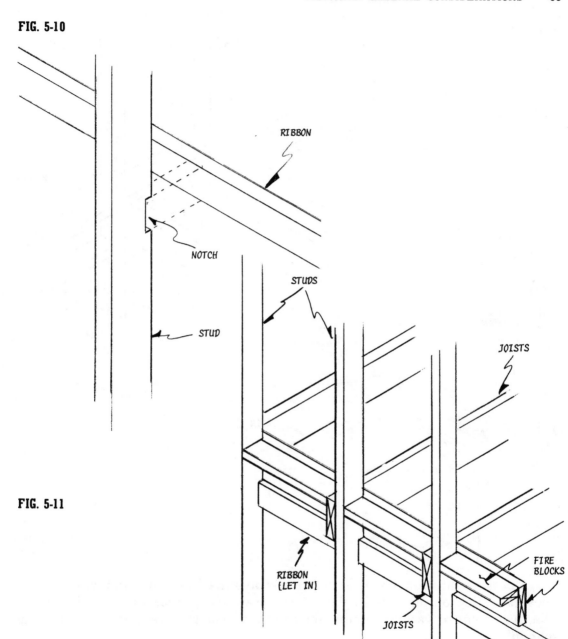

FIG. 5-11

The studs are notched at the second-floor level to receive a ribbon on which the joists will rest (**Figure 5-10**). The joists are placed so they can be nailed to the sides of the studs, and fire blocking is introduced as shown in **Figure 5-11**.

FIG. 5-12

FIG. 5-13

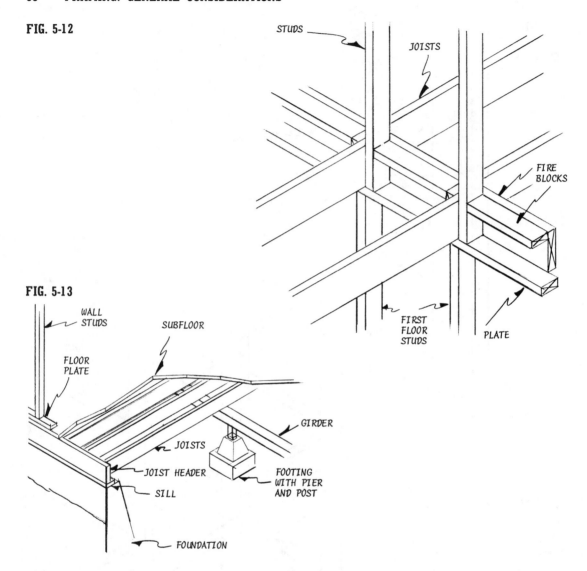

A bearing partition in a balloon frame is organized as shown in **Figure 5-12**. It does not differ too much from the exterior wall except that a plate is included. The placement of various components permits a fairly straightforward nailing schedule.

Framing over a crawl space, shown for a platform in **Figure 5-13**, does not differ from the methods already described; only the pier-and-post supports for beams or girders differ from similar components that may be needed when a basement is part of the structure.

JOISTS

Joists are horizontal, structural components, placed on edge, which carry the house loads to girders and sills. They also must be strong and stiff enough to carry loads over the span areas. The thickness, width, and spacing of joists together with sizes and spacing of subsupports (beams or girders) are determined by the live load placed on them. The live load, of course, has everything to do with the design of the house and the materials used in its construction. Structures with tile roofs, plastered walls and ceilings, and the like are heavier than structures done with shingles, plywood walls, and acoustical ceilings.

Joists that are 2×6s, spaced 16 inches on centers, can span about 12 feet under a 20-pound live load. If the load is increased to 60 pounds the span is decreased to between 8 and 10 feet, depending on whether plaster ceilings are involved. If the joists are 2×8s, a 16-foot span may carry a 20-pound load while 11 to 13 feet might do for the 60-pound load. If they are 3×10s, spans may increase to 23 feet for the light load, and 16 to 20 feet for the heavier one. Another factor is the wood used. Some species are stiffer and stronger than others.

A check of current construction methods reveals much flexibility in the combinations of joist and girder sizes and spacing. A very common practice over crawl spaces uses 4×6 girders, 6 feet O.C. (on centers), with 2×6 joists spaced 16 inches O.C.

Other common joist sizes are 2×8s, 2×10s, and 2×12s, with spans of 8 to 10 feet, 10 to 12 feet, and 12 to 14 feet respectively. Girder sizes commonly run 4×6, 4×8, 4×10, and 4×12, with respective spans of 6 to 8 feet, 8 to 10 feet, 10 to 12 feet, and 12 to 14 feet.

Many of the variables are caused by local conditions, which, of course, are the prime factors that often result in changes and different interpretations of national codes by building authorities in the area. The spacing of joists is pretty standard at 16 inches; span lengths are affected considerably by whether a basement is involved. Here, it is desirable to hold substructures to a minimum to achieve open space, and so heavier joists and longer spans are apt to be used. This consideration is not so critical when building over a crawl space, and so lighter joists and shorter spans are apt to be used.

The chances are excellent that you can get vital information concerning joists, girders, posts, and spacings out of local building-code literature. Technical information is available from various sources, including the government. You can get a slide-rule computer that even gets into grades and species of lumber for $1 from the Western Wood Products Association. (See Appendix for address.)

GIRDERS, BEAMS, AND POSTS

Joists are supported where necessary between foundation walls by girders (also called "beams"). The ends of the girders rest on the foundation walls in the pockets provided for them, and they are supported at intermediate points with posts. The girders should bear a minimum of 4 inches on the foundation and should clear the walls of the pocket by ½ inch on both sides and the end.

The girders can be solid lumber, or they can be assembled by using pieces of 2× material. **Figure 5-14** shows a 40-foot girder built up from six 20-foot pieces and a 24-foot girder built up from six 12-foot pieces. Both are three-piece girders

FIG. 5-14

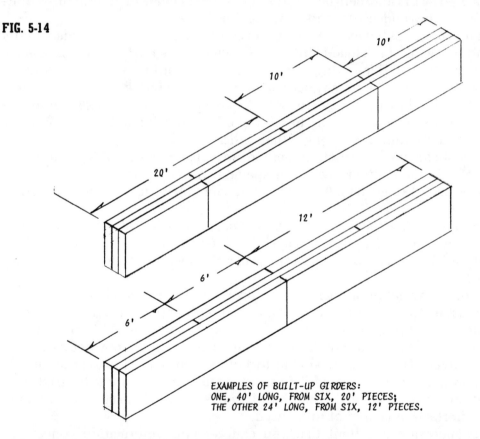

EXAMPLES OF BUILT-UP GIRDERS:
ONE, 40' LONG, FROM SIX, 20' PIECES;
THE OTHER 24' LONG, FROM SIX, 12' PIECES.

—that is, they have three layers of 2× boards. There is no reason why you can't work with other boards of standard lengths, but the joint design should follow what is shown in the examples. Drive 20d nails from each side (for a three-piece girder), positioning two near each end of each piece and spacing others between about 30 inches apart in a staggered pattern. A two-piece girder is nailed the same way but with 10d nails. For a four-piece girder, add a fourth board to the three-piece design with 20d nails. In all situations, it's wise for the joint in a built-up girder to occur over a support post. Often a bolster, with a cross section that matches the girder's, is used between post and girder (**Figure 5-15**). It does help somewhat to distribute loads, but it is used mostly to provide a decorative detail.

FIG. 5-15

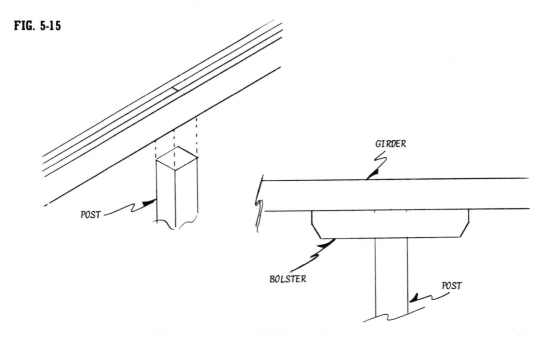

In most situations, the cross-sectional dimensions of a wood post should match the thickness of the girder. Thus, 4×4 posts are okay for a 4-inch girder, 6×6s for a 6-inch girder, 8×8s (sometimes 6×8s) for an 8-inch girder. In some areas, this general rule applies only when the post is not longer than 9 feet or smaller than a 6×6, so check.

Steel I-beams are often used as girders but, like their wooden counterparts, sizes — cross-sectional dimensions and web thicknesses — must be based on the load they will carry and the span. Supports for I-beams often take the form of Lally columns, many of which have a built-in adjustment for height (**Figure 5-16**). Obviously, this can be an advantage when doing initial leveling, and even later should any sag occur. Lally columns may also be used under wood girders. They are locked to steel girders with nuts and bolts and to wooden girders with lag screws. The top of a steel beam is usually set level with the foundation wall and a wooden pad is used on its surface as shown in **Figure 5-17** for levelness with the sill. Wood beams are organized so that top surfaces are level with the sill to start with.

FIG. 5-16

FIG. 5-17

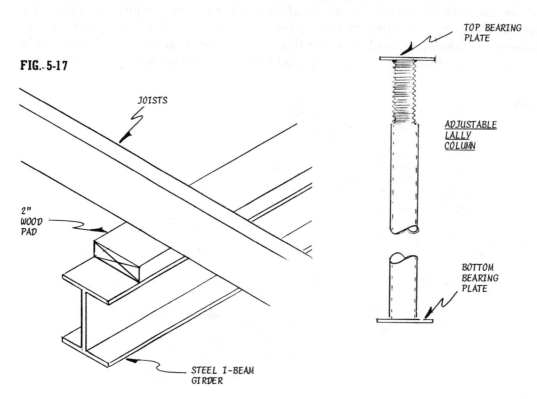

JOISTS

2"
WOOD
PAD

STEEL I-BEAM
GIRDER

TOP BEARING
PLATE

ADJUSTABLE
LALLY
COLUMN

BOTTOM
BEARING
PLATE

Joists that must bear directly against steel beams can be attached as shown in **Figure 5-18**. It is not recommended that you notch the joists so they can rest on the beam's lower flange, since the typical slope of the flange does not provide a good bearing. A common practice is to weld or bolt a steel plate to the bottom of the beam and to use a 2×4 connector across the top. Some clearance between the two materials is usually advisable since wood is more likely to undergo physical changes than steel.

Since this system places the I-beam in the floor space rather than below the joists, it provides more headroom below, which may be an advantage in some situations.

FIG. 5-18

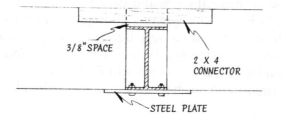

3/8" SPACE

2 X 4
CONNECTOR

STEEL PLATE

POST SUPPORTS

Posts of steel or wood which support girders must have adequate footings so loads will be distributed over a broad, strong area. The method of securing the post to the footing should prevent any lateral motion.

Precast concrete piers, with wood blocks inserted at the top end, are commonly used in a crawl space. The piers are set on a cast-in-place footing with a strong mortar used as a bond (**Figure 5-19**). Both the footing and the pier should be placed carefully to achieve levelness. The post is secured by toenailing through it into the wood block.

FIG. 5-19

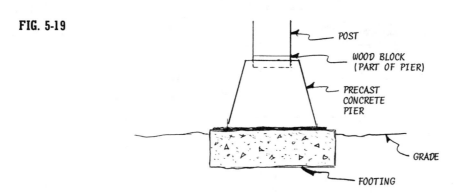

POST
WOOD BLOCK
(PART OF PIER)
PRECAST
CONCRETE
PIER
GRADE
FOOTING

The same type of pier may be used when a concrete floor is placed, but the pier-footing requirements remain the same (**Figure 5-20**). The objection here is the appearance of the pier when a usable basement is part of the design.

FIG. 5-20

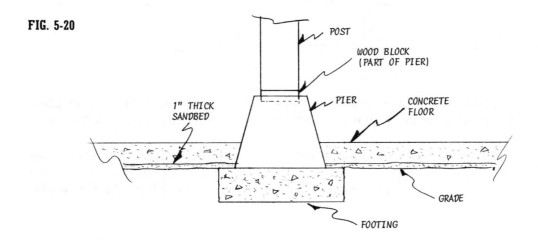

POST
WOOD BLOCK
(PART OF PIER)
1" THICK SANDBED
PIER
CONCRETE FLOOR
GRADE
FOOTING

A similar construction, with cast masonry you can do yourself, often with readymade forms, includes an anchor pin that seats in a hole drilled in the post (**Figure 5-21**). The pin can be a length of ½-inch, or larger, reinforcement rod or a bolt. In either case, it should penetrate the post a minimum of 3 inches. Forms you can buy for round concrete piers (or pedestals) look like large thick-walled mailing tubes. The pins are installed in the concrete as if they were anchor bolts.

FIG. 5-21

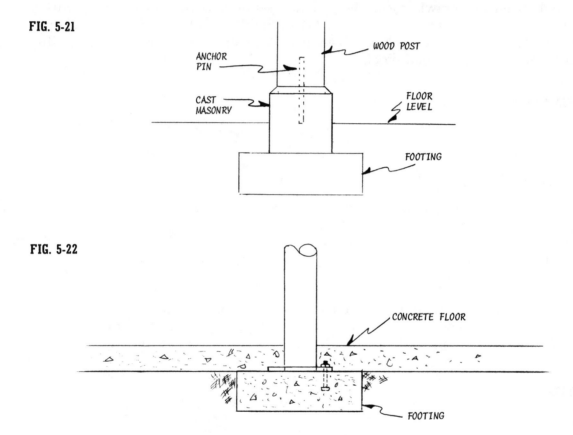

FIG. 5-22

Steel posts and Lally columns are also set on footings. The bottom plate should be placed over bolts that are embedded in the footing pour, as shown in **Figure 5-22.**

Posts that rest directly on a concrete floor are not uncommon, but the chances are they were an afterthought, installed when it became apparent that extra girder support was required. If that is not the reason, there is probably a footing under the slab in that area.

Special anchors, such as shown in **Figure 5-23**, are often used to secure posts to an existing floor, although they are just as usable as inserts in a pour. In the latter case, the anchor bolt is embedded in the concrete. When done on an existing floor, an anchor-bolt hole is formed with a carbide-tipped bit and the bolt is embedded in a special epoxy compound. Such anchors have built-in adjustments that provide some leeway if the anchor bolt is not placed precisely. Note that the anchor is designed to elevate the post above the concrete.

FIG. 5-23

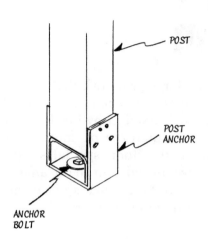

6

FRAMING THE FLOOR

Picture the floor frame as a large gridwork that is supported on the perimeter by the foundation and by girders, if needed, at intermediate points. The joists run parallel to each other and are boxed in by headers, which run across the ends of the joists, and by stringers, which are no more than terminal joists. Considerations for openings through the floor, special supports for partitions, whether joists run continuously or are pieced, and so on, affect the basic picture. Study the typical, complete floor frame in **Figure 6-1**, since it shows the relationship of various components and explains nomenclature.

FIG. 6-1   EXAMPLE OF COMPLETED FLOOR FRAME

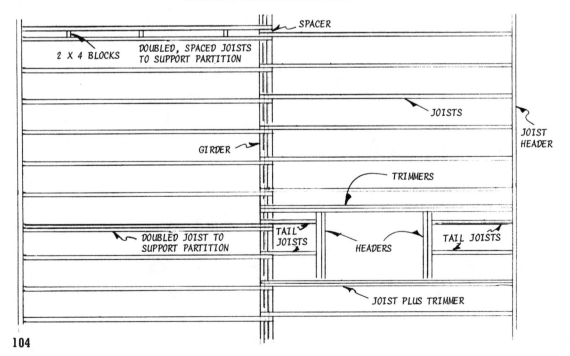

Two types of partition supports are shown. The spaced-joists design is used when passages are required for plumbing, heating, or electrical systems.

The layout for the joists can be done directly on the sills, but in platform construction, it's better to work by marking joist placement directly on the joist headers after they have been cut to correct length. Opposite headers can be butted edge to edge so that lines can be drawn across both at the same time. Place an X-mark on one side of each line to show where the joists must go. When the joists are continuous, the X will be on the same side of each line, as shown in **Figure 6-2**; when joists are spliced, the X is marked as shown in **Figure 6-3**.

FIG. 6-2   CONTINUOUS JOISTS

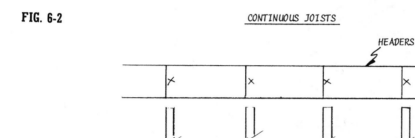

FIG. 6-3   SPLICED JOISTS

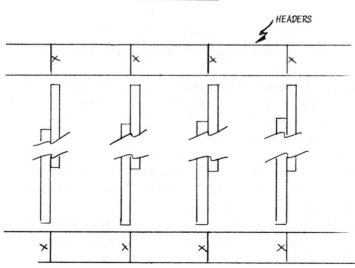

Spacing can be marked off with a measuring tape, but the job will be easier to do accurately if you take a few minutes to make a marking jig such as shown in **Figure 6-4**. The pieces used between the 1× stock should match the thickness of the joists (usually 2×).

FIG. 6-4

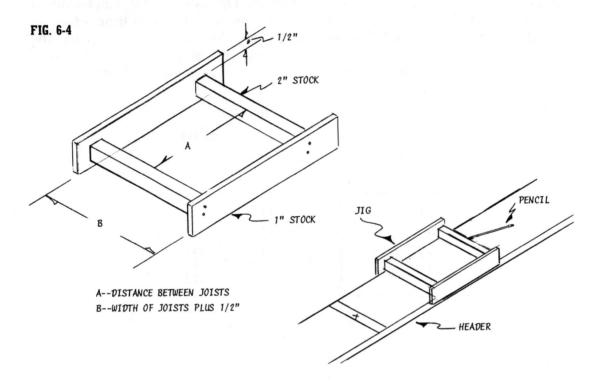

A--DISTANCE BETWEEN JOISTS
B--WIDTH OF JOISTS PLUS 1/2"

The regular joist spacing—usually 16 inches O.C.—should not be interrupted when other components are introduced. These pieces should be viewed as *additions* to the joists required.

Always choose the straightest pieces of wood you have as joist headers. Sight along an edge of each joist before you place it so you can discover any bow or crook. If there is a bow, place it so the high side is up, so that the weight it carries will tend to straighten it rather than bow it further.

When it is necessary to double up on joists or headers or trimmers, work with 16d nails, at least at the start. Drive them staight through and then clinch them where they protrude. This will serve to pull the two pieces tightly together. Space the nails about 12 inches apart and maintain at least a 1-inch edge distance. Assembly may also be done with 12d nails. If these are driven at an angle, holding power will be increased and the nails will not protrude. Both systems are shown in **Figure 6-5**.

FIG. 6-5

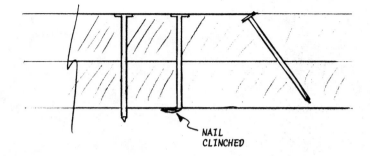

NAIL
CLINCHED

NAILING JOISTS

A good procedure is to start with one joist header, toenailing it to the sill with 10d nails spaced 16 inches apart as shown in **Figure 6-6**. Keep the header as vertical as possible, but don't worry about absolute plumbness since it will be plumbed when you add the joists, the ends of which must be square. Place the stringer joists, toenailing them to the sill as you did the header and securing the stringer-to-header joint with 20d nails.

FIG. 6-6

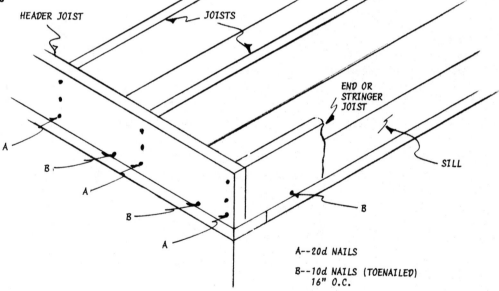

HEADER JOIST

JOISTS

END OR
STRINGER
JOIST

SILL

A
B
A
B
A
B

A--20d NAILS

B--10d NAILS (TOENAILED)
16" O.C.

Place the joists in position, lock them in position with 20d nails, and then add the opposite joist header by following the same nailing schedule. Two critical considerations are the length of the joists and the squareness of their edges. If you

FIG. 6-7

allow variations to occur, contact points will not mate as they should, vertical surfaces will not be plumb, gaps and irregularities will result in poor construction, and you will have problems later.

Be careful when you saw. Much of the preliminary work in housebuilding is called "rough" framing, but the term should not make you careless. If you work with a portable saw, use a crosscut guide that you can make or buy (**Figure 6-7**). If you use a radial-arm saw, be sure it is in correct alignment and that you provide adequate supports on each side of the main table for pieces of wood.

Good nailing means good strength. The sizes and spacings specified here are common practice, but since the nailing patterns are so important and may vary depending on codes, it pays to do some checking. If you find that smaller or fewer nails are okay, you can save some time and money. Don't add more nails as a safety factor. Excessive nailing can cause splitting, now or when the frame is under stress, so too many nails may be worse than too few.

JOISTS OVER GIRDERS

Pieced joists are overlapped where they cross the girder, and solid bridging is installed between them (**Figure 6-8**). Minimum nailing calls for the joist ends to be tied together with two 10d nails and for the joists to be tacked to the girder by toe-

FIG. 6-8

FIG. 6-9

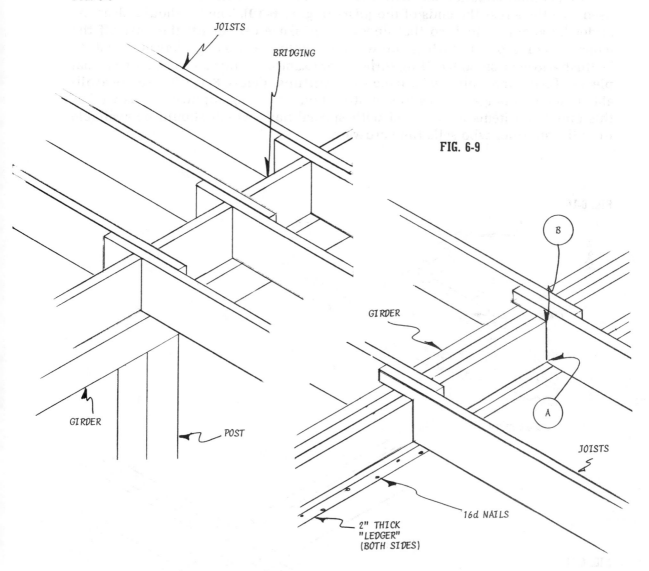

nailing on each side with a 10d nail. The bridging, with nice square ends, is secured by toenailing along the ends into joists and along the base into the girder.

When the girder has been set higher than the sills to gain headroom below, the joists can be organized as shown in **Figure 6-9**. The ledger is a very important component here, so choose good, knot-free material and attach it securely with 16d nails. Cut the notch in the joist so there will be clearance over the girder. The place marked A is where bearing should occur — not at point B. The joists are toenailed to the girder and surface-nailed to each other with 10d nails.

In a similar situation but with the joists in line, a strip of 1× or 2× wood is used as a tie across the ends of the joists (**Figure 6-10**). The tie should clear the girder by about ½ inch so that unequal shrinkage cannot lift the joist off the ledger. Often, especially when you want the bottom edges of girders and joists to be flush — to get a smooth ceiling surface, for example — you can work with special pieces of hardware called joist hangers or stirrups (**Figure 6-11**). These are available in a broad range of sizes to suit other house-framing applications as well as this one. Such items are secured with special nails which should be available from the supplier who sells the hardware.

FIG. 6-10

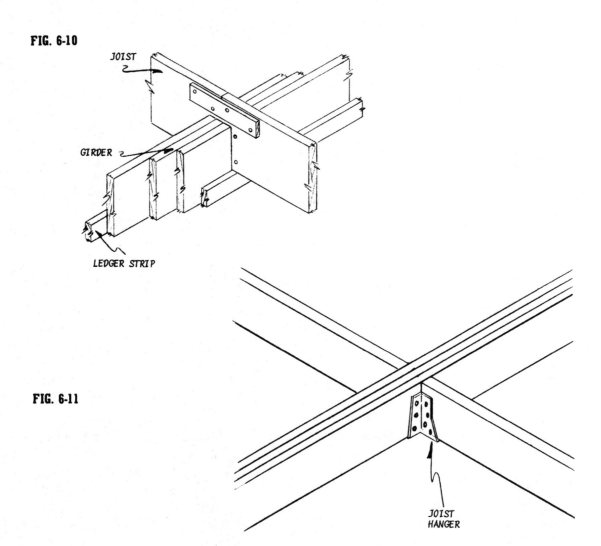

FIG. 6-11

JOISTS UNDER PARTITIONS

Use doubled joists to provide support for any parallel-running partition wall (**Figure 6-12**). The joists can be spaced if you will need room in that area for plumbing or heating requirements. If so, use 2×4 blocks, as in **Figure 6-13**, spaced about 4 feet apart, as bracing for the joists. If you find that a block will interfere with a future installation, you may move it, but don't reduce the number of blocks the 4-foot spacing calls for. In some situations—for example, an opening through the floor for a heating duct—you can move a block to one side of the opening and add another on the other side. It's okay to add but not to reduce.

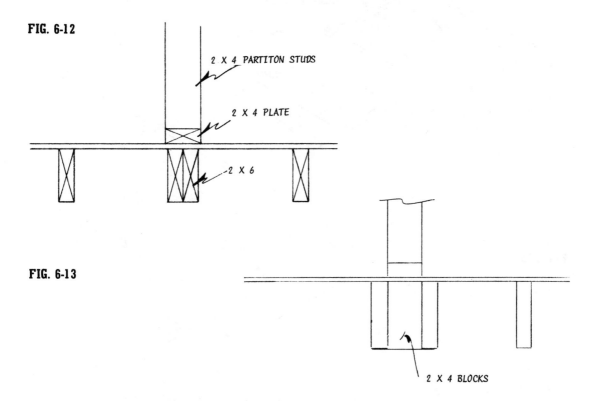

FIG. 6-12

2 X 4 PARTITON STUDS

2 X 4 PLATE

2 X 6

FIG. 6-13

2 X 4 BLOCKS

OPENINGS THROUGH FLOORS

Any opening through a floor frame must be organized to adjust for the strength that is lost when a regular joist is cut. A good system, together with nomenclature, is shown in **Figure 6-14**. The nailing schedule for the assembly can become confusing unless you anticipate it and approach the job step-by-step.

FIG. 6-14

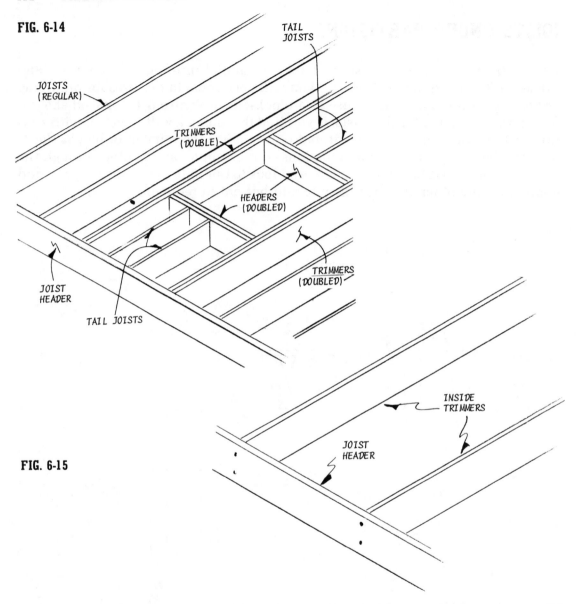

FIG. 6-15

Start by installing the inside trimmers, nailing them in place as you would regular joists (**Figure 6-15**). Set them carefully to be sure they are parallel and correctly spaced. Cut the tail joists to correct length, making sure to allow for the thickness of the double headers. Clamp or tack-nail two of the tail joists to the trimmers as shown in **Figure 6-16** so they become gauges that position the first header correctly. Nail the first header in place with three 20d nails. Position and secure the tail joists by driving three 20d nails through the joist header and

FIG. 6-16

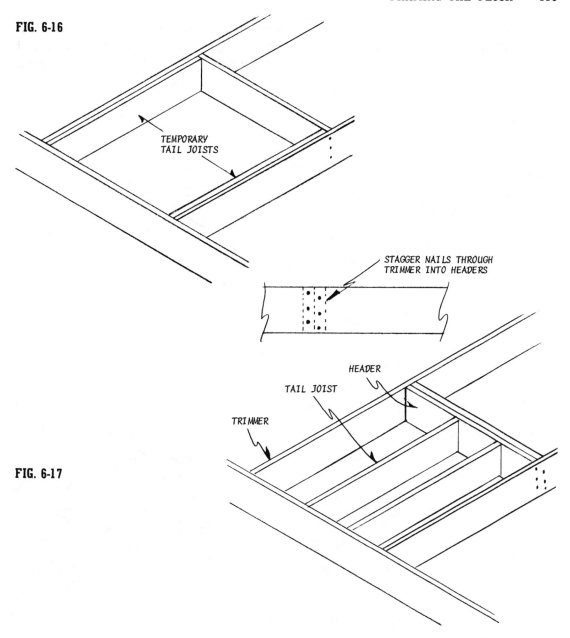

TEMPORARY
TAIL JOISTS

STAGGER NAILS THROUGH
TRIMMER INTO HEADERS

HEADER

TAIL JOIST

TRIMMER

FIG. 6-17

through the opening header. Add the second opening header by nailing through the trimmers as you did for the first one. Bond the two headers together with 16d nails, staggering them and spacing them about 16 inches apart. Keep the nails a minimum of 1 inch away from edges. So far the assembly should look like that shown in **Figure 6-17**.

The other side of the rough opening is framed in the same way. Then the last step is to add the second trimmers. Use 16d nails spaced about 12 inches apart along the edges and staggered as shown in **Figure 6-18**. Maintain the usual 1-inch minimum edge distance.

FIG. 6-18

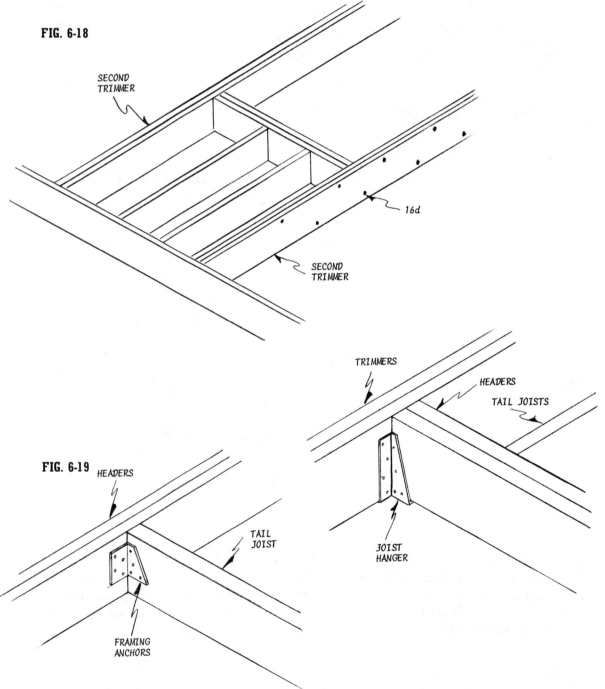

FIG. 6-19

Special joist hangers and framing anchors may be used when assembling trimmers, tail joists, and headers (**Figure 6-19**). They can make the job easier and are often recommended when tail joists are more than 12 feet long.

PROJECTIONS OUTSIDE THE WALL

Special considerations are in order when the house design calls for floor extensions that are outside the basic perimeter of the walls. They may be needed for such things as porch floors, second-story overhangs or decks, or bay windows. If the run of the floor joists permits, they can simply be longer and cantilevered beyond the foundation as in **Figure 6-20**. A small overhang, as for a bay window, will probably not require additional support. If the overhang is wide, as for a porch or veranda, then supports outside the foundation wall will be needed. If the extensions are considered at the start, then substructures could be part of the foundation. If not, they can follow the pattern for joist support in a crawl space, as shown in **Figure 6-21**.

FIG. 6-20

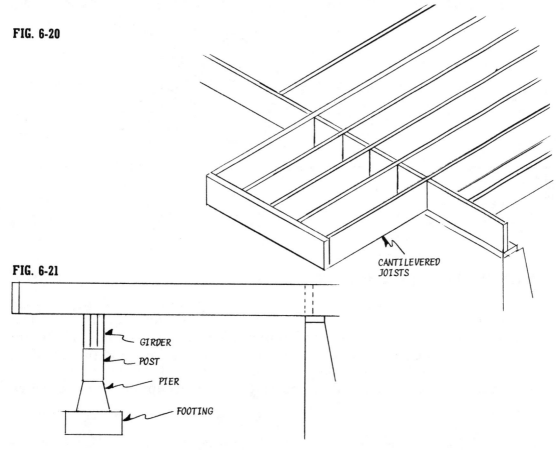

CANTILEVERED
JOISTS

FIG. 6-21

GIRDER

POST

PIER

FOOTING

If the projection joists run at right angles to the floor-frame joists, then the regular joist against which their ends terminate must be doubled (**Figure 6-22**). A general rule for establishing the length of cantilevered joists is that the distance from the foundation wall to the doubled joist should be twice the length of the overhang. Framing anchors are often used on such assemblies and should be considered if only for their convenience. Ledgers can be incorporated, as in **Figure 6-23**. But since the load (A) causes pressure upward against the joists, ledgers should be positioned along the top edges.

FIG. 6-22

FIG. 6-23

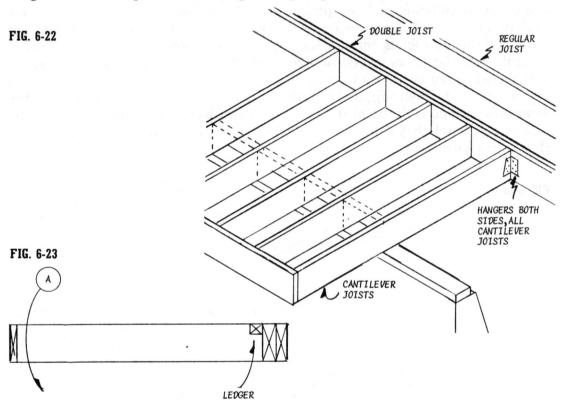

DOUBLE JOIST

REGULAR JOIST

HANGERS BOTH SIDES, ALL CANTILEVER JOISTS

CANTILEVER JOISTS

A

LEDGER

BRACING BETWEEN JOISTS

This is called "bridging," and currently there is some question about its need. The arguments against say that good floor framing, adequately nailed, doesn't require more than the subflooring that follows to keep members fixed and plumb. Most existing codes prescribe it, so if only for that reason, it's included in most structures. A personal observation is that it doesn't hurt and it supplies an extra degree of rigidity.

The most common system uses pieces of 1×3 or 1×4 lumber set between joists as shown in **Figure 6-24**. These should be spaced a maximum of 8 feet and should follow a straight line you can establish easily by snapping a chalk line across the top edges of the joists. In situations where you can't follow the 8-foot spacing, for example if the joists span 12 feet between foundation and girder, just place the bridging along a midpoint line.

FIG. 6-24   CROSS BRIDGING

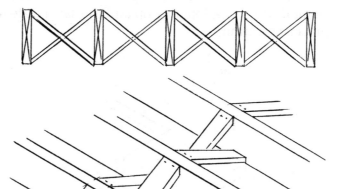

Since many pieces of bridging are required, it pays to organize a system so you can cut them speedily and accurately. First, use a piece of the material as shown in **Figure 6-25** to mark the correct cut angles and length. Use this as a pattern for others if you are sawing by hand, or as a gauge to set up a radial arm saw

FIG. 6-25

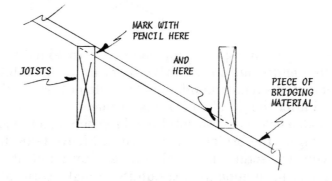

FIG. 6-26

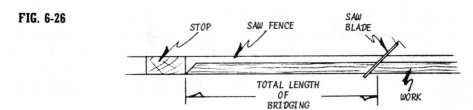

as shown in **Figure 6-26**. If you work with a portable saw, make a jig as shown in **Figure 6-27**. If you make the work-length stop and the saw guide long enough, you can cut several pieces at a time.

FIG. 6-27

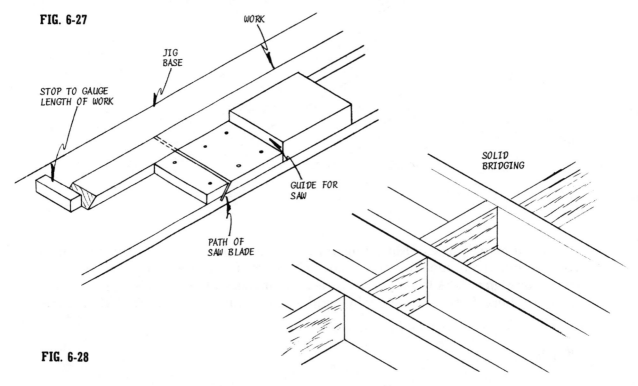

FIG. 6-28

Nail each piece of bridging at the top end with two 8d nails. Use the same size nails at the bottom ends, but do not drive them until the subfloor has been installed. If convenient, it's recommended that you do not secure the bottom ends of the bridging until after the *finished* floor is in place.

Bridging can also be done with solid blocking between the joists. Nailing will be easier if you stagger the bridging as shown in **Figure 6-28**. It's okay to base a choice on the types of "scrap" material you have on hand. It's possible, for example, that you may have enough joist cutoffs to make solid blocking feasible.

CUTS THROUGH JOISTS

The load on a joist, regardless of the wood species, becomes more effective the closer it bears on the center of a span. Thus, this area may be regarded as the most critical point when it is necessary to cut a joist for plumbing or any other reason. The midpoint, across the width, of the joist itself is a neutral axis where compression stresses end and tension stresses begin. The two are interrelated, but in essence compression will bend the joist, tension will crack it (**Figure 6-29**). This should be remembered whenever you cut a joist so you can compensate adequately with additional pieces. A hole bored on the neutral axis will have little effect on the strength of the joist as long as its diameter is not more than ¼ of the joist's width.

FIG. 6-29

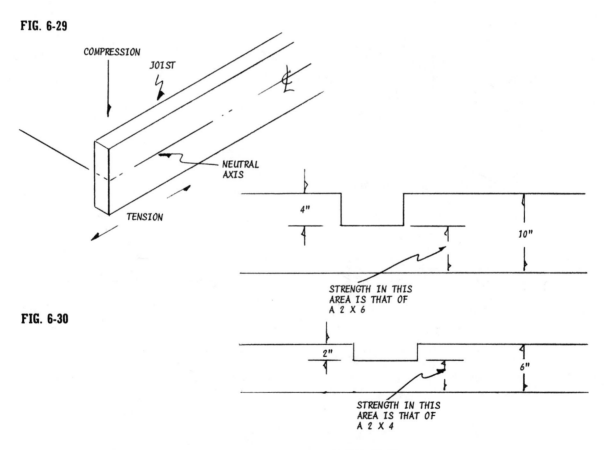

FIG. 6-30

A 4-inch-deep cutout in a 2×10 will reduce the strength of the joist below the cut to that of a 2×6; in a 2×6, a 2-inch-deep cutout leaves the strength of a 2×4 (**Figure 6-30**). A special, well-fitted piece, let into such areas, does the job of taking the compression strains (**Figure 6-31**). This is a good way to work when large plumbing pipes are involved. In all situations, no matter where the cut is located, it makes sense to compensate for lost strength by adding ties, or let-in pieces, or whatever. It's always best to make cuts from the top of the joist. If the joist must be cut through, then compensations take the form of headers, or trimmers, or even additional joists.

FIG. 6-31

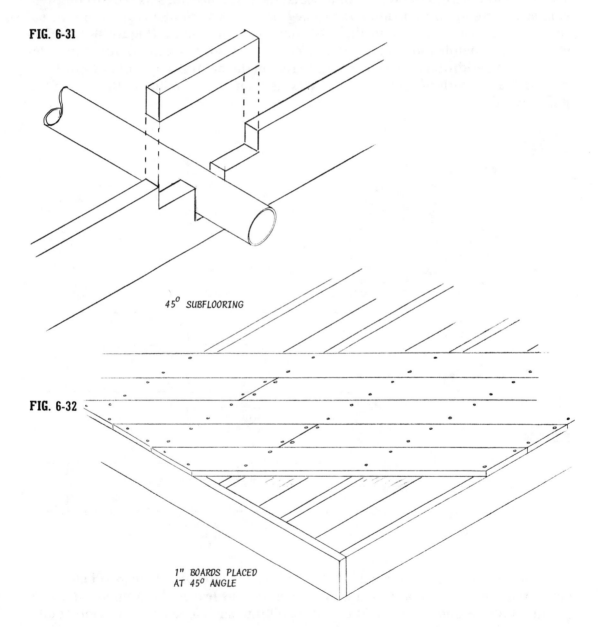

45° SUBFLOORING

FIG. 6-32

1" BOARDS PLACED
AT 45° ANGLE

ADDING THE SUBFLOOR

The subfloor is placed immediately after the frame for the floor is complete. Ordinary boards—often culled from foundation formwork—or special material like tongue-and-groove, shiplap, or end-matched boards or plywoods can be used, but all do the same job, which is to add rigidity to the structure and act as a base for the final, finished floor.

Common 1-inch boards and shiplap are usually placed so they form a 45-degree angle with the joists (**Figure 6-32**). The angular placement provides bracing and allows you to lay T&G (tongue-and-groove) finish flooring on top either parallel or at right angles to the joists.

In both situations it's best to start away from a corner, selecting pieces that can run continuously or, at least, end close to a joist. Let all ends extend beyond the floor-frame perimeter and the edges of openings. The overhangs can be trimmed flush after they are nailed. Use cutoffs and short lengths to work back to corners.

Nail the boards at each joist crossing with two 8d nails if the boards are 6 inches or less in width. Secure wider boards with three 8d nails. End joints for both boards and shiplap should occur over a joist (**Figure 6-33**). If you use

FIG. 6-33

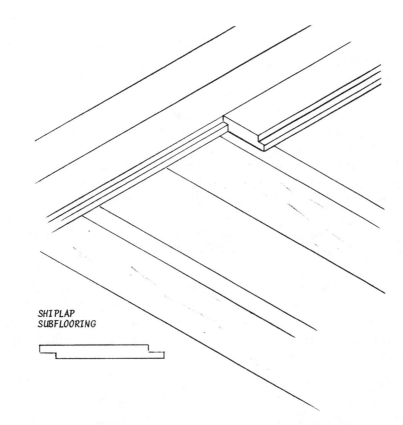

SHIPLAP
SUBFLOORING

FIG. 6-34

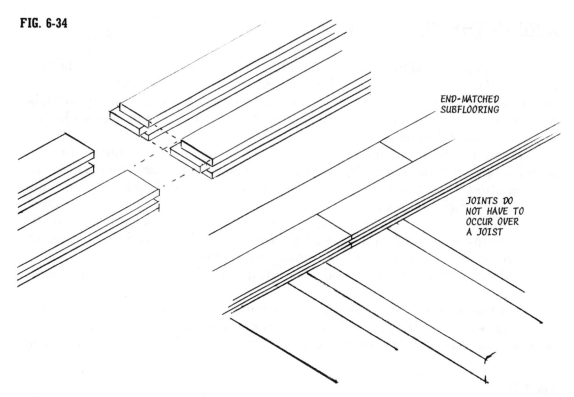

END-MATCHED
SUBFLOORING

JOINTS DO
NOT HAVE TO
OCCUR OVER
A JOIST

subflooring that is tongued and grooved on both edges and ends, then the end joints can occur anywhere (**Figure 6-34**).

It's best to lay board subflooring without clearance between joints. Often one is advised to leave gaps between boards if they are apt to be rained on during construction. The gaps are provided for drainage so water won't swell the floor or cause the boards to cup or warp. Another, perhaps better, preventive measure is to cover the subfloor with plastic sheeting.

Plywood has come a long way toward replacing boards as a subfloor material. The large sheets go down rapidly, sawing is reduced to a minimum, and the finished job presents a smooth, even base for the final flooring material. The laminating glue in plywood panels does vary, so you should select a type that is suitable for your own construction conditions. An intermediate glue is okay if only moderate delays are expected before protection is provided. Such a panel can take high humidity and even some water leakage temporarily. However, panels with exterior glue lines can withstand more adverse conditions. Since such panels are recommended for use when permanently exposed to weather or moisture, it makes sense to select exterior grades no matter what. It might cost a bit more, but it's cheap insurance. Use ⅝-inch-thick sheets even though ½-inch sheets are accepted by many codes.

Place the sheets so the long dimension runs at right angles to the joists and so the joints will be staggered as shown in **Figure 6-35**. Note the gaps recommended between joints. Don't try to measure this kind of thing; just use a strip of wood or heavy cardboard or some such thing as a gauge.

Nailing is done with 8d nails and follows the pattern shown in **Figure 6-36**.

FIG. 6-35

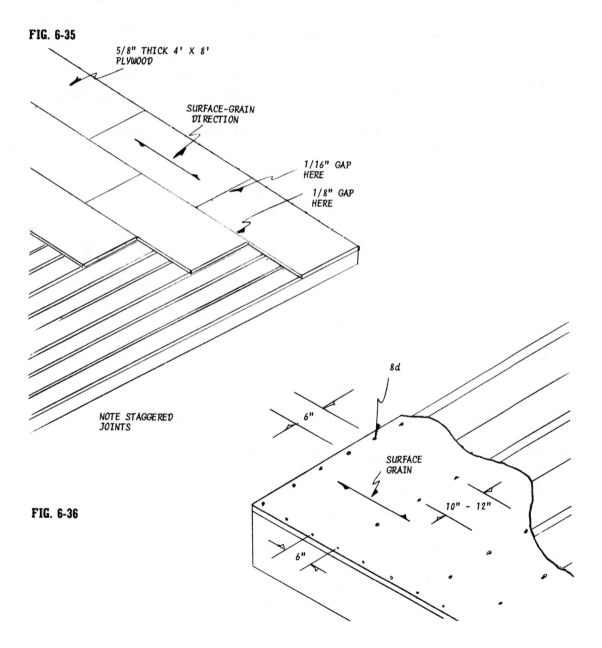

5/8" THICK 4' X 8' PLYWOOD

SURFACE-GRAIN DIRECTION

1/16" GAP HERE

1/8" GAP HERE

NOTE STAGGERED JOINTS

8d

6"

SURFACE GRAIN

10" - 12"

6"

FIG. 6-36

Drive the nails so they will be centered in the joists and headers. Snapping chalk lines will help you achieve good nailing patterns. Drive nails only until they are flush with the plywood. Hammering beyond that will only cause dents in the subfloor. It's a good idea to plan the placement of sheets so that no piece will span less than three or four joists.

THE 2-4-1 PLYWOOD FLOOR SYSTEM

This system calls for T&G plywood panels that are usually 1⅛ inches thick and function as both structural subflooring and underlayment. Because the thickness of the panels provides considerable strength, codes often permit floor-frame designs with 2× joists spaced 32 inches O.C., as shown in **Figure 6-37**, or application directly to 4× girders spaced 4 feet apart, as shown in **Figure 6-38**.

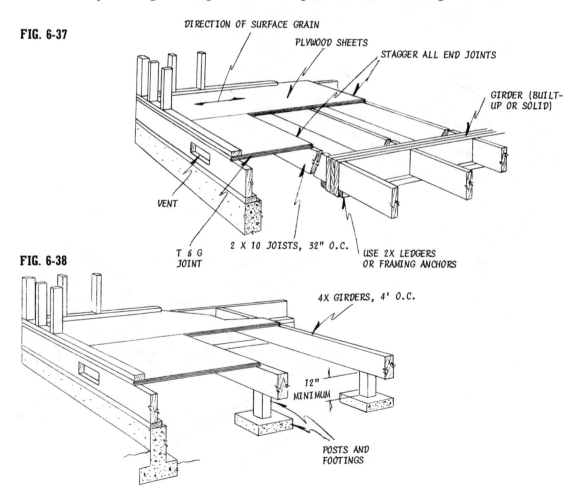

FIG. 6-37

DIRECTION OF SURFACE GRAIN

PLYWOOD SHEETS

STAGGER ALL END JOINTS

GIRDER (BUILT-UP OR SOLID)

VENT

T & G JOINT

2 X 10 JOISTS, 32" O.C.

USE 2X LEDGERS OR FRAMING ANCHORS

FIG. 6-38

4X GIRDERS, 4' O.C.

12" MINIMUM

POSTS AND FOOTINGS

The system can be modified to gain additional stiffness generally or in selected areas where heavy traffic is anticipated or where a very thin vinyl finish floor, for example, will be used.

2×4 blocking, placed 24 inches O.C. and toenailed between girders, will increase stiffness as much as 43 percent (**Figure 6-39**). This is reduced quite a bit if the blocking is spaced 48 inches.

Blocking that is placed diagonally at an angle of 45 degrees will be easier to nail and will provide even more stiffness than straight blocking (**Figure 6-40**). The blocking will be most effective if it crosses panel joints at midspan.

A third method uses "strongbacks," which are 2×4s placed flat and nailed or screwed to the bottom of the plywood sheets midway between the beams, as shown in **Figure 6-41**.

FIG. 6-39

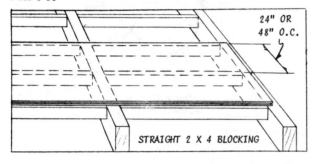

FIG. 6-40

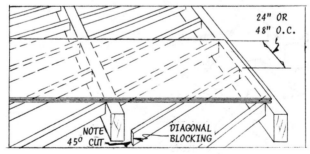

FIG. 6-41

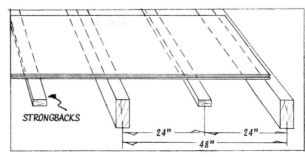

For a low-profile floor that permits the 2-4-1 plywood to be attached directly to sills, the girders can be set in pockets in a concrete wall as in **Figure 6-42**, or on posts as in **Figure 6-43**, which might be easier to do with a block wall.

This is just a brief presentation of an intriguing floor system. In-depth technical data is easily available; see the back of the book for sources of information.

FIG. 6-42

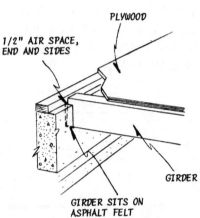

1/2" AIR SPACE, END AND SIDES

PLYWOOD

GIRDER

GIRDER SITS ON ASPHALT FELT

FIG. 6-43

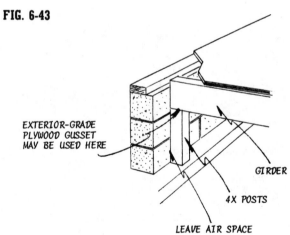

EXTERIOR-GRADE PLYWOOD GUSSET MAY BE USED HERE

GIRDER

4X POSTS

LEAVE AIR SPACE

7

FRAMING THE WALLS

View wall frames as strong skeletons consisting of vertical and horizontal components which lock together physically and with nails. The framing supports upper structures like ceilings and roofs and also acts as a base on which you can nail outside and inside coverings, and in which you can place other essentials like electrical wiring, pipes, heating ducts, insulation, and the like. The term "wall frames" also includes room partitions that are erected inside the perimeter walls.

A complete wall frame, with nomenclature, is shown in **Figure 7-1**.

FIG. 7-1

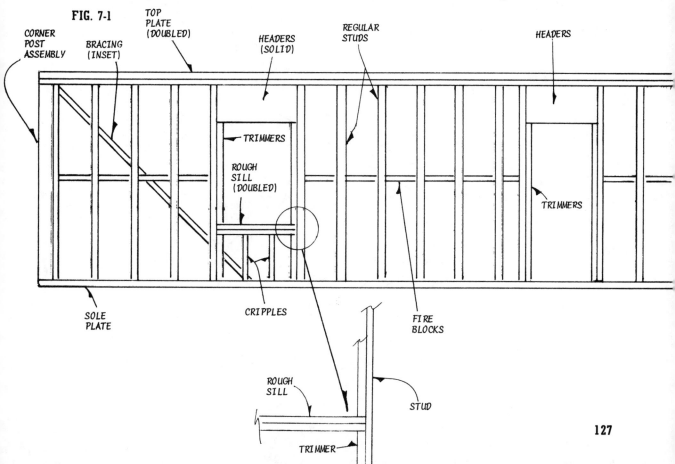

The one-piece header, usually 4×12 stock, is in common usage today because it reduces installation time, as compared to other methods we will show, and because it automatically establishes a common height for all openings with a minimum amount of measuring. Another important fact is that the 4×12 has enough strength to span just about any opening so the chore of having to determine individual header sizes is eliminated.

The bracing may not be needed when types of sheathing that supply lateral reinforcement are used under sidings.

Whether trimmers run full length or break at the rough sill seems to be an arbitrary decision. Arguments for the latter procedure are that shorter pieces of material may be used and some amount of toenailing is eliminated.

START WITH THE SOLE PLATE

Long pieces of 2×4 material are placed flat and nailed along the perimeter of the subflooring as in **Figure 7-2**. Note that the staggered pattern will place alternate

FIG. 7-2

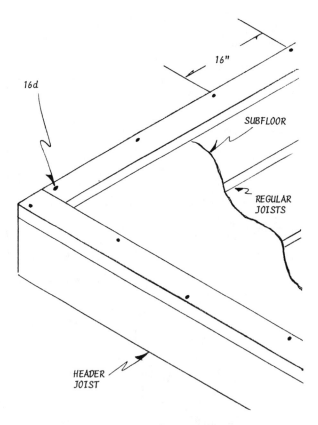

16"

16d

SUBFLOOR

REGULAR
JOISTS

HEADER
JOIST

nails in either the joist header or a joist. Nails through plates that follow a stringer joist are not staggered but should follow the same 16-inch spacing.

Do not attempt, at this point, to make cuts in the plate where door openings occur. Instead, make the plates continuous and do the sawing-out after the wall framing is complete (**Figure 7-3**).

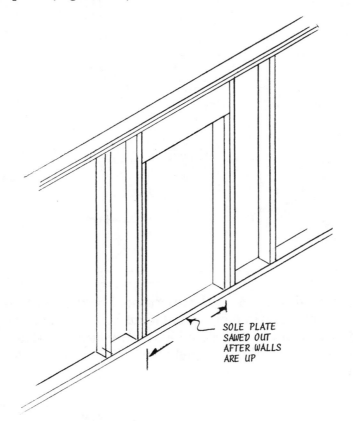

SOLE PLATE
SAWED OUT
AFTER WALLS
ARE UP

STUD SPACING AND LAYOUT

Studs are usually spaced 16 inches O.C., although there are designs (and in some areas, codes that permit it) that call for 24-inch spacings. Actually, some would say that considerably greater spacing would still provide adequate strength, and the spacing of 16 inches (or 24 inches) is established more for accommodating sheathing materials than it is for support. The abbreviation "O.C.," of course, means "on centers"; it is the distance, in this case, between the center of one stud and the center of the next one (**Figure 7-4**).

FIG. 7-4

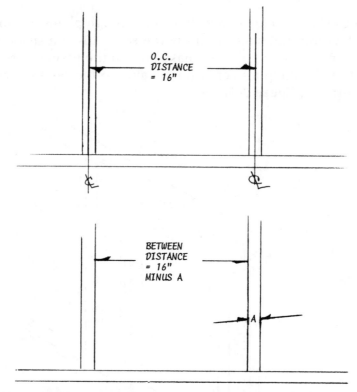

The spacing of studs must never be interrupted by the addition of other components such as cripples or trimmers. Regular studs may become cripples or even trimmers, or additional studs may be added to the basic structure, as shown in **Figure 7-5**, but when the wall is complete, you should be able to find a nailing surface every 16 inches horizontally.

FIG. 7-5

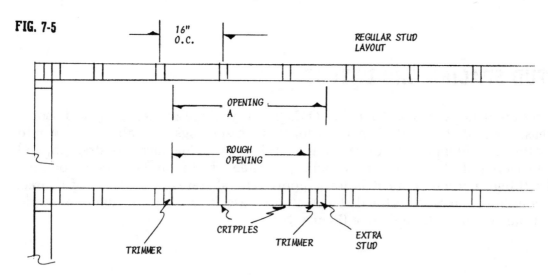

POSTS

Posts are assemblies of 2×4 material that are used at corners of wall frames and where interior partitions intersect or abut a perimeter wall. The assembly of the pieces, even though designs can vary, pretty much indicates the direction of the walls. The posts that are erected at all outside corners may be as in **Figure 7-6**. Areas A and B, which face the inside of the structure, provide nailing surfaces for interior wall coverings.

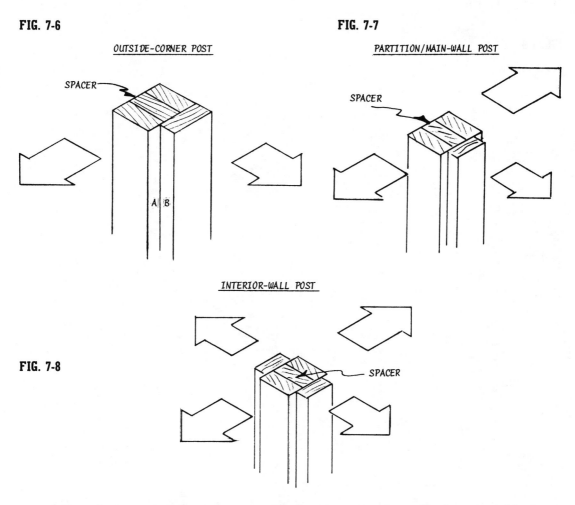

FIG. 7-6

OUTSIDE-CORNER POST

SPACER

A B

FIG. 7-7

PARTITION/MAIN-WALL POST

SPACER

INTERIOR-WALL POST

FIG. 7-8

SPACER

When the post involves a partition abutting a main wall, the assembly is as in **Figure 7-7**. When interior walls intersect, then the post assembly is as in **Figure 7-8**. In each case, you can see that the post assembly, together with plates, ties walls together and supplies nailing surfaces for other materials.

ASSEMBLING POSTS

Be selective when you choose the material for post assemblies or for studs. A slight bow isn't bad, since nailing will pull it into line when it is used for posts, and blocking will straighten it when it is used as a stud, but a crook is something else (**Figure 7-9**). When such pieces are encountered during wall-covering procedures, the crown must often be removed with a plane to eliminate the high spot. It's better to cut such pieces into the shorter lengths required for blocking, braces, ties, and so on.

FIG. 7-9

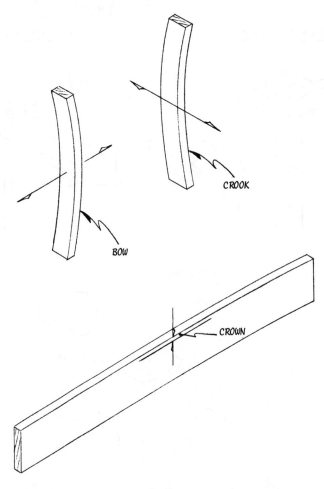

A basic post design starts by assembling two studs and three spacers (**Figure 7-10**). These pieces are nailed together to form a solid unit. Use 10d nails throughout, spacing them about 12 inches apart and following a staggered pattern.

FIG. 7-10

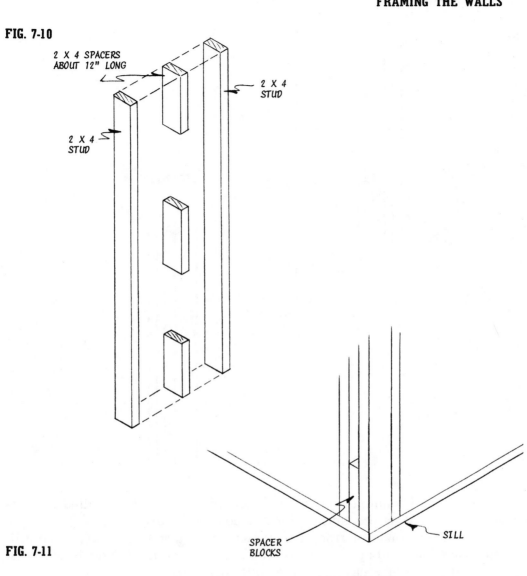

2 X 4 SPACERS
ABOUT 12" LONG

2 X 4
STUD

2 X 4
STUD

FIG. 7-11

SPACER
BLOCKS

SILL

SECTION

When the post is for a corner, add a third stud as in **Figure 7-11**, driving 10d nails so the addition will be locked to both studs and spacers. This design is usable for both platform and balloon framing.

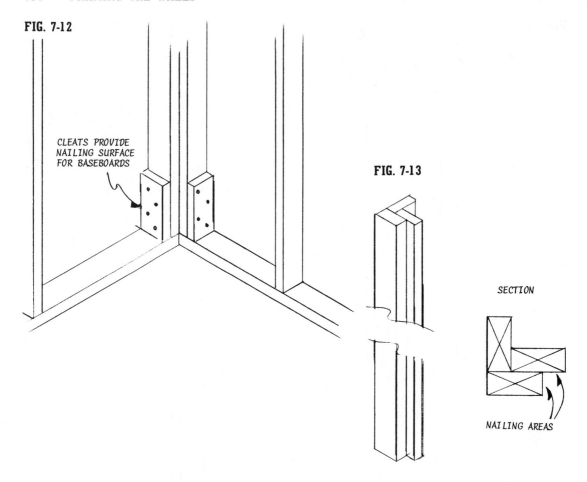

FIG. 7-12

CLEATS PROVIDE
NAILING SURFACE
FOR BASEBOARDS

FIG. 7-13

SECTION

NAILING AREAS

Often, after wall framing is complete, short pieces of 2×4 are nailed to posts to provide additional nailing surface for baseboards (**Figure 7-12**).

Another design for corner posts is shown in **Figure 7-13**. Here, three full-length studs are used and no spacers (or blocking) are required. Note how the assembly provides nailing areas for inside wall coverings.

Solid connections must occur where partitions abut outside walls, and, typical of all posts, the assembly must provide nailing surfaces. This can be done by using extra studs in the outside wall, spacing them with a third full-length stud and then adding a fourth as shown in **Figure 7-14**. The regular studs in the wall may be used as part of this assembly if their location happens to be okay. Partitions can terminate between regular studs if the opening is organized as shown in **Figure 7-15**. Attach the 1×8 board to each of the 2×4 blocks between the studs with 8d nails. This idea can be used anyplace between studs. It is not necessary for the partition to fall at the midpoint.

FIG. 7-14

FIG. 7-15

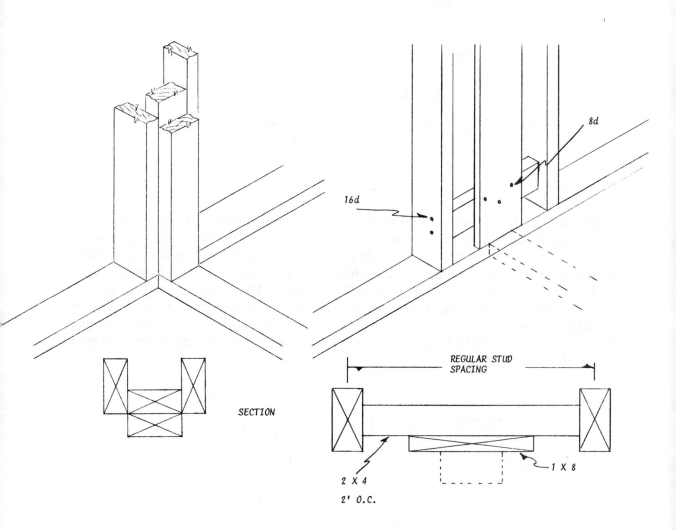

8d

16d

REGULAR STUD
SPACING

SECTION

2 X 4

2' O.C.

1 X 8

MARKING THE SOLE PLATE

Stud locations can be marked off with a steel measuring tape, many of which are specially marked for correct spacing of studs. This can lead to human error, so a better idea is to make a marking jig similar to the one we described for locating

FIG. 7-16

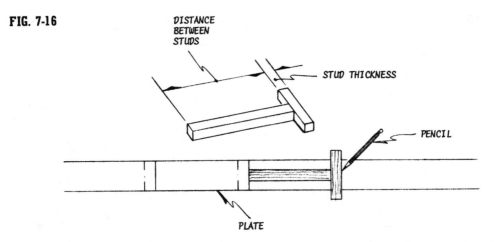

joists or to make a T-gauge (**Figure 7-16**). The T-gauge indicates spacing between studs and actual stud position. At any rate, the center of the first stud is 16 inches from a corner; other studs run 16 inches O.C. continuously. When you do the sole plate, place top-plate material alongside so layout marks can be carried across both pieces. Indicate studs by marking Xs (**Figure 7-17**). The next step is to show the positions of trimmers and cripples. The chances are that some studs will be eliminated because of door openings and others will become cripples and, maybe, trimmers. As shown in **Figure 7-18**, mark the layout so you will know the role each piece will play.

FIG. 7-17

FIG. 7-18

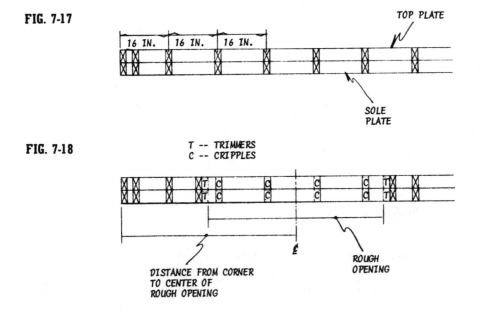

The sizes of rough openings depend, of course, on the units you plan to install. In modern construction, doors and windows are purchased as complete assemblies which are just slipped into place. You can, for example, buy prehung doors which arrive already hinged, with holes drilled for hardware and with jambs in place. The manufacturer's literature will tell how large the rough openings must be—so, it makes sense to know what you are going to use before you do the wall framing. It isn't necessary to have the units delivered before you are ready for them, but you should know how to prepare for them.

Since wall frames will contain extra studs, trimmers, cripples, and headers, it's a good idea to mark their location on the subfloor so you can locate them later, if necessary, after the walls are covered. If you don't have a plan on paper now, make a rough sketch, at least, and file it for future reference.

ERECTING THE STUDS

Each stud is toenailed to the plate with two 8d nails on each side (**Figure 7-19**). Often, if diagonal boards or plywood will be used as sheathing, the carpenter will drive just one nail on each side. I have also seen workers drive two *extra* nails, one into each narrow edge of the stud. Housebuilders have various systems, but the four-nail procedure is pretty standard and safe.

FIG. 7-19

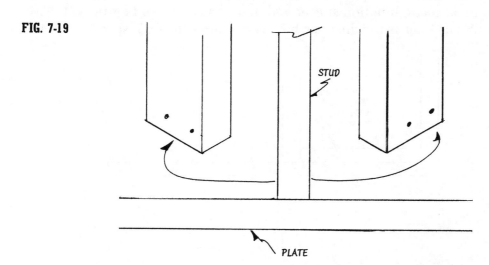

Toenailing can be a problem only if you don't take steps to keep the stud firmly in place as you drive. The angle of the nail to the surface of the plate should be about 60 degrees, while the head-height should be about 1 inch (**Figure 7-20**).

FIG. 7-20

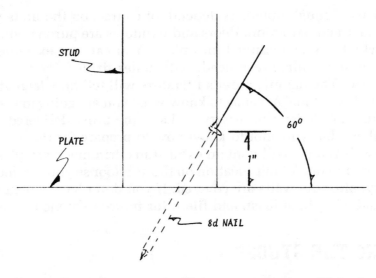

This makes driving easier and assures sufficient penetration. This doesn't mean you must work with a protractor and a scale to place each nail, but you should strive for what is structurally correct even if you must practice a bit.

Professionals do so much of this thing that they work by bracing the stud against a toe. That's okay but chancy — it's not likely that studs that twist a bit will be removed and reset.

You can make a combination spacer and brace as shown in **Figure 7-21**. This is a piece of 2×4 as long as the distance between studs — the O.C. spacing less the

FIG. 7-21

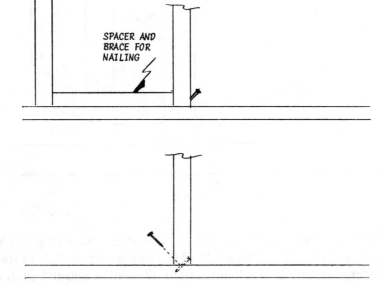

thickness of one stud. Put it in place and step on it as you drive two of the nails. The nails on the opposite side will not be a problem, since the stud will be pretty firmly anchored.

A more elaborate jig is shown in **Figure 7-22**. It does the job of the spacer-brace described above but includes two side pieces which work to keep the stud edges square to the plate. The clearance bevels at each end of the 2×4 are required so the jig can be tilted up after the first two stud nails are driven (**Figure 7-23**).

The jig, or the more simple spacer, may be used at the top of the studs when upper plates are being installed, as well as at the bottom ends.

FIG. 7-22

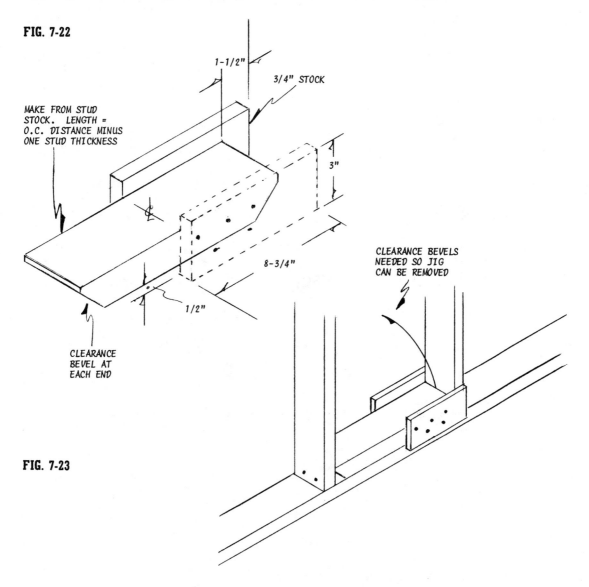

FIG. 7-23

TOP PLATES

The top plates are two pieces of 2×4, arranged so they form types of lap joints over corners, as in **Figure 7-24**; where partition walls abut main walls, as in **Figure 7-25**; and where partitions intersect, as in **Figure 7-26**.

Work with long, straight pieces of lumber. End joints in the bottom member should always occur over a stud. End joints in the top member should be planned so they will be about 4 feet away from those in the bottom piece.

FIG. 7-24

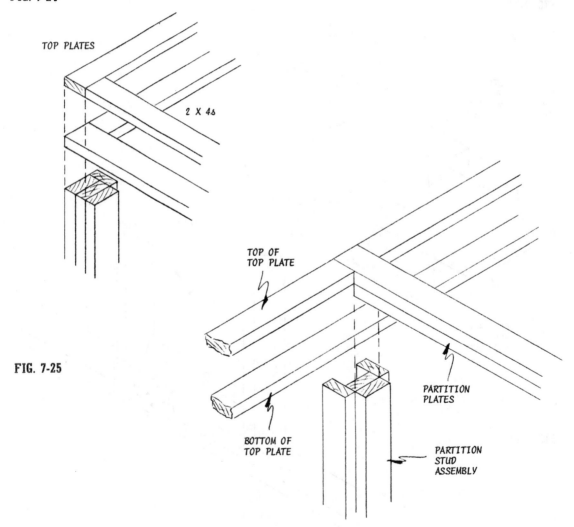

FIG. 7-25

FIG. 7-26

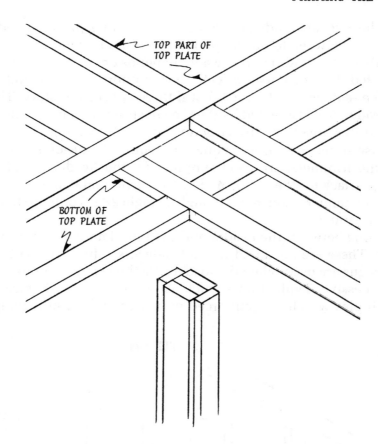

TOP PART OF
TOP PLATE

BOTTOM OF
TOP PLATE

Surface-nail the bottom part of the plate to each post and stud with two 16d nails. The top piece is secured to the lower one with 10d nails. Use two at each end and space those between about 16 inches apart in a staggered pattern.

The assembly of the top plates should result in solid connections to all posts and studs and should be further strengthened by the interlocking action of the various pieces.

HEADERS

Headers span across the top of openings in a wall to supply support for the weight of upper structures. Most times, and especially in bearing walls, when a stud is cut or removed, a header is installed to compensate. All true headers rest on trimmers, so their total length must equal the width of the rough opening plus the thickness of two trimmers, usually 3 inches.

Built-up headers are made of two pieces of 2× stock placed on edge and with plywood spacers between to build up total thickness to match the width of other framing members. The width of the header depends on the span and the load to be carried. Doubled 2×8s can span 8 feet when supporting a roof and ceiling. The span should be reduced by at least 1 foot if there is a second story. This is typical of how the header size is affected by the type of structure. Local codes can supply information in this area to assure structurally sound framing.

Always use studs, even if extra ones are required, at each end of the header. These, plus the trimmers, are good supports for the header but also provide a broad nailing surface for the inside and outside trim that comes later. Double the rough sill for the same reason, even though a single 2×4 would do the job as far as strength is concerned.

Any opening between the header and the top plate is filled in with cripples (**Figure 7-27**). These are toenailed to the header like studs but with 10d nails.

Truss designs are used when the load above the header is unusually heavy or the span is excessive. Typical ones, such as those shown in **Figure 7-28**, are no more than cripples which are reinforced with diagonal, inset bracing.

FIG. 7-27   **FIG. 7-28**

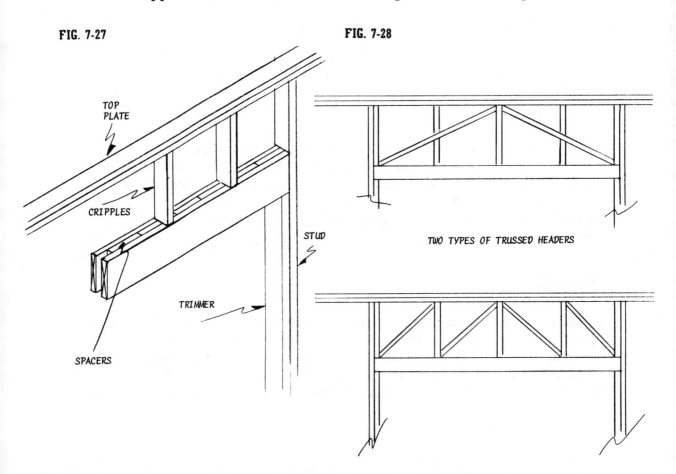

TOP PLATE

CRIPPLES

SPACERS

TRIMMER

STUD

TWO TYPES OF TRUSSED HEADERS

A typical nailing pattern for a rough window opening is this. Nail the studs at each end of the header with four or more 10d nails depending on the header's width. Trimmers and studs are nailed together with 10d nails spaced about 16 inches apart and following a staggered pattern. Toenail at the bottom end of the trimmers as if they were studs. Nail through the lower part of the sill into the cripples below with two 10d nails per cripple. Attach the upper part of the sill with 10d nails spaced about 8 inches apart and staggered. **Figure 7-29** shows a completed opening. The ends of the sill pieces may also be nailed from the outside of the trimmers, but this would require that the studs be placed last or that the rough opening (or at least the trimmers and sill) be done as a subassembly and then slipped into place. A way to work when the frame is not being done flat, to be tilted into place later, is to break the trimmers at the sill line so the sills can be end-nailed through the stud (**Figure 7-30**).

FIG. 7-29   **FIG. 7-30**

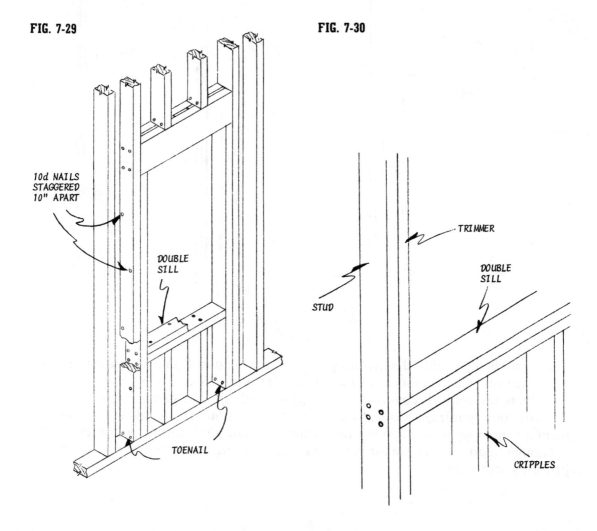

An advantage of using full 4×12 headers throughout, as shown in **Figure 7-31**, becomes apparent when a span requires a header that is so large there is little room between it and the top plate to install cripples. Flat blocking is often used in such cases but it too can be a nuisance to install. Many builders will then use full headers even though they might use other designs elsewhere.

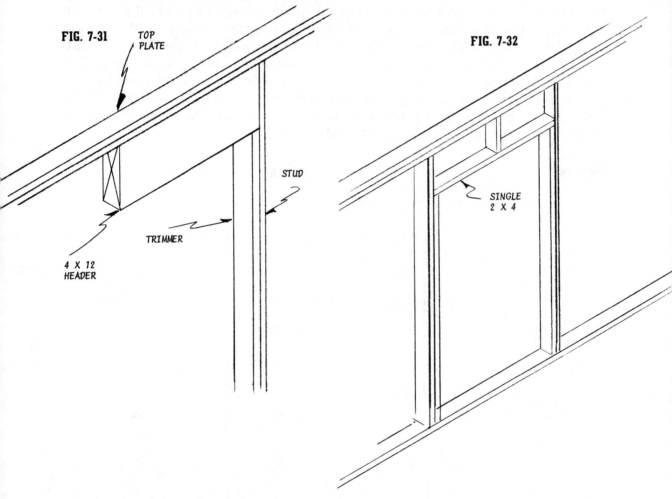

FIG. 7-31   TOP PLATE   STUD   TRIMMER   4 X 12 HEADER

FIG. 7-32   SINGLE 2 X 4

There is a possible disadvantage to the full header. Some designers say that excessive shrinkage can occur in a full header and cause cracks above windows and doors unless wall coverings are applied with special precautions.

All door openings through bearing partitions should have headers. When the header is not required, the opening is framed in simple fashion, as in **Figure 7-32** —enough to provide a base for jambs if a door is hung, or for trim if the opening is just a passthrough.

Walls should not be supported by fireplace masonry. Use headers as you would for any opening and provide at least 2 inches of clearance on all sides, as shown in **Figure 7-33.**

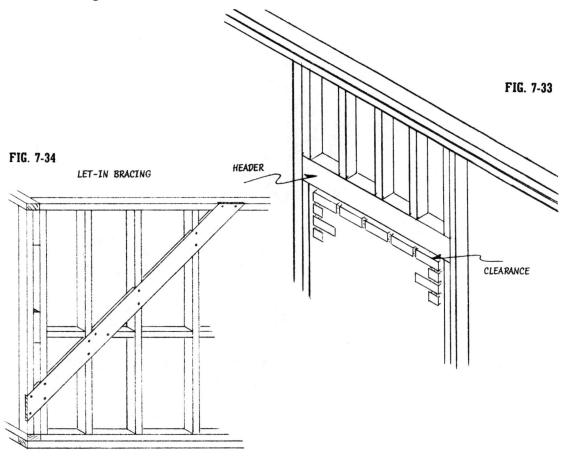

FIG. 7-33

FIG. 7-34

LET-IN BRACING

HEADER

CLEARANCE

BRACING

Bracing, used on exterior walls to resist lateral stresses, can be "let in" or "inset." The let-in type is usually done with 1×4 or 1×6 boards that fit snugly in notches cut in each crossed member of the frame (**Figure 7-34**). The notches must be cut carefully for the brace to function as it should. The best method is to position the brace and draw its outline across the frame members. Use a crosscut saw to make the side cuts as deep as the board is thick and then clean out the waste material with a chisel.

Attach the brace with three 10d nails at each end and two 10d nails at each crossing. Bracing application may vary to accommodate openings, as shown in **Figure 7-35**, and it can even take the form of a let-in ribbon (**Figure 7-36**).

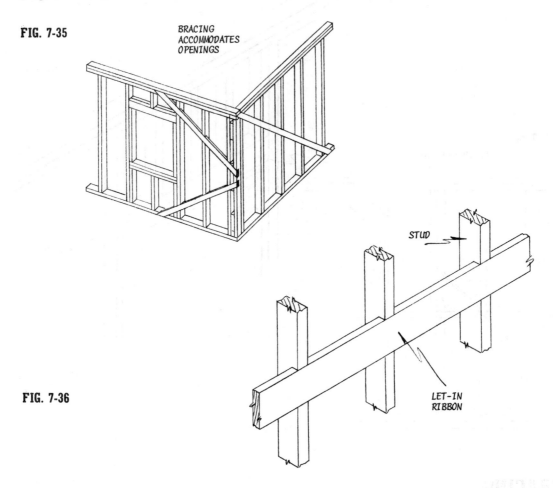

FIG. 7-35

BRACING ACCOMMODATES OPENINGS

FIG. 7-36

STUD

LET-IN RIBBON

Inset bracing is done with material that matches the cross-sectional dimensions of the studs (**Figure 7-37**). This calls for a lot of angular cutting and fitting, but most people feel it is easier to do than the many notches required for the let-in design.

Much depends on the design of the house, the severity of local conditions, and the type of exterior wall covering. In some situations—for example, if a heavy plywood is used as sheathing—it may not be needed at all. Some checking of local standards is in order.

FIG. 7-37

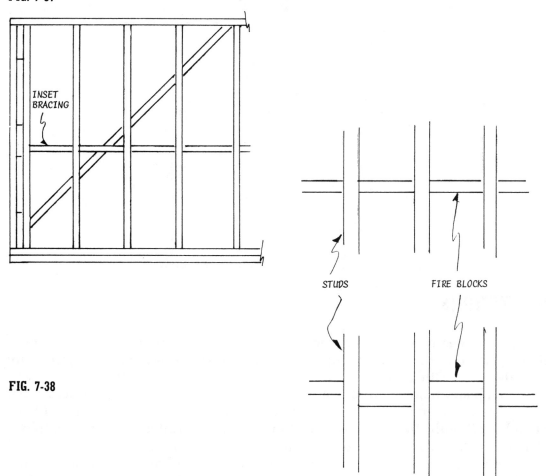

INSET
BRACING

FIG. 7-38

STUDS   FIRE BLOCKS

FIRE BLOCKING

These are 2× pieces running horizontally in a wall frame to break air spaces and so prevent drafts that would encourage fire to spread. They are safety factors but also do much to stiffen studs. The blocks may be installed on a line midway up the wall, but no law says you can't stagger them so they will be easier to nail (**Figure 7-38**).

Blocking can serve two purposes. A double line might be used, for example, to provide nailing surfaces for board-and-batten siding (**Figure 7-39**). In such cases, do not stagger the blocking since it would result in an unsightly nail pattern on the outside of the wall when siding is up.

FIG. 7-39

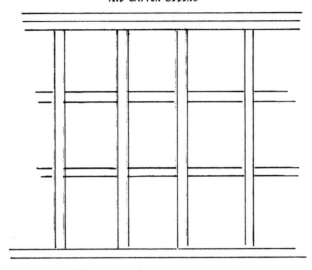

BLOCKING FOR BOARD-
AND-BATTON SIDING

PARTITIONS

Partition frames are erected after perimeter framing is complete. Most times, only those partitions that supply support for upper structures (bearing walls) are included now. Others that merely divide areas can be done after outside coverings are up and there is protection from the weather. An exception occurs when the design of the roof frame (certain types of roof trusses) requires support from outside walls only. In such cases, partitions can be done after the roof is complete. The idea is to make the structure rainproof as soon as possible.

Partitions are organized just like outside walls except that headers and, maybe, bracing are needed for bearing partitions only. Like all walls, the designs must provide nailing surfaces for the wall coverings to be attached later. Although headers are not usually required in nonbearing walls, it's a good idea to include trimmers around all openings if only because of the broader nailing surfaces they provide for any casing and trim that will be used to finish the openings after walls are covered.

Like the sole plates in exterior walls, those for partitions are run continuously and sawed away at door openings after framing is complete.

Framing that is done for nooks, closet walls, and the like is often of 2×2 material or of standard 2×4 stock set with the 4-inch side parallel to the plates. The advantages here are savings in space and money, but don't do either unless the thinner sections abut conventional walls and are relatively short. You don't want shaky walls even if they are just for a closet.

SOME SPECIAL CONSIDERATIONS

Walls can be made thicker by using oversize material for plates and studs (**Figure 7-40**). You might wish to consider this for all walls so that extra insulation can be used, and for bathroom and utility-room partitions that must enclose considerable plumbing. Often, thicker inside walls are designed as in **Figure 7-41**, with standard 2×4s staggered but still maintaining the 16-inch-O.C. distance on each side of the wall. This will give you more room for pipes or you can weave

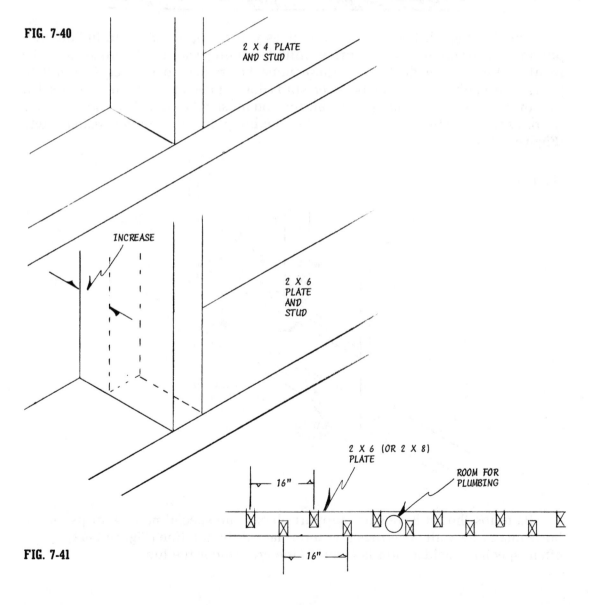

FIG. 7-40

2 X 4 PLATE
AND STUD

INCREASE

2 X 6
PLATE
AND
STUD

2 X 6 (OR 2 X 8)
PLATE

ROOM FOR
PLUMBING

16"

16"

FIG. 7-41

soundproofing material in such a wall (**Figure 7-42**). Such soundproofing may be desirable when kitchen and living room, or kitchen and utility room, for example, have a common wall.

FIG. 7-42

INSULATING MATERIAL

Anticipating what will come later gives you the opportunity to include important construction details during initial framing stages. Openings may be required for heating ducts, and these must be framed to correct size and to provide strength. An opening between studs doesn't need more than a block set at the correct height. A wider opening means you must cut a stud. The code may say it's okay to use a single 2×4 "header" but doubling up will provide greater safety (**Figure 7-43**).

FIG. 7-43

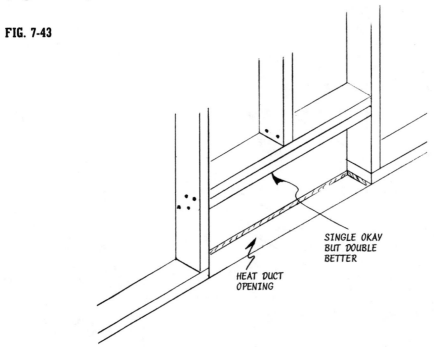

SINGLE OKAY BUT DOUBLE BETTER

HEAT DUCT OPENING

Bathtubs should have extra support blocks and special nailing strips placed horizontally for wall coverings that terminate at the tub line (**Figure 7-44**). Quite often, special consideration is given to the area under the tub.

FIG. 7-44

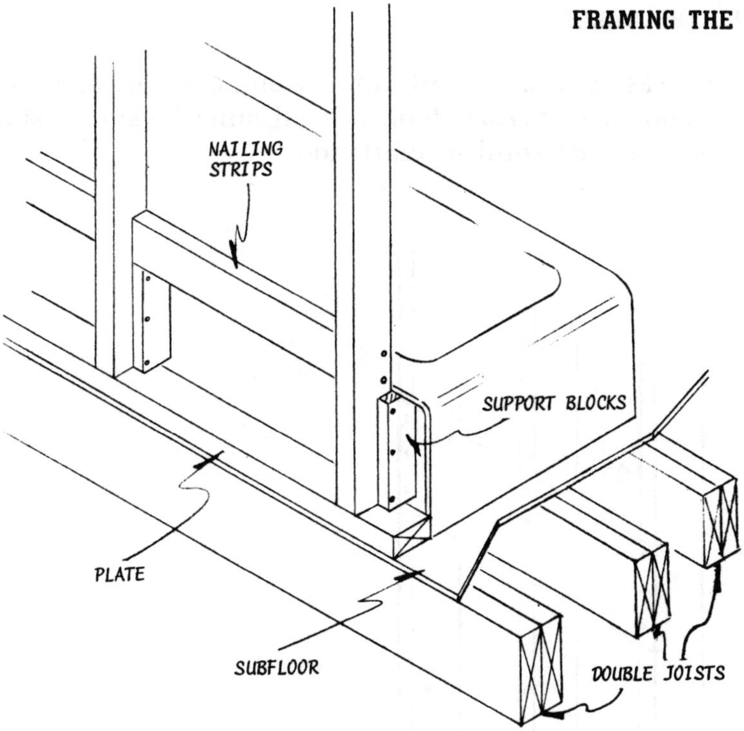

NAILING
STRIPS

SUPPORT BLOCKS

PLATE

SUBFLOOR

DOUBLE JOISTS

FIG. 7-45

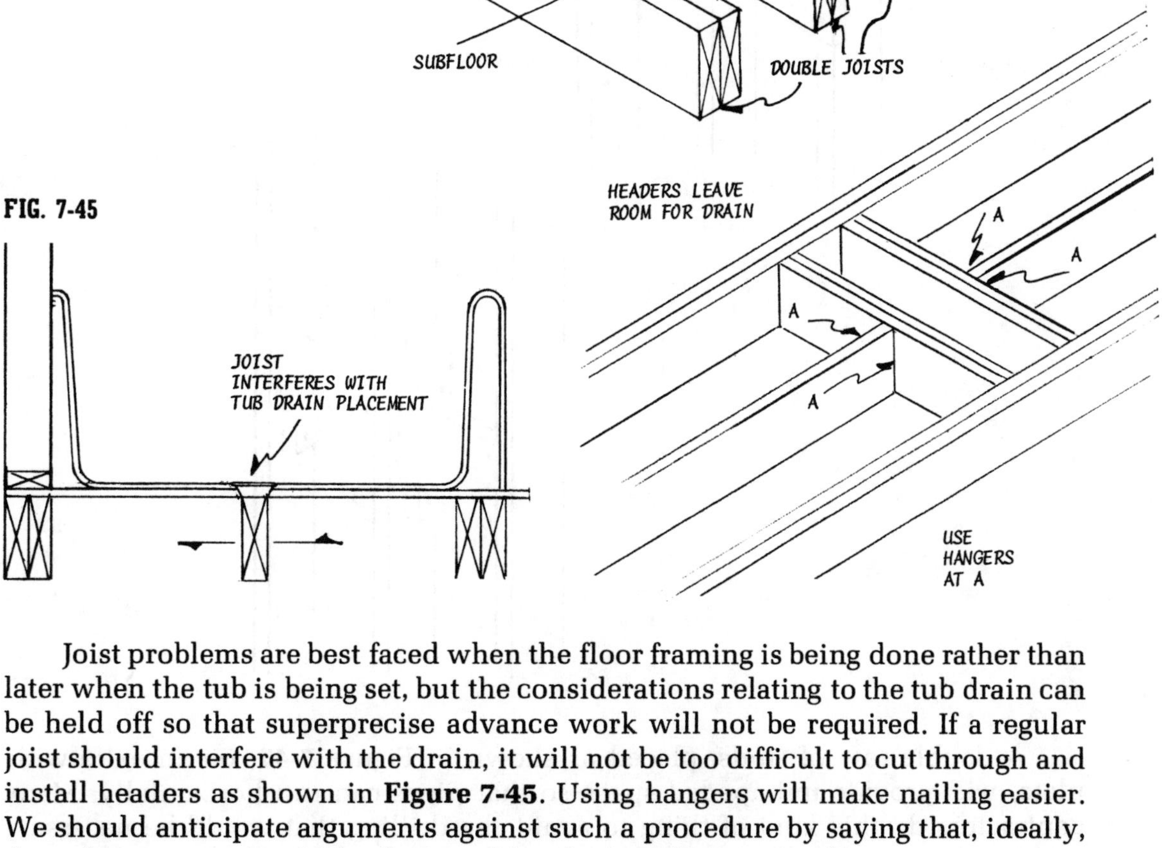

JOIST
INTERFERES WITH
TUB DRAIN PLACEMENT

HEADERS LEAVE
ROOM FOR DRAIN

A

A

A

A

USE
HANGERS
AT A

Joist problems are best faced when the floor framing is being done rather than later when the tub is being set, but the considerations relating to the tub drain can be held off so that superprecise advance work will not be required. If a regular joist should interfere with the drain, it will not be too difficult to cut through and install headers as shown in **Figure 7-45**. Using hangers will make nailing easier. We should anticipate arguments against such a procedure by saying that, ideally, the substructure should be designed for the installation in advance.

Backings for wash basins or wall cabinets and even towel rods and such can be provided as shown in **Figure 7-46**. A wall-hung lavatory usually hangs on brackets and these require solid supports too.

FIG. 7-46

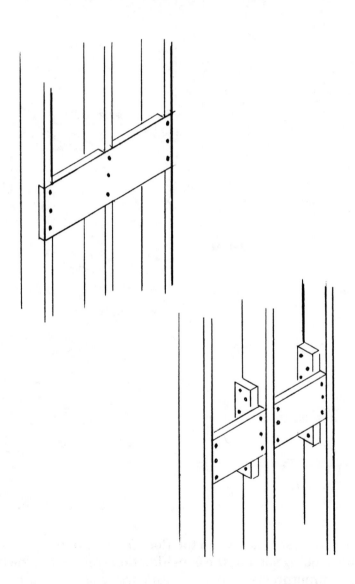

A double row of studs, placed as shown in **Figure 7-47**, on an extra-wide plate, will provide unimpeded space so that pipes can run horizontally. Many contractors do this kind of thing to avoid the drilling that would be necessary with conventional stud placement.

FIG. 7-47

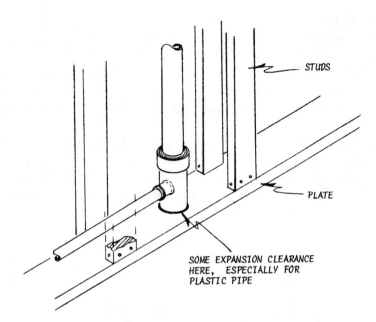

STUDS

PLATE

SOME EXPANSION CLEARANCE
HERE, ESPECIALLY FOR
PLASTIC PIPE

Large vents that pass through plates can weaken the area considerably; so reinforcement, in the form of "scabs," is used around the opening (**Figure 7-48**). Make the scabs from 2×4 stock and cut them long enough so they span across at least two studs.

FIG. 7-48

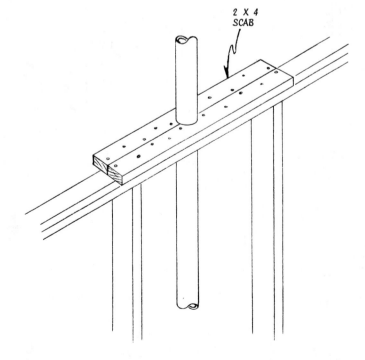

2 X 4
SCAB

Now is the time to provide for some extra storage by using some of the wasted space in the partition walls. Ideas like that in **Figure 7-49** are usable in kitchen and utility rooms for storing canned goods and the like, and even in the bedroom for things like shoes. All you need at the framing stage are two pieces of 2×4 placed horizontally to tell the top and bottom of the cabinet. You can line the interior and add the shelves later. Do the frame and hang the door after the wall covering is complete.

Prefab fireplaces require enclosure frames that provide clearance specified by the manufacturer between wood and metal. Note that a header, when required, is supported by trimmers (**Figure 7-50**). In some situations, depending on the design of the fireplace, it makes sense to actually do the installation while rough framing is going on.

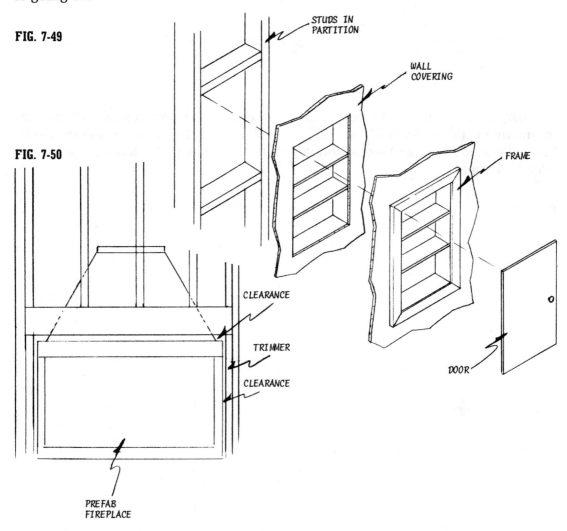

FIG. 7-49

FIG. 7-50

STUDS IN PARTITION

WALL COVERING

FRAME

CLEARANCE

TRIMMER

CLEARANCE

DOOR

PREFAB FIREPLACE

FRAMING ANCHORS AND HANGERS

By using readymade pieces of hardware like those shown in **Figure 7-51**, you can probably frame an entire house without having to drive a single nail at an angle, and the house will be stronger. They are more costly than nails but you may find the convenience is worth it, at least for some nailing jobs.

You'll find that the name of the piece often identifies its application. "Rafter anchor," "joist anchor," "post base," and the like are typical. Some types come flat so you can bend them to suit; all are secured with special nails.

The examples shown here are just a few of the many available. Best bet is to visit a local building-supply yard or request a catalog from the address noted at the back of the book.

FIG. 7-51   FRAMING ANCHORS AND HANGERS

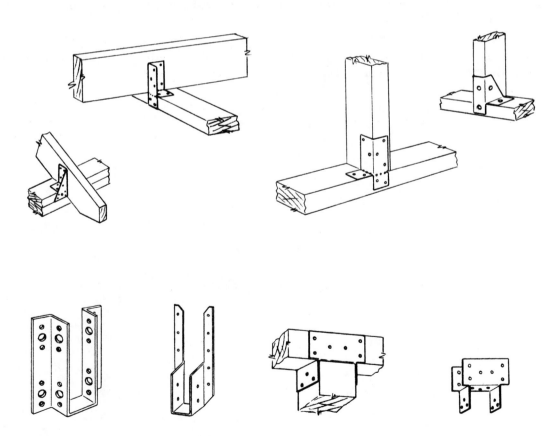

BUILD UP OR BUILD FLAT?

Most professionals build wall frames—even to the point of applying sheathing—as subassemblies which they then tilt into place. Anyone can work this way if there is enough manpower available to lift the frame. Start by marking the positions of studs, trimmers, and so on on the top plate and the sole plate. Place the pieces on edge and space them with a few studs. Incidentally, studs are not cut to length on the job. They are purchased as "studs" already cut to correct length.

Place the remaining studs in correct position and nail through each plate into each stud with two 16d nails. Cut headers and trimmers to correct length; install the trimmer and then the header. You can choose to do the cripples and the rough sill now or wait till after the walls are up. Add partition posts and blocking and any bracing that is used.

Work carefully when doing framing regardless of whether you do it a piece at a time vertically or as an assembly that you tilt. Check often with a level and a square, and use as much temporary bracing as you need to keep the wall plumb. Do not remove the bracing until all walls are up and you are sure of alignment and rigidity.

This is called "rough" framing, true, but plumbness and squareness contribute to strength and will be appreciated later when it is time to install windows and doors and wall coverings.

8

INSTALLING CEILING JOISTS

Ceiling joists are installed after wall and partition frames are up. They pretty much resemble the joists in the floor except that they are not done with perimeter headers and they may differ dimensionally depending on the material used for the ceiling and the load of the upper structure. In a two-story house, the ceiling joists at the second level become floor joists, and it is reasonable to assume they must be as sturdy as those used below, and constructed in similar fashion.

Lighter joists are okay under unused attic space but should bulk a bit more if they must support a plaster ceiling rather than, for example, an acoustical-tile installation. Other factors to consider are whether the attic, if any, will be used for storage and whether you might someday want to make wasted space livable. Joists also act as ties between opposite walls and resist stresses that could cause walls to lean out (**Figure 8-1**). All in all, it would seem to make sense to be generous with

FIG. 8-1

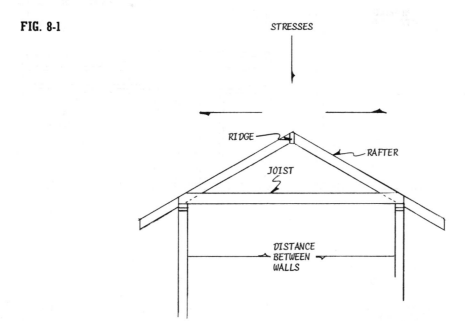

STRESSES

RIDGE — RAFTER

JOIST

DISTANCE BETWEEN WALLS

ceiling joists instead of doing the opposite to save some money on material. At any rate, local codes may make the decision for you. You can also think in terms of preassembled roof trusses, in which the bottom chord of the truss does the job of the joist (**Figure 8-2**).

FIG. 8-2

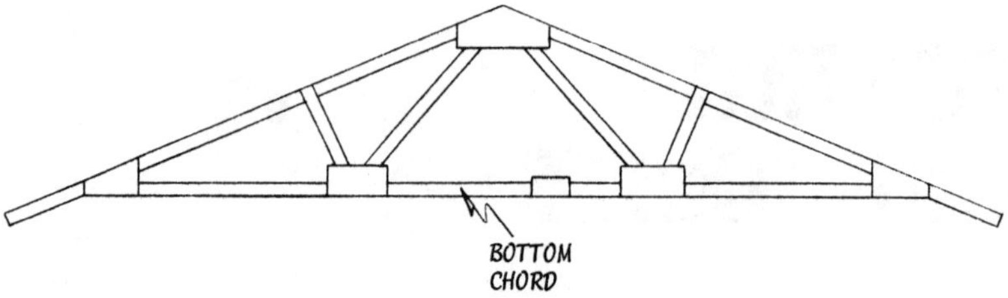

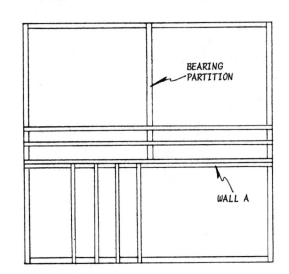

BOTTOM
CHORD

It's a good idea to run joists across the narrower dimensions of the wall frame so spans can be reduced even if it means changing the direction of the runs. In the example in **Figure 8-3**, the long joists receive intermediate support from the bearing partition. Shorter joists are placed at right angles and are supported at each end on wall plates (**Figure 8-4**). Wall A should then be regarded as a bearing wall.

FIG. 8-3

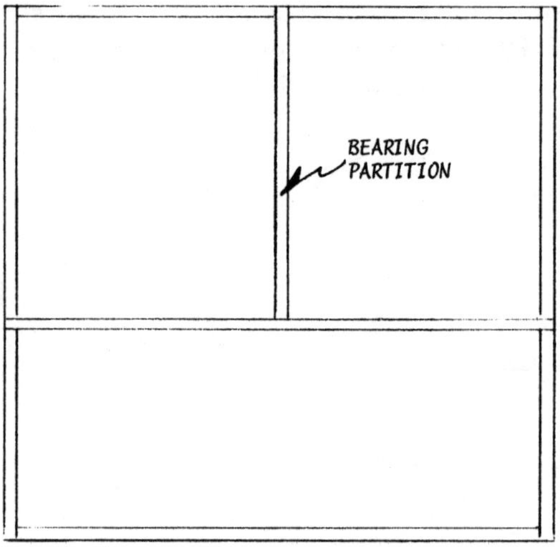

BEARING
PARTITION

FIG. 8-4

BEARING
PARTITION

WALL A

The joists running across the wall can be one piece or made up of shorter pieces if they are overlapped or spliced at the wall (**Figure 8-5**).

FIG. 8-5

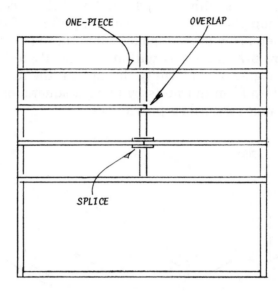

ONE-PIECE   OVERLAP

SPLICE

Another situation that calls for special joist considerations can occur when the roof has little slope (low pitch) and the joist run is parallel to the roof edge. Here, the roof slope may not make it over the outside joists, so shorter members are placed at right angles to the regular joists (**Figure 8-6**).

FIG. 8-6

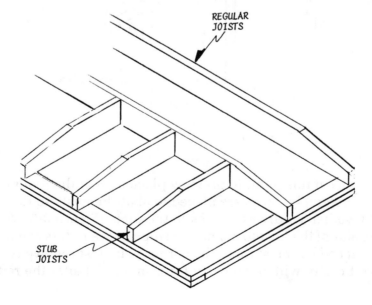

REGULAR
JOISTS

STUB
JOISTS

SPACING JOISTS

The spacing can vary but, as with studs, 16 inches O.C. is accepted as a standard. Because the doubled top plate of the wall frame supplies sufficient strength, joists can be placed anywhere. However, it makes some sense to set them over the stud positions as in **Figure 8-7**, since this will automatically tell the correct spacing and will establish a degree of conformity throughout the framing system. Layout marks can be carried up from the studs by using a square, or you can work with one of the spacing gauges described previously.

 FIG. 8-7

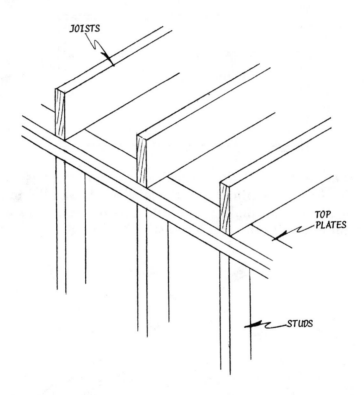

It's a good idea at this point to preview placement of the roof rafters that will be added later. Preferably, these are placed to abut the joists so the two members can be nailed together solidly as well as to the plate (**Figure 8-8**). Any cut that is required at the end of the joist to match the slope of the rafters can be done before the pieces are placed but most times the corner you must remove is small enough so you can do it easily with a handsaw (or even an ax) after the rafters are up.

FIG. 8-8

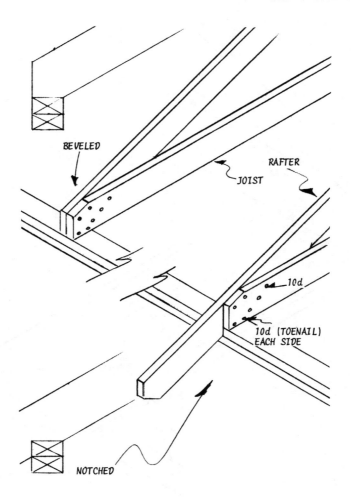

BEVELED

JOIST

RAFTER

10d

10d (TOENAIL)
EACH SIDE

NOTCHED

NAILING THE JOISTS

Toenail the joists to each perimeter plate with two 10d nails on each side. In areas where high winds occur frequently, metal strap anchors are often added as shown in **Figure 8-9**, to provide more strength. Other types of readymade framing anchors may also be used.

Joists that cross a partition are toenailed to the plate and 2×6 blocks are added between them as in **Figure 8-10** to provide nailing surfaces for wall-covering materials.

FIG. 8-9

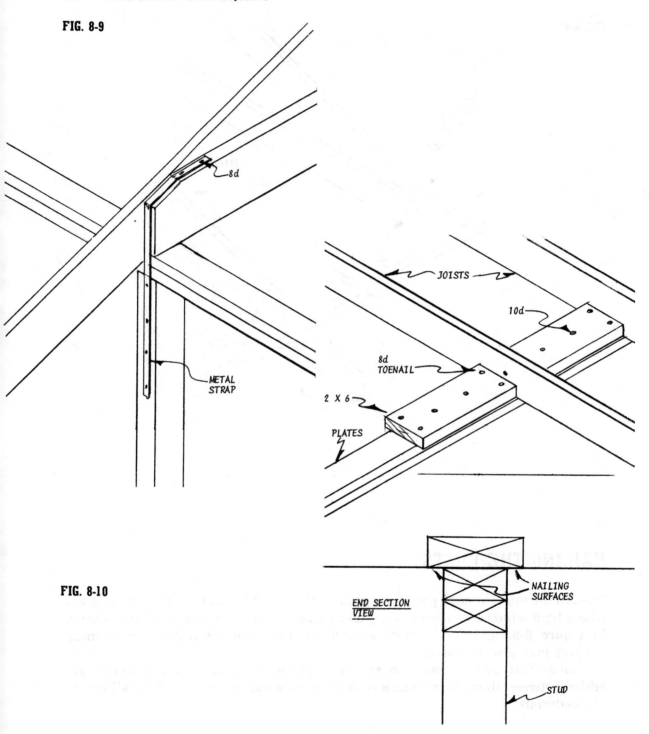

8d

METAL
STRAP

JOISTS

10d

8d
TOENAIL

2 X 6

PLATES

FIG. 8-10

END SECTION
VIEW

NAILING
SURFACES

STUD

Pieced joists that overlap at a crossing must be tied together by surface nailing in addition to being toenailed to the plate (**Figure 8-11**). If the pieces are aligned, splice them at the joint by using 2× material that is at least 24 inches long (**Figure 8-12**).

FIG. 8-11

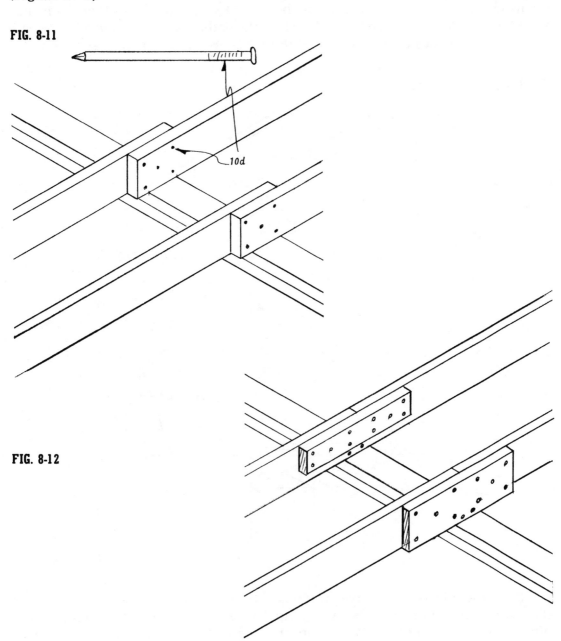

FIG. 8-12

Partitions that are parallel to the joists must be tied in for rigidity and to supply nailing surface for other materials. Various procedures may be followed as long as they comply with the needs, but a fairly straightforward one, used often, is shown in **Figure 8-13**. Place the 1×6 nailer first, centering it over the plate and securing it with 7d or 8d nails. Place the 2×4 blocks and toenail them as if they were joists before you drive the 16d nails at the ends. Work with a level to be sure the interior surfaces of the nailer and the joists are on the same plane.

FIG. 8-13

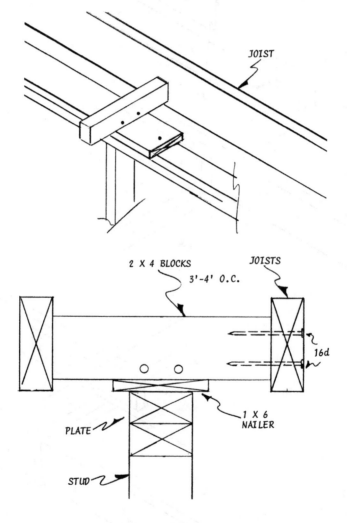

If the room below will have an exposed beam ceiling, the joists are placed in standard fashion with suitable nailing blocks placed between them (**Figure 8-14**). When a hidden beam is used to support joists so there will be unbroken space

FIG. 8-14

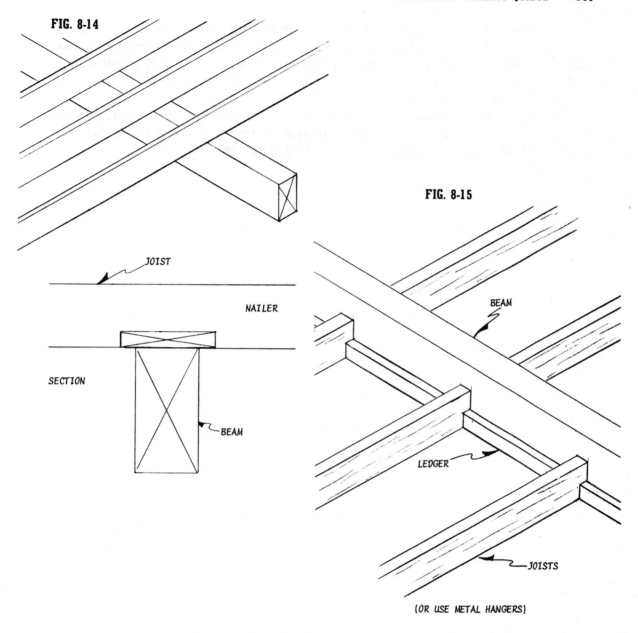

JOIST

NAILER

SECTION

BEAM

FIG. 8-15

BEAM

LEDGER

JOISTS

(OR USE METAL HANGERS)

below, as for a very large living room, the framing can be done with ledger strips, as shown in **Figure 8-15**, or by using metal joist hangers. The beam must be big enough to span the area safely without intermediate supports below. Just figure the beam will be replacing a partition that normally would support the ceiling joists.

Openings through a ceiling frame are handled like those through a floor. If the upper area will be nothing but an unused attic that requires only an access hole that isn't more than 2 or 3 feet square, you can probably skip the trimmers and doubled headers. Local codes that relate to fire regulations may tell you what size such openings should be.

If you are considering anything like a prefab folding or disappearing stairway, information on the size of the rough opening required and possibly even the design of the framing will be furnished along with the unit.

9

FRAMING THE ROOF

Up to now, we've worked mostly by making square cuts across framing members and by using butt joints at all connection points. The members of a roof frame — unless the roof is flat — get us into angular cutting, the complexities of which are directly related to the design of the roof.

The shed roof, often called a lean-to roof, and the plain gable roof are examples of roofs whose framing members don't require more than simple miter cuts (**Figure 9-1**). In both cases, one angle applies regardless of where a cut is

FIG. 9-1

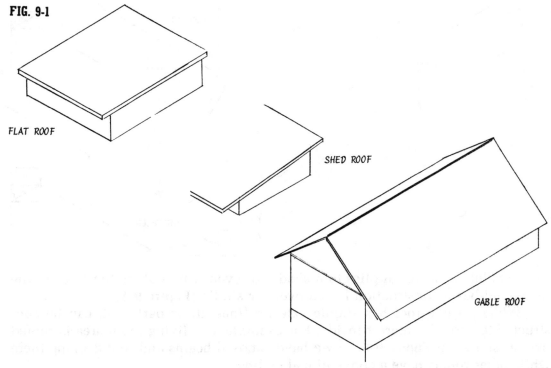

FLAT ROOF

SHED ROOF

GABLE ROOF

made in the rafters. As shown in **Figure 9-2**, on gable roofs with different slopes, the cuts at A, called "plumb cuts," are parallel, and this may even be carried to the overhang end of the rafters if the design you choose is applicable (**Figure 9-3**).

FIG. 9-2   **FIG. 9-3**

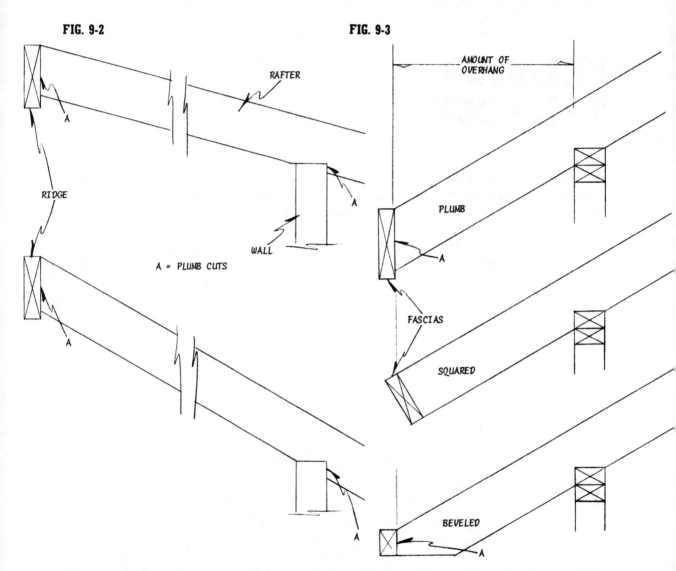

The parallel-cut factor applies to a shed roof even if it is along the lines of the plank-and-beam construction I used over my studio (**Figure 9-4**).

While a gable roof has simple exterior lines, it, or part of it, can be constructed to provide interest inside. For example, our living-room area is roofed over as shown in **Figure 9-5**, so we have exposed beams and roof decking there while other rooms have a conventional ceiling.

FIG. 9-4

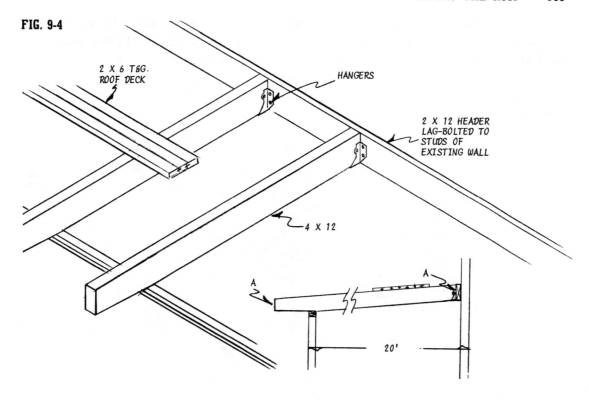

2 X 6 T&G.
ROOF DECK

HANGERS

2 X 12 HEADER
LAG-BOLTED TO
STUDS OF
EXISTING WALL

4 X 12

A

A

20'

FIG. 9-5

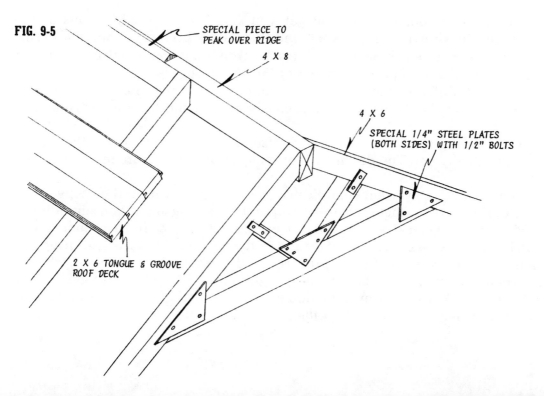

SPECIAL PIECE TO
PEAK OVER RIDGE

4 X 8

4 X 6

SPECIAL 1/4" STEEL PLATES
(BOTH SIDES) WITH 1/2" BOLTS

2 X 6 TONGUE & GROOVE
ROOF DECK

Gable or shed dormers are often added to plain gable roofs to break monotonous lines, as in **Figure 9-6**, or for the more practical purpose of providing light and air and additional space to make attic areas more usable.

FIG. 9-6

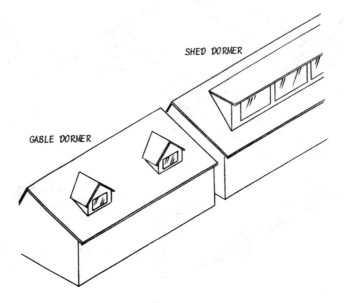

SHED DORMER

GABLE DORMER

Examples of roof designs that get you into more complex construction procedures are shown in **figures 9-7** and **9-8**. The hip roof is a popular design because it can have a silhouette that hugs the ground and automatically provides protective overhangs on the perimeter of the structure.

The main difference between the gambrel roof and the gable is the break in the slope, which can be used advantageously at a second level to provide more headroom.

The mansard roof, like the gambrel, also has a double slope, but the second one is practically flat. The design will provide additional headroom for a second story.

Roof framing may seem complicated, but it is well within the scope of the amateur. If it weren't, there would be no point in talking about housebuilding at all. Much of the geometry involved, for example, has already been done and is available in the form of printed tables and in the framing square, a measuring tool that is a math book in itself when properly used. Also, there is no reason why the shape of components can't be picked up right on the job—the wall frames being the base to work from—or even by doing a large scaled drawing. It's often even possible to make a layout right on the subfloor.

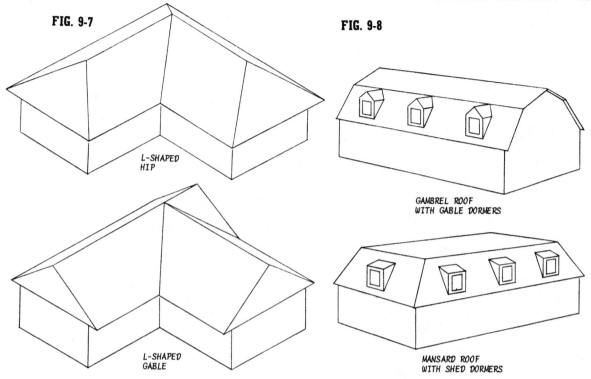

FIG. 9-7

L-SHAPED
HIP

L-SHAPED
GABLE

FIG. 9-8

GAMBREL ROOF
WITH GABLE DORMERS

MANSARD ROOF
WITH SHED DORMERS

With most pieces you can establish cut angles on one and then use it as a pattern for others even when the *length* of companion pieces changes. Some designs require more thought than others, but the actual doing in any case relies more on the builder's dedication than on encyclopedic knowledge or mysterious skills.

KNOW THE NAMES AND THE TERMS

Most roof-framing pieces are "rafters," but they are further identified by the part they play and placement in the structure. The simplified top view of a roof frame for a T-shaped house in **Figure 9-9** includes enough roof designs so principal rafters can be named. Note that the common rafters, which play a part in most sloped-roof framing, run continuously from the ridge to the wall, and beyond if there is an overhang. Common rafters are used exclusively in a plain gable roof and are the easiest to form, since they need only simple miter cuts. You can see why a jack rafter, for example, needs a compound angle cut, at one end at least, if you take a close look at how it connects to the hip rafter, which is already set at an angle (**Figure 9-10**). This also applies to cripples where they connect to valley raf-

FIG. 9-9

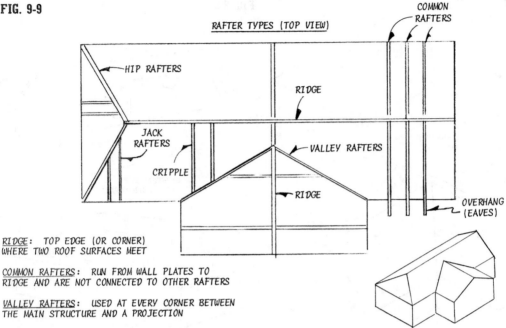

RAFTER TYPES (TOP VIEW)

COMMON RAFTERS

HIP RAFTERS

RIDGE

JACK RAFTERS

VALLEY RAFTERS

CRIPPLE

RIDGE

OVERHANG (EAVES)

RIDGE: TOP EDGE (OR CORNER) WHERE TWO ROOF SURFACES MEET

COMMON RAFTERS: RUN FROM WALL PLATES TO RIDGE AND ARE NOT CONNECTED TO OTHER RAFTERS

VALLEY RAFTERS: USED AT EVERY CORNER BETWEEN THE MAIN STRUCTURE AND A PROJECTION

HIP RAFTERS: SIMILAR TO VALLEY RAFTERS

JACK RAFTERS: REST ON WALL PLATES LIKE COMMON RAFTERS BUT CONNECT AT OTHER END TO HIP RAFTERS

CRIPPLE RAFTERS: CALLED SO BECAUSE THEY DO NOT BEAR ON THE WALL PLATE AT ALL

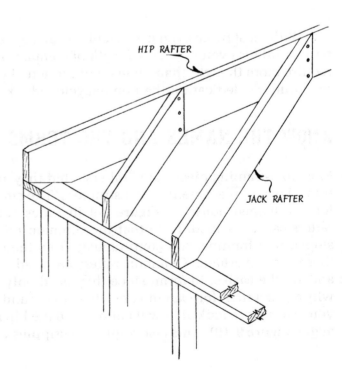

HIP RAFTER

JACK RAFTER

FIG. 9-10

ters, to the valley rafters themselves, and to hip rafters. Compare this with the way common rafters join the ridge board in **Figure 9-11** and you can quickly see the difference in the sawing required. (We'll talk about the collar beam later.)

FIG. 9-11

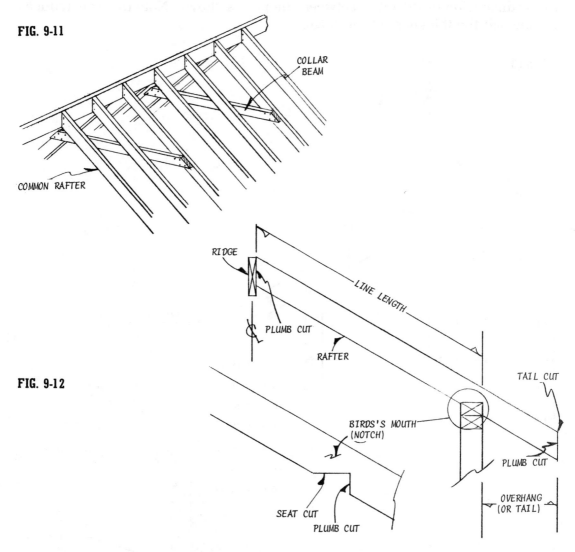

FIG. 9-12

The nomenclature and the various cuts needed in a common rafter are shown in **Figure 9-12**. While the cuts are simple, they should be done carefully so the rafter will fit solidly against the ridge and snugly over the plate. The bird's-mouth is needed only when the rafter extends to form an overhang. You still need the seat cut if the rafter ends at the plate, but the plumb cut is done so the end of the rafter will be flush with the outside edge of the plate.

The run of a rafter is the distance from the centerline of the ridge to the outside of the wall. If the ridge is centered, as in **Figure 9-13**, then the run of all rafters is the same and equals one half the total span. The *line length* is the true linear dimension of the rafter between the points shown. Note that it is reduced by one-half the thickness of the ridge.

FIG. 9-13

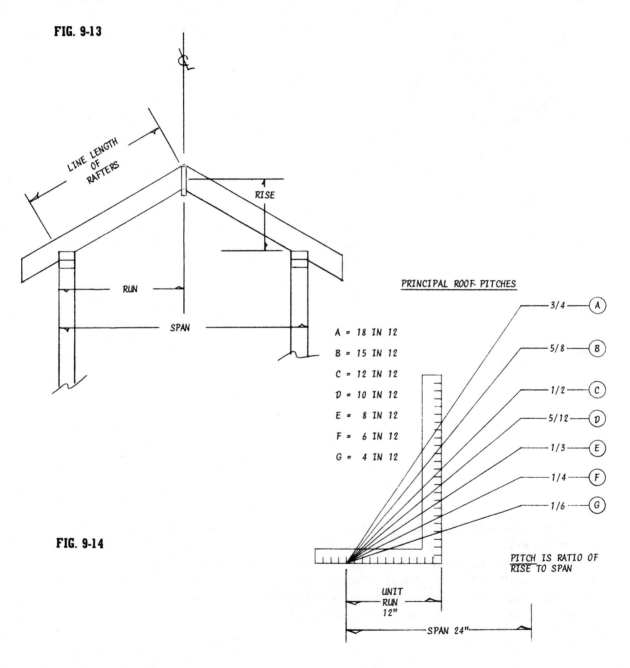

A = 18 IN 12

B = 15 IN 12

C = 12 IN 12

D = 10 IN 12

E = 8 IN 12

F = 6 IN 12

G = 4 IN 12

PRINCIPAL ROOF PITCHES

3/4 —— A

5/8 —— B

1/2 —— C

5/12 —— D

1/3 —— E

1/4 —— F

1/6 —— G

PITCH IS RATIO OF RISE TO SPAN

UNIT RUN 12"

SPAN 24"

FIG. 9-14

FIG. 9-15 *EXAMPLES OF PITCHES*

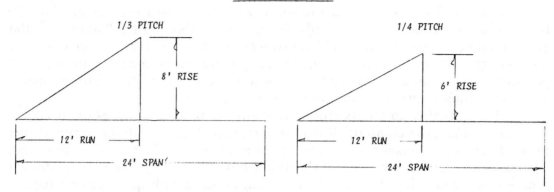

THE PITCH AND THE SLOPE

Both terms indicate the incline of a sloping roof. *Pitch* expresses it as a ratio of the rise to the span — a fraction arrived at by dividing the rise by the span. Remember, the span is twice the run (**Figure 9-14**). A 6-foot rise divided by a 24-foot span will have a ¼ pitch. If the span is the same but the rise is 4 feet, the pitch would be 1/6 (**Figure 9-15**).

Slope is the incline as a ratio of the rise to the run expressed in inches per foot. If the incline increases 6 inches for every foot of run, the slope is said to be 6 in 12. If the increase is 4 inches, then the slope is 4 in 12.

Roofs can range from no slope for a flat design, through slight slope for a shed, intermediate slope for a ranch or rambler-type house, comparatively sharp slope for a Cape Cod, to the extreme slope of an A-frame. While the slope of the roof is usually an architectural consideration, it can also be affected by the type of roofing to be used. This is not as critical today because modern roofing materials and newer methods of installation provide a great deal of leeway. For example, if you *increase* the thickness of the underlay material and *decrease* the exposure-to-the-weather distance of shingles, you can often get by with a lesser slope than might normally be required. By being very careful when you select asphalt and aggregate surfacing materials, you can do a built-up roof on a slope. At one time, the technique was specified only for roofs that had a very small slope or were flat.

RAFTER SIZES AND SPACING

Rafters work for a roof as joists do for a floor in that they provide a skeleton to which you can nail covering materials — and they must support live and dead loads. Since the latter, having to do with the weight of the structure itself, snow loads, wind pressures, and the like, can vary, so can the cross-sectional dimensions of rafters as specified in building codes. Since rafters do not normally carry a heavy ceiling load, it's possible to use a wider spacing than the standard 16 inches prescribed for joists and studs. Do remember that "rafters" applies to flat and shed roofs as well as those with considerable slope and that in such cases the rafters might well be joists requiring similar spacing. The same thought applies if you think you might someday finish the attic to provide additional living space. The rafters then do double duty by taking on joist chores.

Although typical rafter spacing ranges anywhere from 16 to 24 inches, with 20 inches being quite common, it would seem that staying with the spacing used for joists and studs makes sense even if a few more pieces of material are required. Also consider now the type of roof sheathing you will use. Appropriate spacing of rafters can mean a lot of convenience when sheet sheathing, plywood for example, will be used.

When rafters are spaced like the joists, each one can be tied to its companion joist as well as to the plate, as in **Figure 9-16**. If the spacing differs, some attempt should be made to tie in as many rafters as possible in similar fashion. This calls for some compatibility in the spacing, so adjustment may be in order. This should not be done on the plus side and never so that spacing between rafters is irregular. This top view of a roof frame in **Figure 9-17** shows how the system works when joists are 16 inches O.C. and rafters are 24 inches O.C.

FIG. 9-17

FIG. 9-16

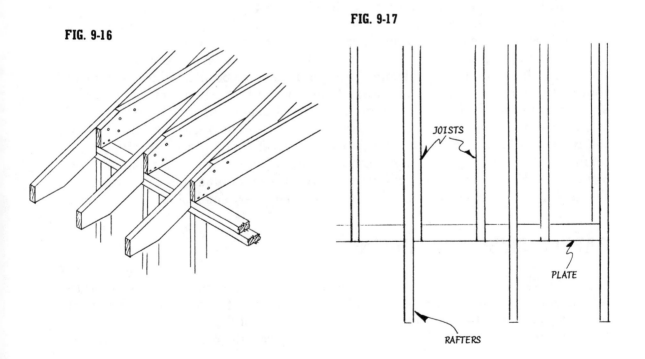

JOISTS

PLATE

RAFTERS

MAKING A COMMON RAFTER

If you view a half cross section of a sloped roof you will see a right triangle whose altitude, base, and hypotenuse relate specifically to the rise, run, and rafter length of the structure (**Figure 9-18**). This relationship never changes, which is why you can work with a small right triangle that has a 12-inch base to represent the unit run and whose altitude equals the slope's rise per foot of run, as in **Figure 9-19**, to step off the length of the rafter. This is also why the framing square is so useful in rafter layout. Quality squares have stamped tables so you can actually calculate rafter lengths by reading the square, but that system is not used too often even by pros.

FIG. 9-18

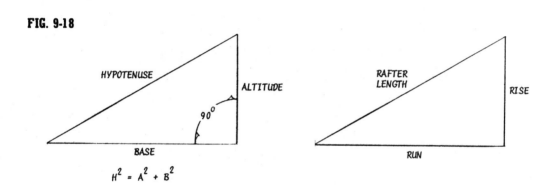

$$H^2 = A^2 + B^2$$

FIG. 9-19

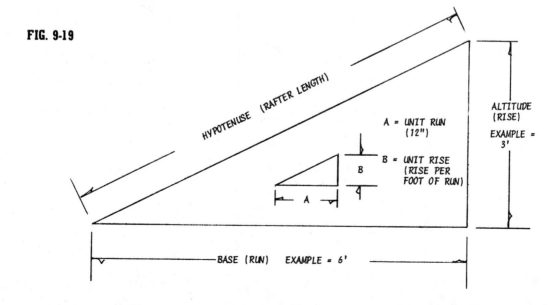

The squares are L-shaped and have a body (or blade) and a tongue (**Figure 9-20**). To step off the length of the pattern rafter, place the square as shown in **Figure 9-21** so the blade indicates the unit run (12 inches) and the tongue shows whatever the unit rise may be. Mark around the square with a pencil and repeat the procedure for each foot of run (**Figure 9-22**). The system works regardless of the amount of slope in the roof. The unit run dimension is constant; the unit rise depends, of course, on what the slope must be (**Figure 9-23**).

FIG. 9-20

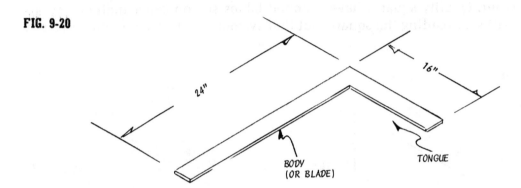

FIG. 9-21

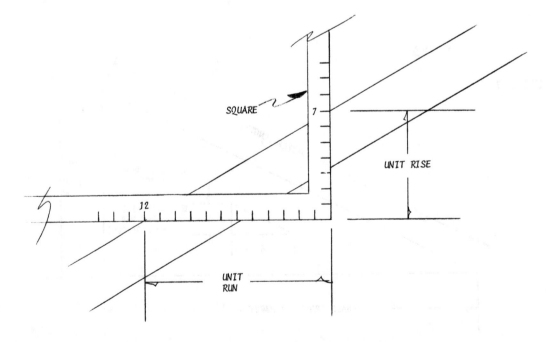

FIG. 9-22

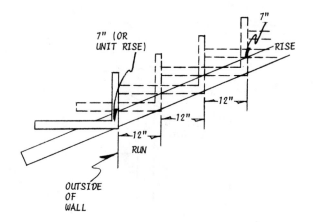

FIG. 9-23

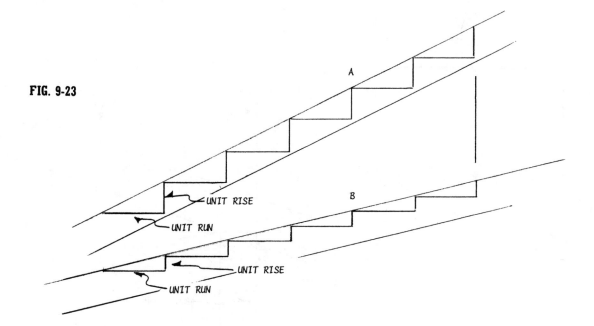

If there is an odd unit in the run, for example, if the total run is 10 feet 6 inches, lay out the six inches at the ridge end of the rafter as shown in **Figure 9-24**. Note that the position of the square in relation to the rafter remains as if you were marking for the full run and rise units but that the tool is moved toward the last layout mark "A" to set it for the odd 6 inches. All the vertical lines, including the last one marked, will be parallel.

FIG. 9-24

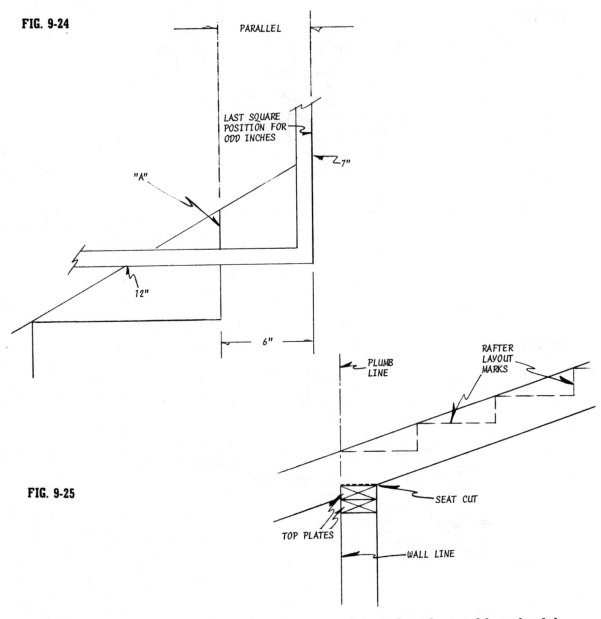

FIG. 9-25

Since the layout is made to the centerline of the ridge, the total length of the rafter must be reduced by one half the thickness of the ridge board.

The seat for the bird's-mouth is a horizontal cut that is about as long as the plate is wide (**Figure 9-25**). Lay out the overhang length of the rafter by starting from the line of the plumb cut. This may also be done with the square, but many carpenters just take a linear measurement with a tape.

WORKING WITH A TEMPLATE

The template shown in **Figure 9-26** simulates a framing square and is used the same way to step off the length of the pattern rafter. Make it from ¼-inch plywood, so it is a right triangle with a base equal to the unit run, and an altitude that equals the unit rise. If you make the triangle oversize to begin with, you can position the edge guide to provide the correct rise and run units. If there is an odd unit, mark it on the base of the template.

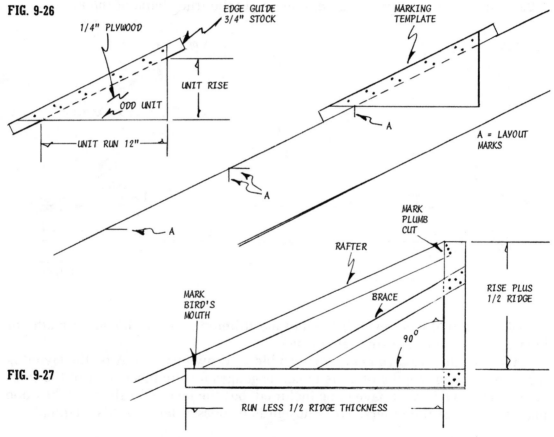

FIG. 9-26

1/4" PLYWOOD
EDGE GUIDE 3/4" STOCK
MARKING TEMPLATE
UNIT RISE
ODD UNIT
UNIT RUN 12"
A
A = LAYOUT MARKS
A
A

MARK PLUMB CUT
RAFTER
BRACE
RISE PLUS 1/2 RIDGE
MARK BIRD'S MOUTH
90°
RUN LESS 1/2 RIDGE THICKNESS

FIG. 9-27

You can also make a template that is full-size. Here too, the project is a right triangle but its base and altitude are equal to the full dimensions of the rise and run. Make it with 1× stock, braced for sturdiness, and use it as shown in **Figure 9-27** for a pattern rafter, or to mark all rafters. If you set it up sturdily across sawhorses, it can even be used as a cut guide. If you do the latter, it might be wise to make the altitude piece longer and just mark it for correct positioning of the rafter.

WORKING FULL-SIZE

This is a procedure that makes the most sense if it is done right after the floor frame and subfloor are complete. The idea is to use the platform as a giant drawing board on which you can place actual roof-framing material to mark lengths and cuts.

Snap a chalk line across the platform at right angles to the edges and then mark its center. From that point, erect a perpendicular and mark it to indicate the rise. Nail down thin sections of ridge-board and plate material as shown in **Figure 9-28**, and you can see that it is possible to get the true shape of the rafters.

FIG. 9-28

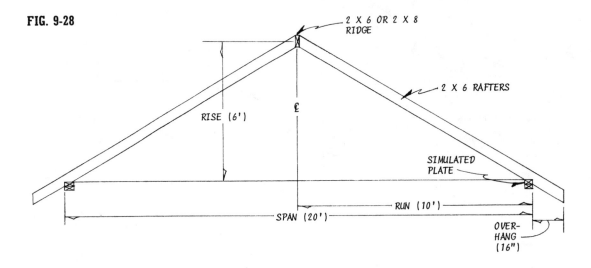

The idea will work even if the roof has different slopes — it's just a matter of where you position the simulated ridge.

Working this way, you can preassemble your own trusses. After the layout is made, nail down lengths of 2×4s to act as gauges for the placement of the truss components. Webs will have to be included, but the layout for them can be done just as it was for the rafters. Truss design is discussed later in this chapter.

CUTTING THE RAFTERS

How you cut the rafters depends, of course, on the tools you have available. The work can be done with a handsaw, but it will be tedious and time-consuming. It's better to work with a portable cutoff saw even if you must rent one. A radial-arm saw, set up with side extensions to support long pieces, is ideal. Special cutters are available for the saws so you can gang pieces and form the bird's-mouth in a single pass. However you work, be sure the pattern rafter, which you should double-check, is used to mark all other pieces. This is the way to avoid cumulative errors.

ERECTING THE COMMON RAFTERS

These are all you will need if the roof is a plain gable, but even with other designs, the common rafters should be organized first so you can work from them to set up other components. The top ends of the rafters abut the ridge board, which, in a gable design, runs the full length of the structure. Since it's not likely you can do this with one piece, plywood splices are used to join sections. It's a good idea to plan the length of the ridge pieces so the splice occurs midway between two rafters (**Figure 9-29**). Since the ridge board is used mostly as an aid in the erection and alignment of rafters (it is not a true structural member), its thickness is not critical. Usually 1× lumber is used, although 2× stock is not out of line if the house is relatively large. At any rate, the width of the ridge board should be at least equal to the length of the cut in the rafter. **Figure 9-30** shows how a 2×6 rafter would mate with a 2×8 ridge; here actual planed dimensions are shown.

FIG. 9-29

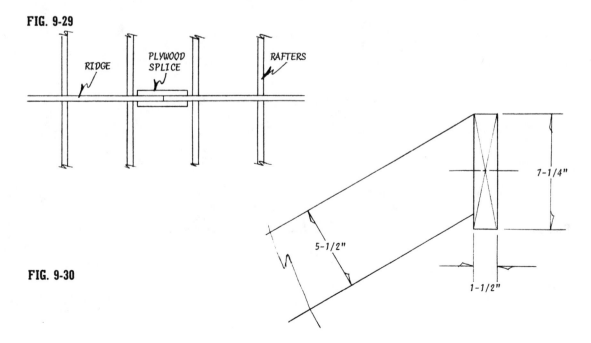

FIG. 9-30

A good way to start setting up rafters is to make a few special supports that you nail temporarily to joists to hold the ridge board in correct position (**Figure 9-31**). They must, of course, be set on a common centerline and sized so the ridge height is correct.

FIG. 9-31

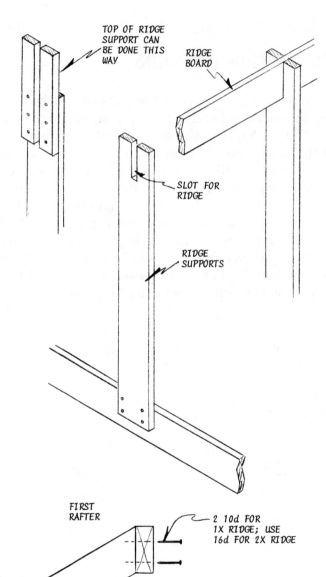

TOP OF RIDGE
SUPPORT CAN
BE DONE THIS
WAY

RIDGE
BOARD

SLOT FOR
RIDGE

RIDGE
SUPPORTS

FIG. 9-32

FIRST
RAFTER

2 10d FOR
1X RIDGE; USE
16d FOR 2X RIDGE

Start by placing two end rafters so they are flush with the outside of the wall. Nail the first one through the ridge as shown in **Figure 9-32**, and attach its mate by toenailing as shown in **Figure 9-33**.

FIG. 9-33

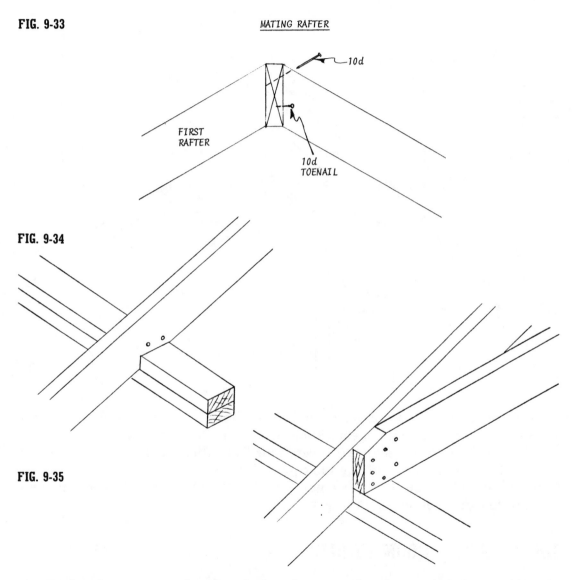

MATING RAFTER

10d

FIRST
RAFTER

10d
TOENAIL

FIG. 9-34

FIG. 9-35

Each rafter is secured at the plate with two 10d nails (toenailed) on each side (**Figure 9-34**). If the rafter abuts a joist, the toenailing is done on each side of the assembly and five additional 10d nails are driven through the joist into the rafter (**Figure 9-35**). Driving the nails through the rafter into the joist is also okay, or, as

FIG. 9-36   RAFTER FRAMING ANCHORS

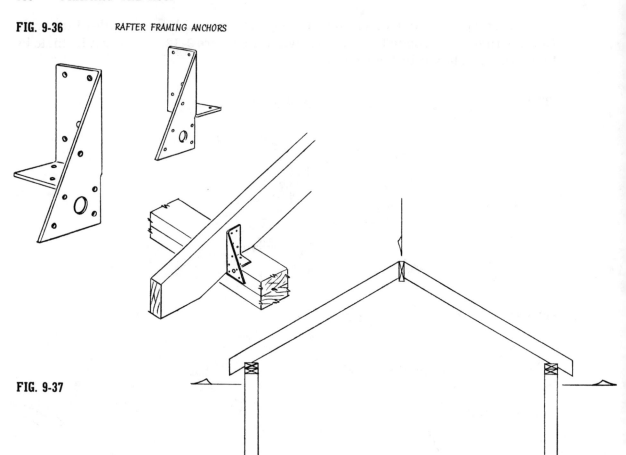

FIG. 9-37

some carpenters do, you can drive two nails on one side and three on the other —
or you can do the job with special framing anchors (**Figure 9-36**).

Follow the same procedure on every fourth or fifth rafter and then come back
and fill in the ones between. Another procedure is to set up rafters at each end
and then stretch a line across as a guide for the ridge. After that, work on every
third or fourth rafter, and then fill in.

BRACING THE ROOM FRAME

Rafters are connected solidly at the wall plates and they lean against each other at
the ridge to provide mutual support, but they impose an outward thrust at the
plate line (**Figure 9-37**). To counter this thrust, you can include collar beams in
the structure (**Figure 9-38**).

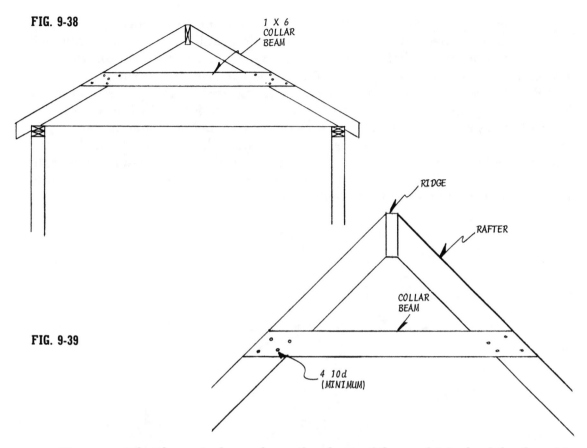

FIG. 9-38

1 X 6
COLLAR
BEAM

FIG. 9-39

RIDGE

RAFTER

COLLAR
BEAM

4 10d
(MINIMUM)

How great the thrust is depends on the slope of the roof. Much of the thrust is opposed by the joists, but nevertheless, especially in high wind areas, collar beams placed at the midpoint of the rafters are desirable. Also consider that the beams will oppose the sag that can develop when the rafters are very long. In addition, since they reduce the span of the rafters, collar beams might make it possible to use lighter rafter material. This, of course, should be checked before roof framing is started and, as always, with local codes in mind.

A common method is to use collar beams of 1×6 material and to place one across every third or fourth pair of rafters. A better method is to use one on every rafter, even though it takes a little more time and material. The reason is that the stiffening effect of collar beams is negligible except in relation to the rafters to which they are secured.

Collar beams are often located close to the ridge so they can be used as ceiling joists later when the attic space is made livable (**Figure 9-39**). In such situations the beams should definitely be placed across all rafters and should be 2×4 or 2×6 material depending on the span.

The nearer a collar beam is to the ridge, the greater the leverage action on it, with increased tendency for the collar-beam nails to pull out. In such situations it might pay to use longer nails and to clinch them over where they protrude. Persnickety craftspeople I have known substituted carriage bolts for nails.

Purlins are often included when collar beams are not used across all rafters. These are long pieces of 2×4 attached horizontally across the inside edges of the rafters (**Figure 9-40**). A more basic use for purlins, though, is when the rafter span is excessive and requires additional support. The purlins are attached the same way but are braced with lengths of 2×4 that rest, preferably, on supporting partitions (**Figure 9-41**). The angle of the brace is not critical. The idea is to run from the midpoint of the rafter to a suitable support below.

FIG. 9-40

FIG. 9-41

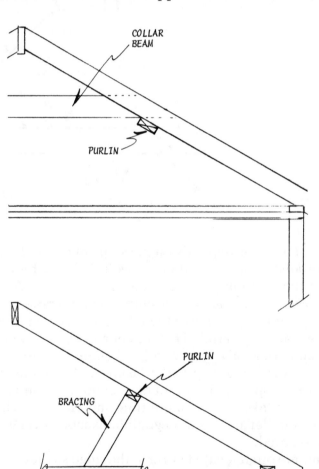

REINFORCING AN END WALL

An end wall that runs parallel to the basic roof structure does not receive the anti-thrust support that side walls get from joists and collar beams. A typical example occurs under sections of a hip roof. When necessary, ample strength is provided by running stub joists at right angles from a regular joist to the wall plate. The stub joists are nailed in place as shown in **Figure 9-42**, and may be reinforced further by using an anchor strap that is long enough to span three regular joists plus a good part of the stub joist. If a subfloor is planned, you can work with framing anchors to secure the stub joist to the regular joist. Regard the short pieces as regular joists when it comes to spacing and nailing at the wall plate.

FIG. 9-42

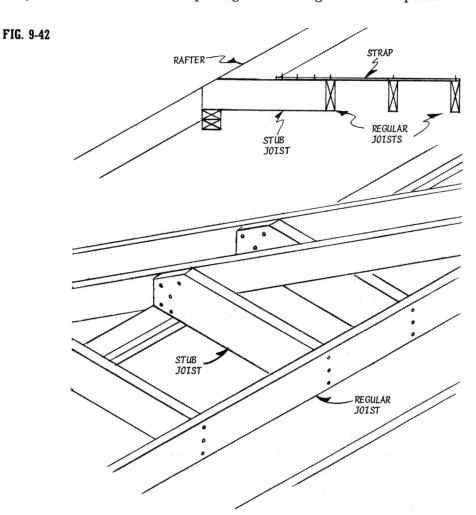

FRAMING AT THE GABLE ENDS

The opening between the plate and the end rafter is framed with studs that are notched at the top to fit the rafter (**Figure 9-43**).

FIG. 9-43

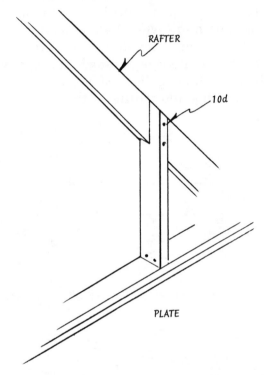

RAFTER

10d

PLATE

The best way to proceed is to drop a plumb bob from the center of the ridge and mark the point on the wall plate. Since this is the place where attic vents are usually installed, you want to locate the positions of the first studs (right and left) to accommodate the opening you need. Mark regular stud spacing from these points to the outer walls. The best way to find the correct stud heights and cuts is to place a length of 2×4 at the point marked as #1 in **Figure 9-44**. Use a level to be sure the wood is plumb and then mark across it to show the top and bottom edges of the rafter. Do the same thing with another piece of stock at point #2. The difference in length between these two pieces will apply to all others and, of course, the shape of the notches will be the same in each piece except that the direction of the angle will change depending on whether they are used on the right or left side of the centerline. These short studs are often installed without a notch, the top end being cut straight across at an angle that matches the slope of the rafter. It's an easier but not a better way.

FIG. 9-44

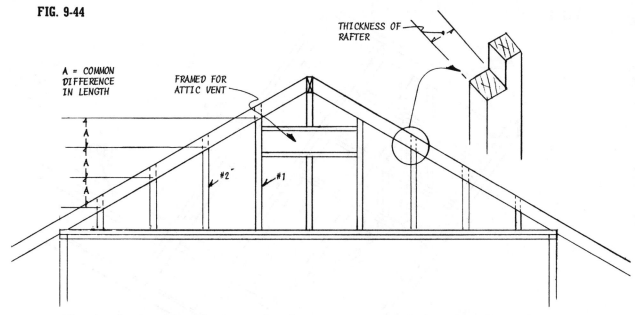

Put in the horizontal pieces for the vent opening before installing the two-piece center stud. Trimmers and headers are not required here.

FRAMING FOR A GABLE OVERHANG

Longer rafters provide for overhangs along side walls, but when the roof must extend at the ends, special framing is called for. There are various ways this can be accomplished but a fairly straightforward one is shown in **Figure 9-45**. The gable end is framed conventionally except that the studs are not notched for a rafter but are angle-cut straight across for a plate. The frame thus constructed provides support for lookout rafters that travel up the slope like the rungs of a ladder. The entire assembly can be done following conventional nailing practices or you can work with framing anchors, which may not be a bad idea where the short rafters meet the regular rafter. Note how the ridge board extends. It is not a good idea to plan a ridge-board splice in this area.

THE HIP RAFTER

If you slice vertically through a gable roof and then slope down from that point to the end wall, you have the form of a hip roof. A side view of the framing will show that a common rafter makes the connection between the wall plate and where the

FIG. 9-45

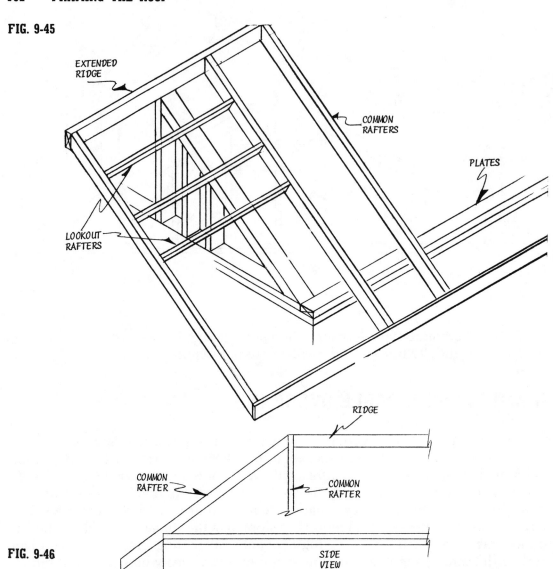

FIG. 9-46

ridge of the gable ends (**Figure 9-46**). If you look down on the framing you will see that the common rafter of the hip and the last common rafter of the gable form a 90-degree angle (**Figure 9-47**). The hip rafter is a diagonal that runs from the intersection to the corner of the walls (**Figure 9-48**). The length of the rafter and the cuts required can be calculated geometrically and by working with a framing square, but they can also be determined by working right on the assembly, a procedure that has much going for it, especially for the first-time builder. It's been tested — it works.

FIG. 9-47

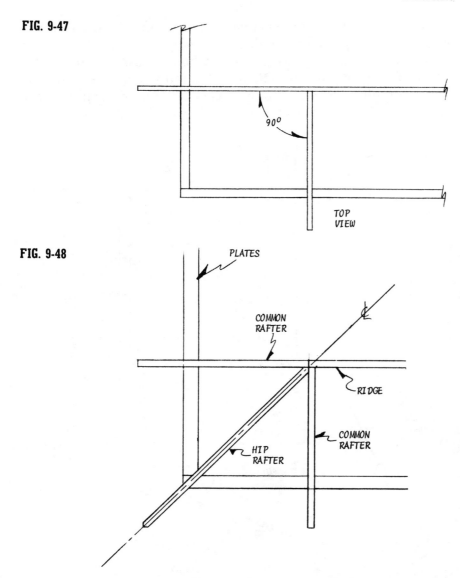

90°

TOP
VIEW

FIG. 9-48

PLATES

COMMON
RAFTER

RIDGE

COMMON
RAFTER

HIP
RAFTER

Tack-nail a block at the corner of the walls to represent the height of the rafter where it crosses the plate. This is a common dimension and is easily picked up from the common rafters already in place. Drive a small nail at the intersection and at the corner of the height block and stretch a line between them. The line represents the center of the hip rafter, so you can measure it to find the line length and then add the amount of overhang. In situations like this it's best to make the overhang longer than necessary and then cut to fit, on assembly, when you get to installing fascia boards.

Use a T-bevel to find the vertical angle between the line and the intersection of the common rafters at the ridge. You can also do this by using a piece of stiff cardboard, pressing one edge into the corner and marking across the top of the line with a pencil. Transfer the angle you pick up to the rafter stock, as shown in **Figure 9-49**, and make a simple miter cut. This angle also applies to the plumb cut of the bird's-mouth.

FIG. 9-49   **FIG. 9-50**

FIG. 9-51

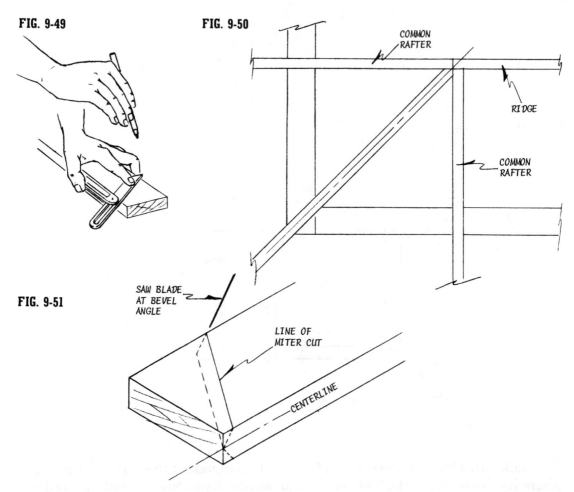

So that the end of the hip rafter will fit snugly in the corner, the miter cut is beveled on each side of the rafter's centerline (**Figure 9-50**). Another way to work would be to mark the line of the miter on both sides of the rafter and then set the blade of a cutoff saw to the correct bevel angle. Saw by following the miter-cut lines and you will have the correct compound angles that are called for (**Figure 9-51**). This, of course, may also be done on a radial-arm saw.

The top edge of the hip rafter will be above those of other rafters. Since this would interfere with placing roof sheathing flat, it's a good idea to bevel those edges as much as necessary by working with a plane or with a saw (**Figure 9-52**). Another method that is often used is to deepen the seat cut of the bird's mouth. This solves the sheathing problem by dropping the rafter lower than it would normally be.

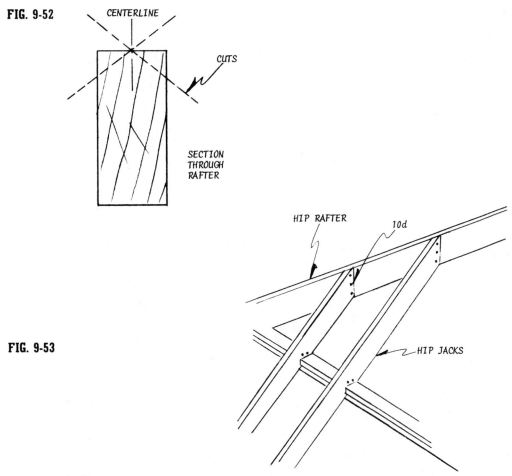

FIG. 9-52

CENTERLINE

CUTS

SECTION THROUGH RAFTER

FIG. 9-53

HIP RAFTER

10d

HIP JACKS

JACK RAFTERS

The jack rafters run parallel to the common rafter and span between the hip rafter and the wall plate. The form of the bird's-mouth is just like that of the common rafters, but the end against the hip rafter requires a compound angle cut (**Figure 9-53**). If you view the assembly from both the top and the side, you will see the side or angle cut and the plumb cut that together form the compound angle (**Figure 9-54**). Follow the same procedure we described for the hip rafter and you will be able to cut the jacks with a minimum of fuss. Stretch a line from the hip rafter to the wall plate to indicate the centerline of the first hip jack (**Figure 9-55**). Hold

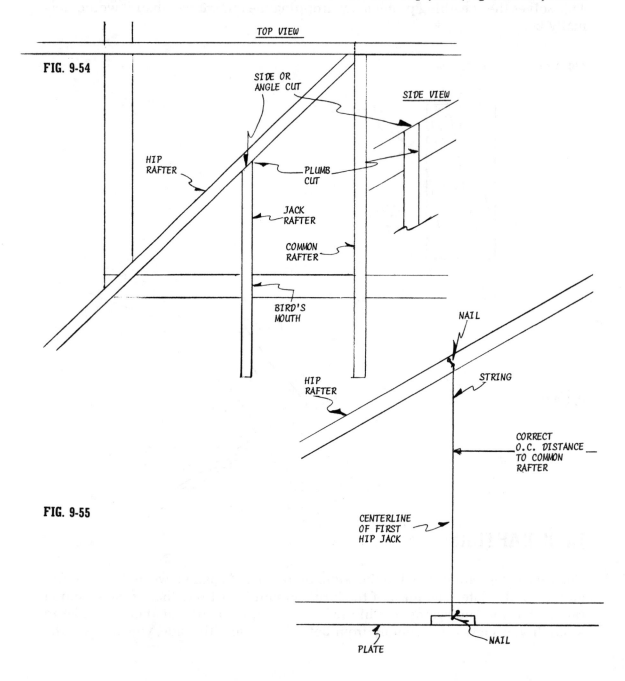

FIG. 9-54

TOP VIEW

SIDE OR
ANGLE CUT

SIDE VIEW

HIP
RAFTER

PLUMB
CUT

JACK
RAFTER

COMMON
RAFTER

BIRD'S
MOUTH

NAIL

HIP
RAFTER

STRING

CORRECT
O.C. DISTANCE
TO COMMON
RAFTER

FIG. 9-55

CENTERLINE
OF FIRST
HIP JACK

PLATE

NAIL

FIG. 9-56

FIG. 9-58

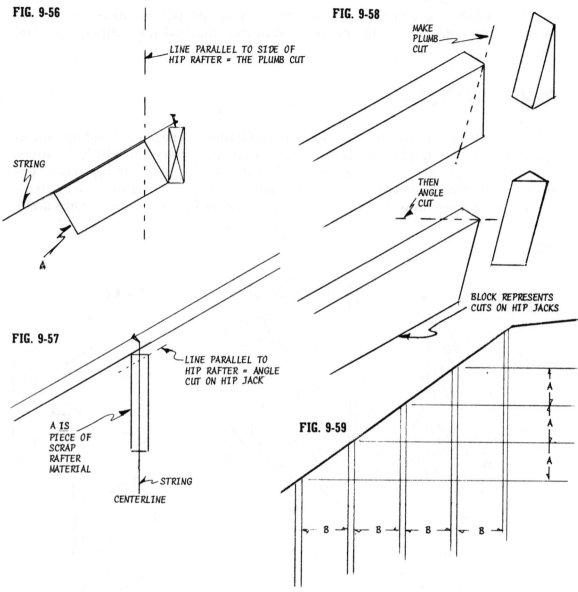

LINE PARALLEL TO SIDE OF
HIP RAFTER = THE PLUMB CUT

STRING

A

MAKE
PLUMB
CUT

THEN
ANGLE
CUT

BLOCK REPRESENTS
CUTS ON HIP JACKS

FIG. 9-57

LINE PARALLEL TO
HIP RAFTER = ANGLE
CUT ON HIP JACK

A IS
PIECE OF
SCRAP
RAFTER
MATERIAL

STRING

CENTERLINE

FIG. 9-59

a short scrap piece of rafter material in the position shown in **Figure 9-56** and mark its end with a line parallel to the face of the hip rafter. Make the cut, and then use the piece of wood in the position shown in **Figure 9-57** so you can mark the line for the angle cut. This time, the line is parallel to the longitudinal run of the rafter. Make this cut, and the sample tells you the shape you need for all the jacks (**Figure 9-58**). The difference in length of all the jack rafters is consistent as long as they are equally spaced. Once you have cut the first two you can apply the change in length to all the others (**Figure 9-59**).

It's a good idea to erect jack rafters in pairs so they will oppose each other to keep the hip rafter from being pushed out of line. Start with a midpoint set, and then fill in with the others.

OTHER RAFTERS

All other rafters that may be required in a roof frame can be cut accurately following the on-job principles we have described for hip rafters and hip jack rafters. The valley rafter, in most situations, is almost a duplicate of the hip rafter. The difference between the two is told by the names. The hip rafter forms an outside corner; the valley rafter works the same way for an inside corner (**Figure 9-60**).

FIG. 9-60

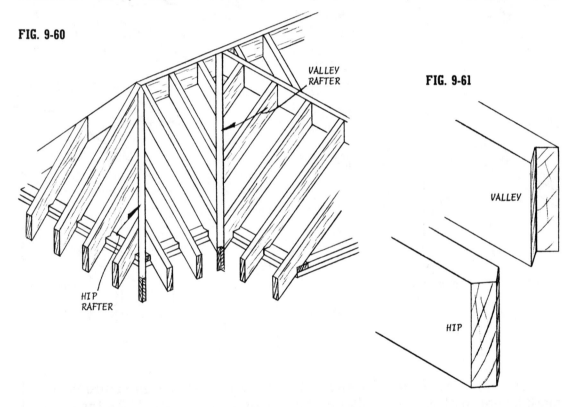

FIG. 9-61

The shapes of the rafters at the end of the overhang are opposed. As you can see here, in **Figure 9-61**, they both form a V, but the one on the valley is inverted and a bit trickier to cut than the other. However, you can simplify the chore by making a plain plumb cut and then adding beveled strips with small nails and an ample amount of waterproof glue (**Figure 9-62**).

FIG. 9-62

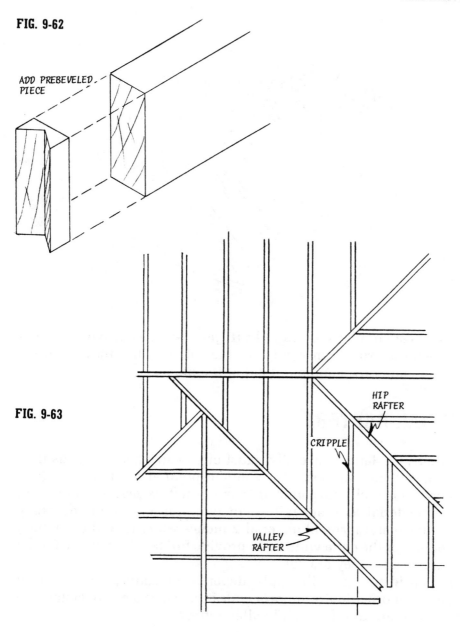

ADD PREBEVELED
PIECE

FIG. 9-63

HIP
RAFTER

CRIPPLE

VALLEY
RAFTER

Cripples are rafters that do not attach to either a ridge or a plate. They can connect, for example, between a hip rafter and a valley rafter, as in **Figure 9-63**, and are named specifically for their position in the frame. For example, they may be called a "valley cripple jack" or a "hip-valley cripple jack." Since they span between sloping rafters, they require compound angle cuts at each end.

FIG. 9-64

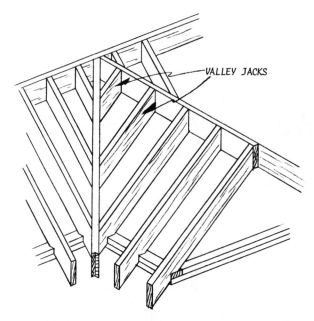

Valley jack rafters are brothers to hip jacks (**Figure 9-64**). They work in similar fashion, but since they run from a ridge to the valley rafter, they require a plumb cut at the top end and a compound cut at the other.

SIZES OF SPECIAL RAFTERS

Hip and valley rafters, the latter especially, must often carry heavier loads than common rafters. Hip rafters that meet at a common point at the top get a good measure of support from each other and, if they are not excessively long, can safely be the same material as other rafters. In premium construction, valley rafters are often doubled and made of material 2 inches wider than the common rafters. Another reason for the extra width is to provide full bearing surface for the ends of the jack rafters after they are cut.

Whether the considerations will apply depends of course on your roof design. You might get information from local codes or, just as a safety factor, use heavier stock or double up for all hip and valley rafters.

DORMERS — SHED AND GABLE

The gable dormer is so called because its roof design is the same as that of a regular gable roof and usually has a matching slope. Sometimes it's purely decorative but may combine its esthetic value with a practical function such as providing attic ventilation. In the latter case, the front of the dormer would have screen-backed louvers and the whole assembly could be viewed as an add-on to the basic roof frame. It looks pretty from the front but from the back you would see that it doesn't do much more than cover a hole through the roof.

Of course, it may also be designed to provide light and air for an attic room. Then its design would be more integrated with the basic roof framing.

The shed dormer is something else, since it can be incorporated to provide a lot more additional space inside. Its construction can be quite simple if you view its front wall as a vertical extension of the house wall and its rafters as common ones that have a different slope than the main roof—the run of the rafters is the same. There is much leeway here in how much room you can pick up, since the width of the shed dormer can run the full length of a gable design.

In both cases, the opening through the roof framing is reinforced with doubled rafters or trimmers and, where applicable, doubled headers between rafters. Study **figures 9-65** and **9-66** and you will see that the framing procedures we have discussed in relation to walls and roofs apply to dormers as well.

FIG. 9-65

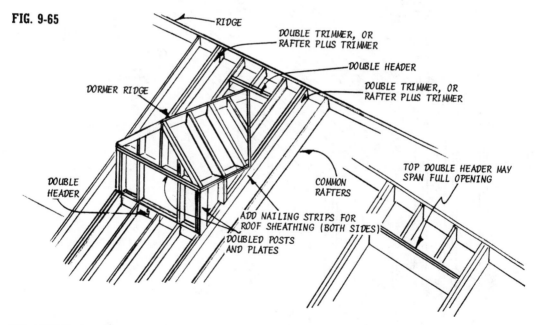

IF THE ROOF IS FLAT

A flat roof, or one with a slight slope, is easier to frame than conventional double-slope roofs. The design has always been with us, but its contemporary lines plus improvements in roof-surfacing materials and methods of application are making it more popular.

FIG. 9-66

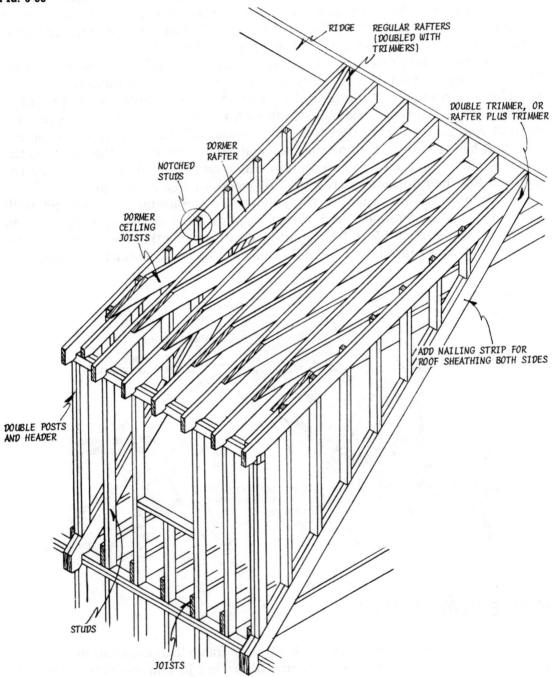

RIDGE

REGULAR RAFTERS
(DOUBLED WITH
TRIMMERS)

DOUBLE TRIMMER, OR
RAFTER PLUS TRIMMER

DORMER
RAFTER

NOTCHED
STUDS

DORMER
CEILING
JOISTS

ADD NAILING STRIP FOR
ROOF SHEATHING BOTH SIDES

DOUBLE POSTS
AND HEADER

STUDS

JOISTS

The horizontal members, which, in this case, are both rafter and joist, are called "roof joists." Since they must support both the finish roof and the ceiling, wider material than you would use for conventional joists and rafters is called for. 2×10s or even 2×12s are not uncommon; 16-inch-O.C. spacing is standard, but the requirements of local conditions are, as always, an important factor to consider before making a decision.

The framing, with due consideration for span support and the like, is done as if for a floor. Overhangs are made by extending roof joists where possible and by using lookout rafters on other sides. The roof joists that support the lookouts are doubled to become headers. Conventional nailing can be used throughout, or you can work with framing anchors. The latter will provide more strength if used where the lookouts abut the header.

Either of the two methods shown in **figures 9-67** and **9-68** can be used to do the overhang framing. The doubled king rafter in **Figure 9-68** is better, structurally, when the overhang gets up to 3 feet wide. Often, as a design feature, the roof joists are tapered in the overhang area (**Figure 9-69**).

FIG. 9-67

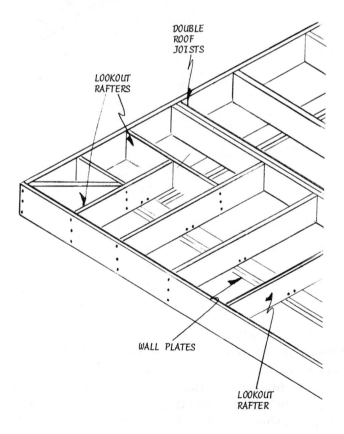

FIG. 9-68

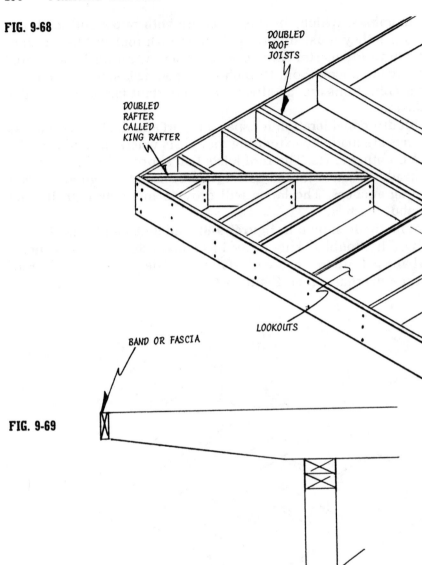

DOUBLED
ROOF
JOISTS

DOUBLED
RAFTER
CALLED
KING RAFTER

LOOKOUTS

BAND OR FASCIA

FIG. 9-69

When you do the roof with heavy solid-wood decking or a similar material, you can substitute beams for roof joists. Many times, this is done so that both the decking and the beams are exposed on the inside. The technique applies specifically to a plank-and-beam construction we will discuss later. Here, rigid insulation should be put down over the decking. When roof joists are used, insulation, and maybe an air space, should be introduced between the ceiling and the roof.

CUTTING THROUGH RAFTERS

A large opening through the roof frame—something you may require for a chimney—should be designed like those you make through a floor (**Figure 9-70**). The doubling up of rafters or the addition of trimmers, plus headers, is necessary wherever you must cut through rafters. If the installation is done after roof fram-

FIG. 9-70

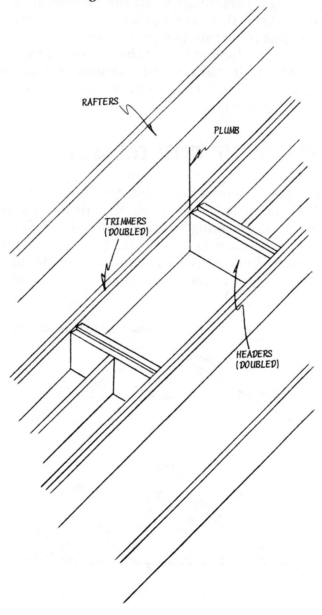

RAFTERS

PLUMB

TRIMMERS
(DOUBLED)

HEADERS
(DOUBLED)

ing is complete, do something to supply temporary support at the cut area until the headers are set. This can be a couple of 2×4s placed across the underside of the rafters and braced from beneath, or just a couple of boards placed across the top of the rafters and tack-nailed to hold components in correct position while you frame the opening.

Small openings that do not require cutting can be framed merely by installing headers between regular rafters. In all cases, the headers are set plumb. That is, their surfaces are on a plane which is perpendicular to the ceiling joists. You can determine this easily just by dropping a plumb bob.

Any opening you make for a chimney should include 2 inches of clearance on all sides. If you plan to install a prefab fireplace together with lightweight, ready-made chimney units, check the specifications that come with the product to find passthrough requirements.

DOING THE ROOF FRAME WITH TRUSSES

A truss is basically a preassembled frame that does the job of conventional rafters and joists and makes it possible to span across outside walls without intermediate supports (**Figure 9-71**). Engineering has been refined and designs developed so applications can vary and spans can run from 20 to more than 30 feet. In most cases, they can save time and material and effort. Since after they are constructed they can be set up fast, you can get a protective covering over the house pretty quickly.

FIG. 9-71

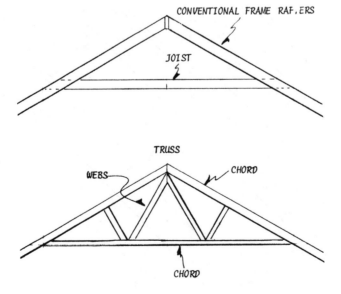

FIG. 9-72

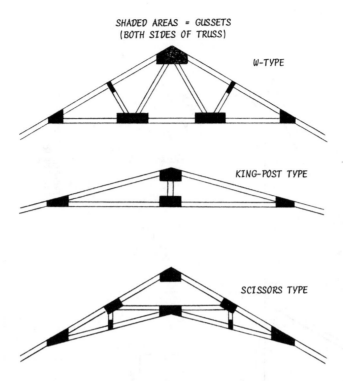

SHADED AREAS = GUSSETS
(BOTH SIDES OF TRUSS)

W-TYPE

KING-POST TYPE

SCISSORS TYPE

There are many types of trusses but the basic ones for residential construction are shown in **Figure 9-72**. The king-post truss is the easiest to do but is also the most limited in terms of span because the chord is supported only at the center. It has fewer parts than other types and so can be assembled more quickly and economically. The W-type truss is used extensively, probably because the extra webs that support the chord permit greater spans than a king-post truss — assuming, of course, that similar materials and member sizes are used in each.

The scissors-type truss is good construction when the house design calls for sloping ceilings. This can be general or limited to one or two rooms if you do the roof frame by, for example, combining scissor-type trusses with W-types.

The design of the truss and the materials used can't be arbitrary. Dead-load and live-load conditions must be considered as well as the span. Local codes can help, and there is much engineering data available from sources listed at the back of the book.

Trusses can be purchased already assembled, made to your specifications if a standard form won't do, or you can make them on the job. What you need is a large platform, which might be a subfloor or something you can set up by covering sawhorses with plywood or boards.

The trusses must be strong, and this calls for structurally sound lumber and

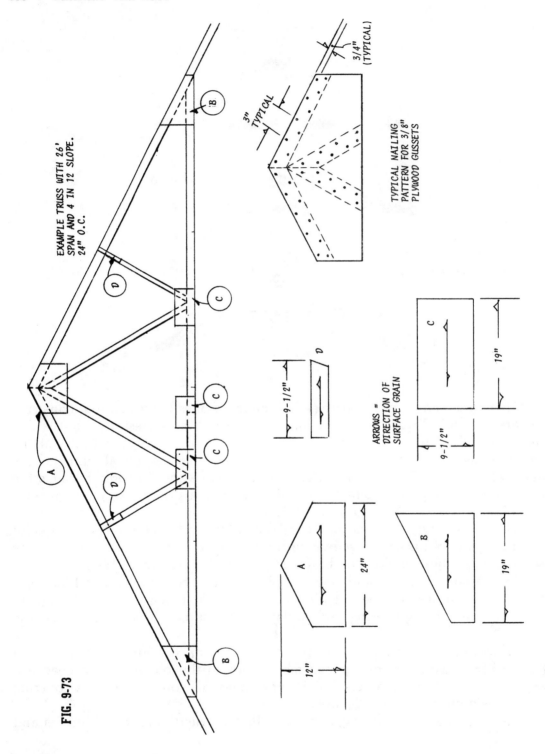

EXAMPLE TRUSS WITH 26' SPAN AND 4 IN 12 SLOPE. 24" O.C.

FIG. 9-73

TYPICAL NAILING PATTERN FOR 3/8" PLYWOOD GUSSETS

3" TYPICAL

3/4" (TYPICAL)

ARROWS = DIRECTION OF SURFACE GRAIN

A   24"   12"

B   19"

C   19"   9-1/2"

D   9-1/2"

careful workmanship. Since the trusses are duplicates, all parts can be precut on a production-line basis after the layout is done and pattern pieces are cut correctly. Do the layout carefully; check it more than once. You can work to chalk lines or you can nail down lengths of 1× stock in appropriate places so the whole will be a master jig for placing and holding components as you do the fastening.

The example shown in **Figure 9-73** is for a W-type truss for a span of 26 feet and a 4-in-12 roof slope. 2×4 stock is used for the chords and webs; the gussets are exterior-grade plywood which can be either ⅜ or ½ inch thick. The gussets might be thinner or thicker depending on strength requirements and the dimensions of chord and web stock. Use 4d nails for gussets up to ⅜ inch thick, and 6d nails for gussets from ½ to ⅞ inch thick.

Gussets are applied on both sides of the truss and are attached with glue as well as nails. Use a waterproof glue even though favorable conditions in some areas might make another type acceptable. Be sure all glue-contact points are clean and that you apply enough to assure some squeeze-out at all lines when you drive the nails. Some waterproof glues achieve strongest bonds when they are applied within a certain temperature range. Check the instructions on the container to be sure what the range is.

Assembled trusses must be handled carefully to avoid stresses that can cause distortion. Their strength is realized only when they are in a vertical position, and this is the way they should be carried and stored. If you must store them flat, be sure they get full-length support. Allowing the center or an end to carry the weight would result in distortion. If you carry them flat, have enough help at both ends and along the center. A common method of erecting them is to place them upside-down across the walls and then use long 2×4s at center points to tilt them upward.

The connections at wall plates are done best with framing anchors. These will supply necessary strength and eliminate the need to toenail through plywood gussets.

You can choose to do the truss assembly with special plates and connectors that are designed for easy application while still providing maximum strength. The plates are available in various shapes and sizes. Some are drilled for nails, others have punched holes so the metal itself drives into the wood (**Figure 9-74**).

FIG. 9-74   TRUSS PLATES

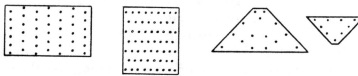

FIG. 9-75

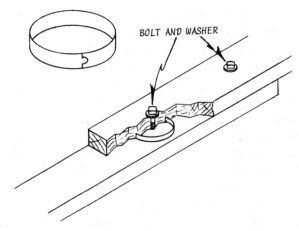

BOLT AND WASHER

Still others have nails that are actually part of the plate. There are also very special ring-type connectors that fit into circular grooves you form with a special tool (**Figure 9-75**). The chances are good that if you buy ready-made trusses, they will have been assembled with the types of hardware described above.

While the types of trusses we have discussed are most adaptable to house designs that require but one truss shape, don't overlook the fact that they can be adapted for L-shapes and that special ones can be designed for hip roofs and for valley areas. It's also possible to combine conventional roof framing with special trusses that take care of problem areas.

GAMBREL ROOFS

The method you use to frame a gambrel roof is quite similar to that for a gable. The big difference is the break in the roof slope that occurs between the wall plate and the ridge (**Figure 9-76**). The angle above the break, in relation to a horizontal plane, is less than 45 degrees, while the one below is more than 45 degrees. Generally, the two work out as 30 degrees and 60 degrees. In the simplest gambrel design, if you bisect the point where upper and lower rafters meet, you'll find that the end cuts on the rafters are the same (**Figure 9-77**). This basic connection can be organized as shown in **Figure 9-78**, when it is the intention to provide a floor area that is free of posts and partitions. Frequently, as you may have seen, the design is used for barn roofs.

Purlins such as shown in **Figure 9-76** can be added in different ways, but since the gambrel design is usually adopted to provide more headspace in what might otherwise be an attic, they are usually made like a doubled top plate and supported by studs that will form a partition (**Figure 9-79**). The upper and lower rafters are notched to fit the plate, the cuts required much like those for a bird's-mouth—you have a plumb cut and a seat cut on each rafter.

FIG. 9-76

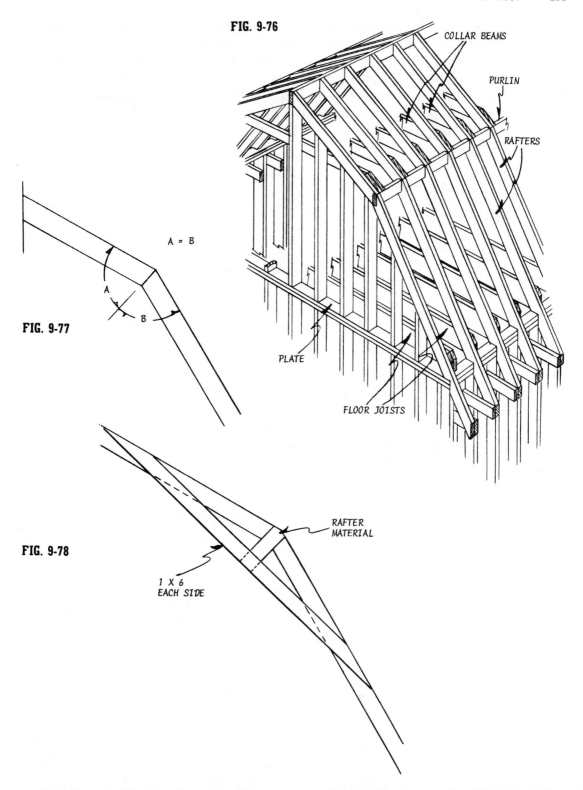

COLLAR BEAMS

PURLIN

RAFTERS

A = B

A

B

FIG. 9-77

PLATE

FLOOR JOISTS

RAFTER
MATERIAL

FIG. 9-78

1 X 6
EACH SIDE

Collar beams can span across the partitions or they can be set higher to increase headroom. If the latter is done, the ceiling would slope from the vertical plane of the partition to the horizontal plane of the collar beams. If you look at **Figure 9-79** again, and forget the word "purlin," you will see that the gambrel can be done like two roofs. One slopes from the wall plates to the "break" partition, the other from the partition to the ridge. You can, if you wish, do the top part of the roof by working with small trusses that span the partitions.

FIG. 9-79

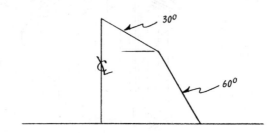

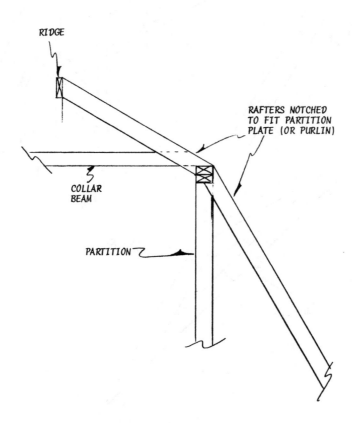

10

ROOFING: GENERAL CONSIDERATIONS

THINK SAFETY

You are on a roof and it slopes—so you can fall off an edge or slide down the incline. These are unpleasant possibilities, but awareness of them is a great safety factor. Knowing what *can* happen should prompt you to take preventive measures.

Shoes with slippery bottoms are out. Good types are of leather, ankle-high, with rubber soles. Nonskid sneakers are a second choice. There is such a thing as a safety harness, an item that is usually available at a marine-supply store, or that you can probably rent from a roofing-materials dealer or a rental establishment. You wear the harness and lock it as you go to a strong line that travels over the ridge and is secured somewhere on the opposite wall frame.

Traveling up and down on a ladder is risky, especially when carrying materials. Professionals erect scaffolds on which they can store materials and on which they can stand when starting roofing work along lower edges. It's something for you to imitate either by making scaffolds or by renting some that can be erected at the site. Dollied scaffolds such as shown in **Figure 10-1** are available. Often, the scaffolds consist of high, sturdy sawhorses, platformed solidly with 2× boards or with plywood that is supported by spaced boards. Whichever way you go, you know the scaffold must support you and maybe others as well as roofing materials. So, sturdiness is important even though the projects are temporary.

When you are actually on the roof, you can use footing scaffolds against which you can brace your feet. These are no more than lengths of 2×4 that are tack-nailed at right angles to the rafters over roofing material already applied (**Figure 10-2**). The tack-nail holes can often be covered by the overlap of the roofing material. If not, tiny dabs of roofing cement will do the job. Another way to secure the footing scaffold is to nail it to vertical pieces which in turn are nailed at the ridge line. Vertical pieces interfere with laying the roofing materials, but this is a minor nuisance worth putting up with for safety.

FIG. 10-1   DOLLIED SCAFFOLD

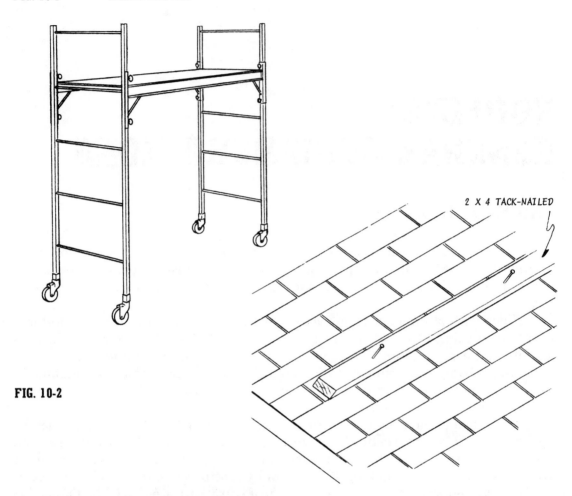

2 X 4 TACK-NAILED

FIG. 10-2

HOW ROOFING SHEDS WATER

Water flows downhill, and on a roof it should do so without interruptions that can cause dams that allow the water to slip *under* roofing materials instead of over them. This is why courses of roofing materials must always be lapped as shown in the lower part of **Figure 10-3**.

Exposure *to the weather* applies to all materials and is a phrase to take literally. It is always specified in inches. In the case of shingles, it is the distance from the edge of one course of shingles to the edge of the next course. As you can see in **Figure 10-4**, a short exposure can increase the thickness of the cover. Less than normal exposures are often done so that a material can be used on a lesser slope than would be possible otherwise.

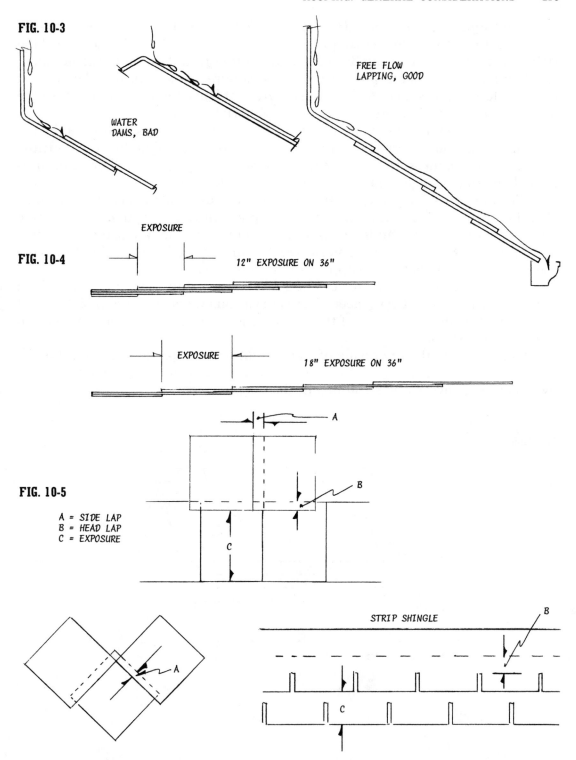

FIG. 10-3

WATER
DAMS, BAD

FREE FLOW
LAPPING, GOOD

EXPOSURE

FIG. 10-4

12" EXPOSURE ON 36"

EXPOSURE

18" EXPOSURE ON 36"

A

B

FIG. 10-5

A = SIDE LAP
B = HEAD LAP
C = EXPOSURE

C

A

STRIP SHINGLE

B

C

Head laps and *side laps* are shown in **Figure 10-5**. In each case, it's the amount of the lower shingle that is covered by the top one. Note, in the case of the strip shingle shown, that the head lap is measured from the top of the notch.

Roofing materials are purchased by the *square*. This indicates the amount required to get 100 square feet of *finished* roof (**Figure 10-6**). Don't confuse the term with the actual square footage of the material itself.

Coverage might be thought of in terms of roofing thickness as affected by the overlapping of the materials. It can be specified as single, double, or even triple coverage and is a measure of the weather protection so provided. Short exposures provide more coverage than long ones.

Flashing is used on a roof wherever the regular placement of covering materials is interrupted. The cause can be a chimney, a valley, a vent pipe, or something else. When there is such an interruption, special precautions are taken in the area to prevent leaks. The flashing material can be similar to what is used to finish the roof, or it can be metals like copper, aluminum, or special alloys.

The *cornice* is a special construction that finishes and trims the underside of the area where roof and wall meet. There are various ways to accomplish this, as I will show. Usually the design of the cornice includes considerations for ventilation of attic spaces.

Soffit refers to the underside of a construction. In relation to a roof, it is the bottom of the cornice, or, if there is no cornice, it can be the bottom of the overhang area covered by sheathing or decking.

FIG. 10-6

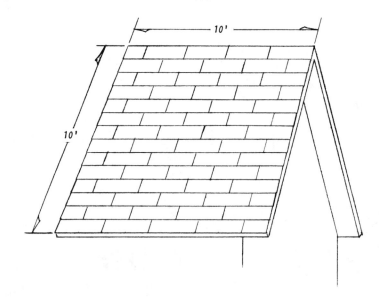

SEEING THE ROOF AS A WHOLE

The roof is what you put down over the roof frame. It consists of sheathing (which can be solid boards, spaced boards, plywood, or special composition materials), maybe underlayment, finish materials, flashing, cornices and soffits, and gutters. All are interrelated, and while there can be a step-by-step procedure, it is wise to check out all the factors before you take the first step. Some of the cornice work might be easier to do if basic pieces are installed along with roof framing. You might consider fascias and even some trim while you do the sheathing. Attachment of gutters might be simplified if you do it along with or immediately after the sheathing process.

A lot depends on the house design and the materials involved, but you can judge the logical next step or what jobs to combine if you see the complete picture beforehand. It's not uncommon, even with professionals, for "remodeling" to become necessary even before the house is finished. Avoid such wasted effort by anticipating what comes next or what can go along with the current chore.

THE FIRST ROOF COVER IS SHEATHING

Sheathing is installed over the roof frame to make it stronger and more rigid and to serve as a nailing base for the roofing material. Most sheathing is done with 1× lumber or with sheathing grades of plywood with a thickness that is suitable for the material to be applied and the rafter spacing. On some types of flat roofs, on those with minimum slopes, and over plank-and-beam constructions, the first cover is often heavy tongue-and-groove boards, or special fiberboard panels that are finished ceiling, sheathing, and insulation all in one. Since even the T&G boards are usually exposed underneath and serve as the ceiling, such materials should be called decking rather than sheathing, even though both are bases for the final cover.

Some new sheathing materials have appeared recently, and there are others being developed that may be available by the time you read this. One you can buy now is composed of solid boards that are sandwiched between veneers of heavy kraft paper. An advantage is that they come in panels as long as 16 feet. Other possible sheathings are composition materials that are light, strong enough to provide rigidity, have good nail-holding power, and have insulating qualities. It's a good idea to check your supply sources to see if such modern materials are available.

BOARD SHEATHING

Strips of solid wood that are as long as possible but not more than 6 inches wide are used to make a solid cover over the roof frame if asphalt shingles or similar materials are used for the final surface. The boards can be common—that is, have square edges and ends—or shiplap, or end-matched (**Figure 10-7**). Run the boards at right angles to the rafters, securing them at each crossing with two 8d nails. Runs may be pieced, but no board should be shorter than the span across three rafters. Use the longest boards you have to cover overhangs, especially if they occur at a gable end. In the latter situation it will pay to use boards that span at least four rafters.

FIG. 10-7

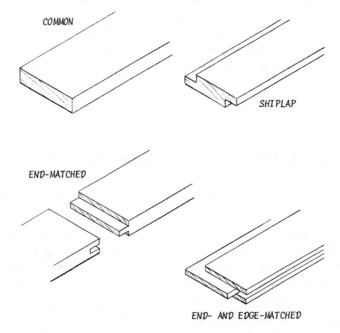

The end joints of common boards must occur over the centerline of a rafter, and it's good practice to work so such joints are staggered (**Figure 10-8**).

End joints can occur between rafters if the boards are end-matched, but place the boards so the joints will not follow a common line. Boards can be set flush at roof edges, or you can allow them to project a bit and then trim them after nailing is done.

Wood shingles and shakes are usually set down over spaced roof boards, especially when climatic conditions involve much dampness. The same rules apply as for solid sheathing, except that the spaced boards can be 1×3s or 1×4s.

FIG. 10-8

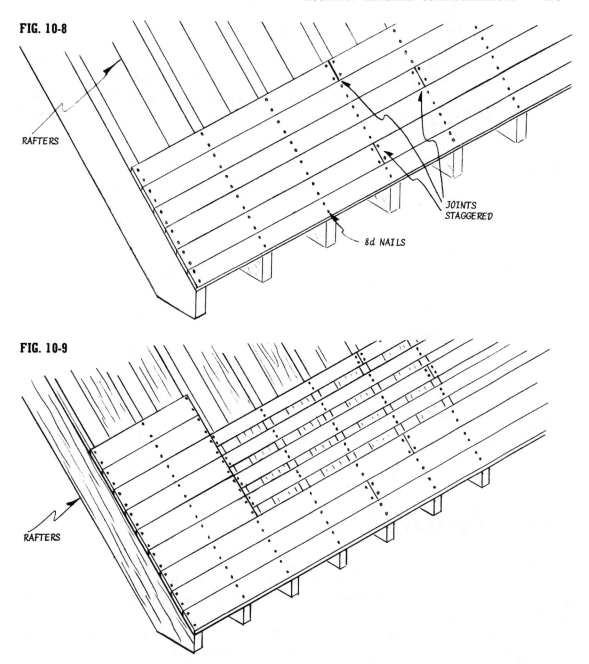

RAFTERS

JOINTS
STAGGERED

8d NAILS

FIG. 10-9

RAFTERS

When open soffits are part of the design, the roof is sheathed with a combination of closed and spaced boards (**Figure 10-9**). Since the underside of the overhangs will be visible, a better-looking grade of material is used to do the covering. Often an exterior-grade plywood is selected, since it can be applied without visible joints.

The O.C. distance of the spaced boards equals the amount that the shingle or shake will be exposed to the weather. As the example in **Figure 10-10** shows, if the exposure is 5 inches, then the O.C. spacing of the boards must also be 5 inches. A always equals B. The open spaces between the boards provide needed ventilation.

FIG. 10-10

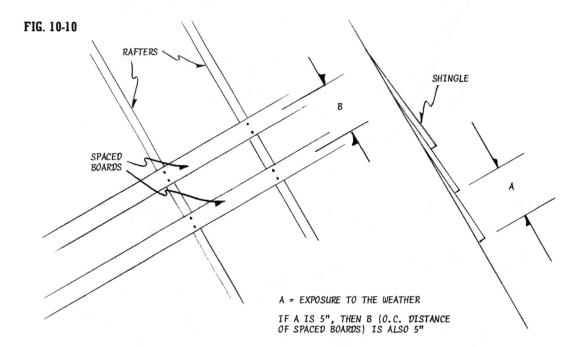

RAFTERS

SHINGLE

B

SPACED
BOARDS

A

A = EXPOSURE TO THE WEATHER

IF A IS 5", THEN B (O.C. DISTANCE
OF SPACED BOARDS) IS ALSO 5"

PLYWOOD SHEATHING

Plywood comes in sheathing grades with or without an exterior glue line. Base your choice on local climatic conditions or simply decide to use the waterproof variety as a safety factor. No matter what, open soffits should always be covered with exterior grade.

Although 5/16-inch-thick panels are often used when rafters are 16 inches O.C., 3/8-inch or even a 1/2-inch panel will contribute much toward a stronger and a smoother roof. The extra thickness also provides better nail penetration when adding the cover. The thicker grades of plywood are a must if rafters are 24 inches O.C., and over *any* spacing if an extra-heavy roofing material like slate or tile or asbestos-cement shingles will be used.

You can use thinner plywood than you could normally if you include blocking between rafters so that all perimeter edges of the sheets can be nailed. There are also plywood systems that include special clips to lock the panel edges together between rafters, and panels with tongue-and-groove edges that increase allowable rafter spans even without blocking. All three systems are illustrated in **Figure 10-11.**

FIG. 10-11

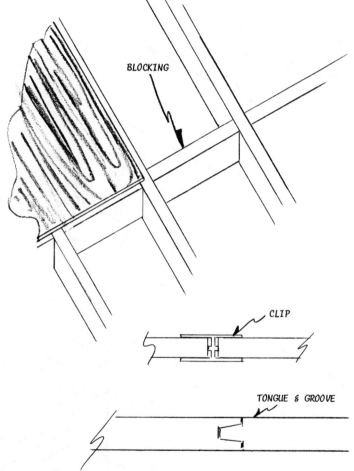

Always place plywood so the face grain runs across the rafters. All vertical joints must fall on the centerline of a rafter and should be staggered. Don't use pieces too short to span across at least three rafters. Separate panels by $\frac{1}{16}$ inch along edges and $\frac{1}{8}$ inch at the ends. Double the spacings in unusually humid areas. Don't attempt to measure the spacings with a tape. Instead, use strips of cardboard or wood as gauges.

Attach sheets that are ½ inch thick or less with 6d nails spaced as shown in **Figure 10-12**. Use 8d nails if sheet thickness is between ½ inch and 1 inch.

Plywood edges, especially if the panel has an interior glue line, should not be exposed to the weather. Usually the finish trim will provide adequate cover, but if you wish to take an extra precaution you can "paint" such edges with a waterproof glue. This particularly applies to edges on the perimeter of the building.

FIG. 10-12

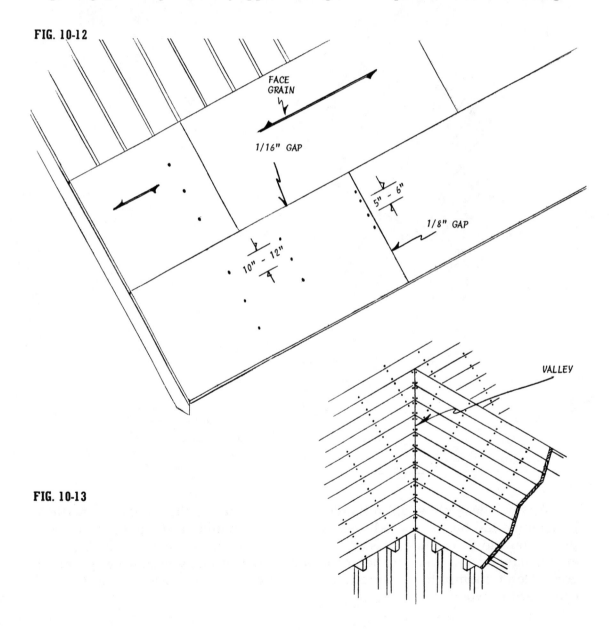

FIG. 10-13

Plywood sheathing is often used under shingles — sometimes under shakes. If you wish to do this and still get by with panels that are less than ½ inch thick, you can do so by adding spaced boards over the sheathing. To increase ventilation area under the cover, use 1×2 boards instead of the usual 1×3s or 1×4s. Spacing, of course, is determined by the shingle or shake exposure. By working with thicker plywood you may be able to attach the cover directly to the sheathing, but check this out against local codes, since some special underlayment may be in order in your area.

AT VALLEYS, HIPS, ETC.

All sheathing should form tight joints wherever interruptions occur. This doesn't mean you must form compound angle joints at, say, valleys, such as shown in **Figure 10-13**, but mating edges should be reasonably snug. You can opt for a persnickety ending at the ridge or just allow the sheathing edges to kiss (**Figure 10-14**). Openings around chimneys should be cut so there will be about ¾ inch clear-

FIG. 10-14

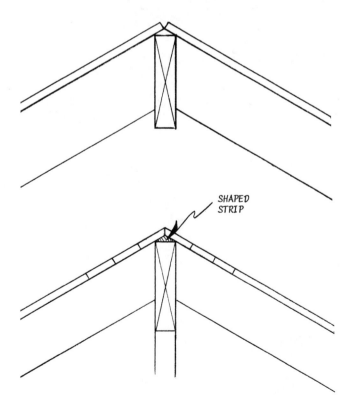

SHAPED
STRIP

ance on all sides between the masonry and the wood. This applies to the sheathing only; specified in Chapter 9, the framing members require a 2-inch clearance (**Figure 10-15**).

FIG. 10-15

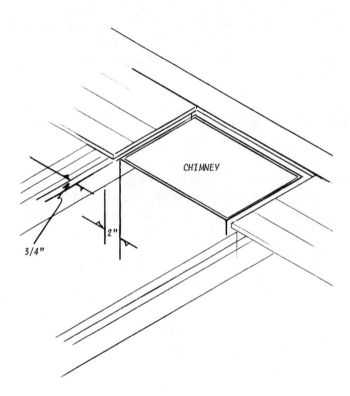

11

BUILDING CORNICES

CORNICES: OPEN, BOXED, OR CLOSED

All cornices may be considered as part of the exterior trim on the house, but they also afford a degree of protection for the walls, which relates to the width of the overhang. When to do the construction depends a good deal on its design. Some components can be attached while roof framing is going on and before roof sheathing is placed. This applies especially to open cornices, which require nothing but the addition of fascia boards, as shown in **Figure 11-1**, or even the fascia boards can be omitted. The fascia is nailed directly to the ends of the rafters. Often it is allowed to project above the rafters an amount equal to the thickness of the sheathing (**Figure 11-2**). The advantages are that the finish roof then covers the joint and the fascia hides and protects the edge of the sheathing.

FIG. 11-1   **FIG. 11-2**

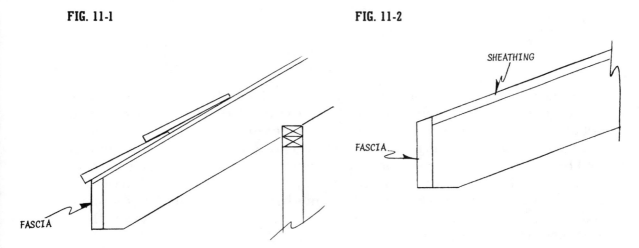

Since the fascia is a visible trim, it should always be a good grade of material and be installed carefully. End joints must meet at a rafter and should be spliced rather than butted. Inside and outside corners are best done with miters (**Figure 11-3**). It's a good idea to coat the mating edges of fascia joints, as well as those in other cornice areas, with a caulking compound. Sometimes, lead paste is used. Apply the material generously so there will be ample squeeze-out, but remove the excess quickly to avoid stains.

FIG. 11-3

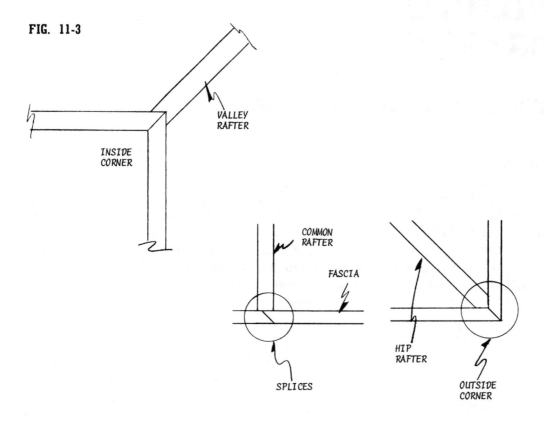

Wide overhangs, which are basically open soffit constructions, can do more than serve as trim for the house. At my place, we extended the roof line for a good part of a large window area as protection from the sun and to make that outdoor area more usable, supporting the overhang with posts and beam as shown in **Figure 11-4**. It became more a veranda than a simple overhang and we made the supports heavier than necessary simply for appearance, but nevertheless, the roof extension was open-soffit construction. The slope had to change a bit for headroom, so the regular roof rafters were stopped at the wall and others took off from there at the new angle.

FIG. 11-4

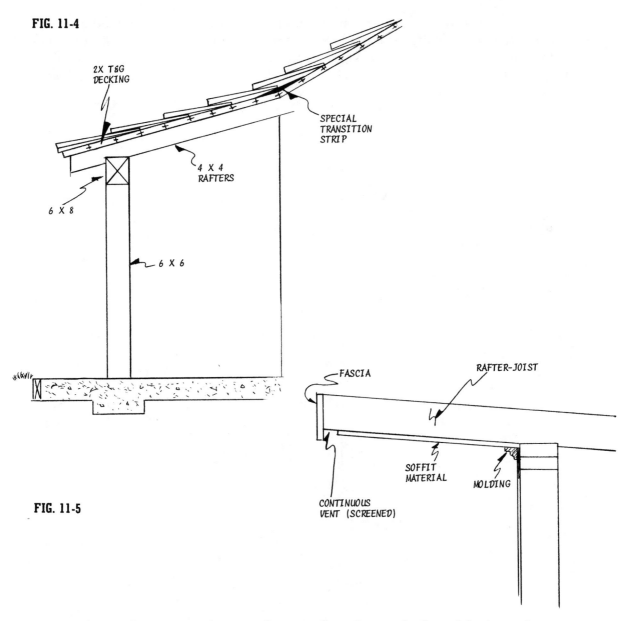

2X T&G DECKING

SPECIAL TRANSITION STRIP

4 X 4 RAFTERS

6 X 8

6 X 6

FASCIA

RAFTER-JOIST

SOFFIT MATERIAL

MOLDING

CONTINUOUS VENT (SCREENED)

FIG. 11-5

Wide overhangs are often used on modern flat or shed roof designs; they are left open or boxed in by adding a soffit to the underside of the rafters (**Figure 11-5**). The latter applies especially when the roof-rafter area is closed in by the finish roof and the inside ceiling. Note that a continuous vent is included and that the soffit-to-wall joint is finished and sealed with a strip of molding, which can be shaped or quarter-round or simple a well-fitted plain board.

The close cornice can be most simple in design and can be added after the roof and the house walls are complete. As shown in **Figure 11-6**, it may consist of just a trim board (called a frieze) and molding, or a plain board that you bevel on both edges to fit the corner. It's obvious that while this serves as a trim, the design doesn't afford much protection for the walls.

FIG. 11-6

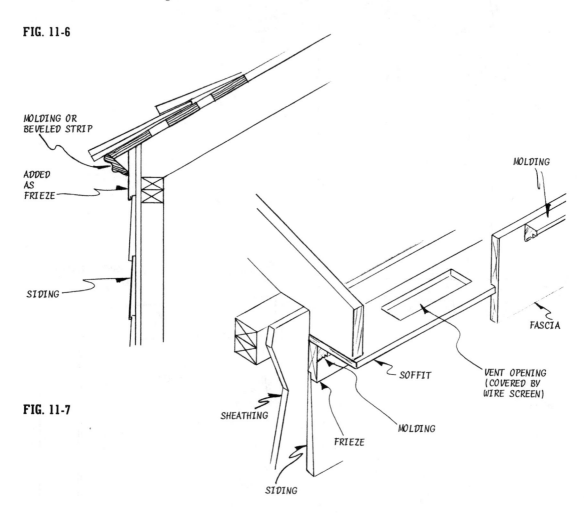

FIG. 11-7

Another type of close cornice (also called a narrow boxed cornice) is shown in **Figure 11-7**. In this design the rafter projections serve as nailing surfaces for both the soffit and the fascia. The amount of overhang available will depend on rafter size and roof slope. Note that wall sheathing has been installed and that the fascia is grooved to receive the soffit. The latter isn't always done but it is premium construction and contributes considerably to weathertightness.

Proper ventilation of attic spaces is necessary, especially when maximum attention is given to insulation. Gabled roofs usually have openings at each end to do the job, but openings through the soffits of cornices will increase air movement greatly. There is a relationship between vent area and ceiling area, but normally, soffit vents have been standardized as shown in **Figure 11-8**. The opening with the round ends will be the easiest to make if you are using a saber saw.

FIG. 11-8

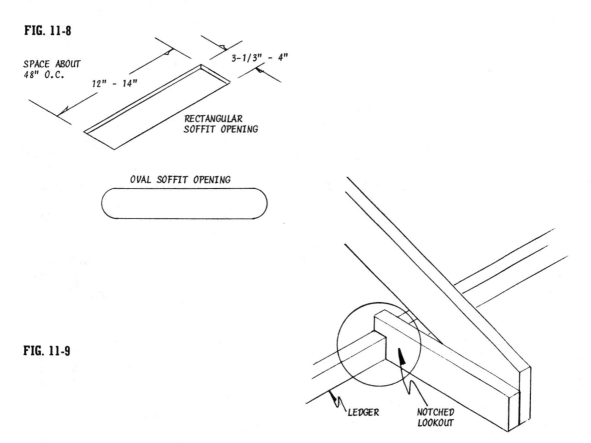

SPACE ABOUT 48" O.C.

3-1/3" - 4"

12" - 14"

RECTANGULAR SOFFIT OPENING

OVAL SOFFIT OPENING

FIG. 11-9

LEDGER

NOTCHED LOOKOUT

THE BOXED CORNICE

The boxed cornice is popular because it is architecturally attractive and provides considerable leeway in terms of design and the width of the overhang. Most of them include 2×4 "lookouts" which span between the wall and the end of the rafters and which serve as nailing surfaces for the soffits. The lookouts tie to ledgers which are nailed to the wall over sheathing. They may be butted and toenailed to the ledger, or they can be notched as in **Figure 11-9**. The width of the overhang is controlled by the extension of the rafters; the depth of the cornice increases in relation to the extension and the roof slope (**Figure 11-10**).

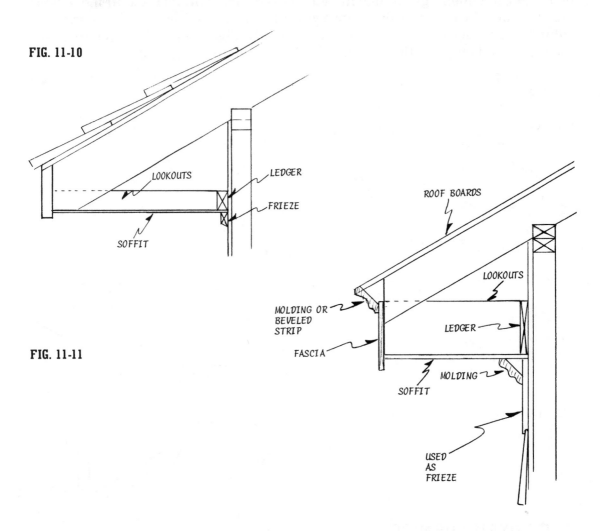

FIG. 11-10

LOOKOUTS

LEDGER

FRIEZE

SOFFIT

ROOF BOARDS

LOOKOUTS

FIG. 11-11

MOLDING OR BEVELED STRIP

FASCIA

LEDGER

MOLDING

SOFFIT

USED AS FRIEZE

Custom designing is possible. The section through an elaborate cornice in **Figure 11-11** shows how you can gain bulk without increasing extension by using wider material for the lookouts and the ledger. Such installations require more time and more material, so be sure they are really desirable before you do them.

In all situations, it's a good idea to install the ledger first, positioning it by using a level between the wall and the end of the rafter and securing it by driving nails that will penetrate the wall studs. Place lookouts at each rafter, toenailing them at the ledger and surface-nailing them to the rafter (**Figure 11-12**).

FIG. 11-12

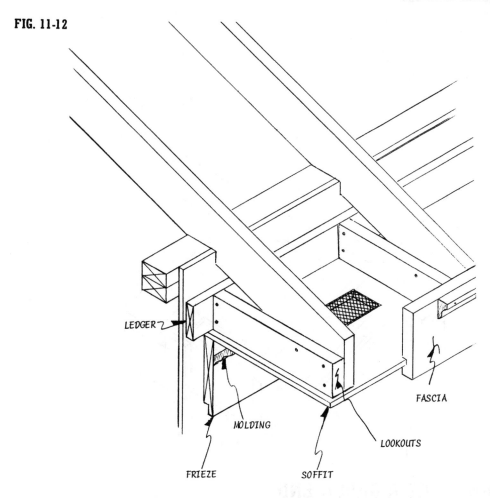

If boards are used for the soffit they should be tongue-and-groove with end-joints occurring at a lookout, and staggered. Plywood is a good material to use, but it must have an exterior glue line. Hardboards and special types of composition boards are also usable. When the soffit material is on the thin side, you can add nailers along the top edge of the ledger and between the ends of the rafters. This will permit complete perimeter nailing. In all cases, the bottom edge of the fascia should be lower than the soffit by at least ½ inch so it will act as a drip to prevent water from getting into the cornice.

Another type of boxed cornice does without lookouts, the soffit being nailed directly to the underside of the rafters (**Figure 11-13**). If you use a grooved fascia, it won't be necessary to angle the cut if you pack the groove well with a caulking compound before you install the fascia. This will provide an effective seal to compensate for the extra-wide groove that will be needed.

FIG. 11-13

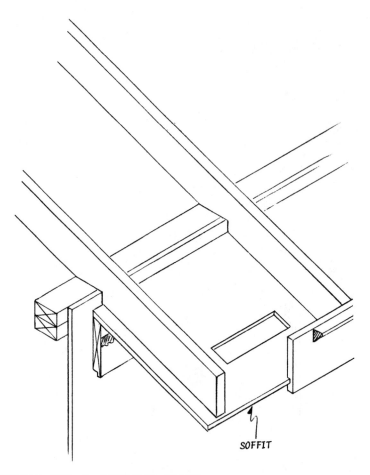

SOFFIT

THE CORNICE AT A GABLE END

A close cornice doesn't require much more than a backup to move the fascia off the house wall. In a sense, the backup is another rafter, but since it substitutes for a frieze board, it should be regarded as trim and so should be better-looking material than you would use for regular rafters. A typical gable-end cornice is shown in **Figure 11-14**. Since there are a lot of joints here, premium construction calls for flashing strips that are placed before the finish roof is added.

A step between the close cornice and an extended one that permits a cornice up to about 8 inches wide with a minimum of fuss can be done by using short lookouts to which the fascia is nailed directly (**Figure 11-15**). If the fascia is placed as shown, it gains strength by the nails driven into it through the roof sheathing. The molding is the final touch, ending the project decoratively and hiding the edges of the sheathing.

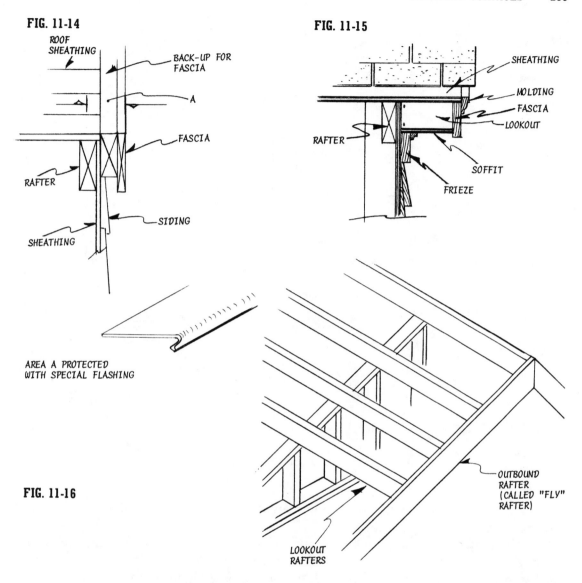

FIG. 11-14

ROOF SHEATHING

BACK-UP FOR FASCIA

A

FASCIA

RAFTER

SIDING

SHEATHING

AREA A PROTECTED WITH SPECIAL FLASHING

FIG. 11-15

SHEATHING

MOLDING

FASCIA

LOOKOUT

RAFTER

SOFFIT

FRIEZE

FIG. 11-16

OUTBOUND RAFTER (CALLED "FLY" RAFTER)

LOOKOUT RAFTERS

An open-soffit construction, if it isn't more than 15 to 20 inches wide, can be done merely by extending solid roof sheathing outboard to a *fly rafter* which ties in with the ridge at the top and the fascia board at the bottom. Since the sheathing carries the fly rafter, the boards or the plywood used should be long enough to span three of the regular rafters plus the extension.

Wider overhangs are anticipated and provided for, as we have already explained, by incorporating lookout rafters in the basic roof frame as shown in **Figure 11-16**. Soffit material is attached directly to the underside of the lookouts

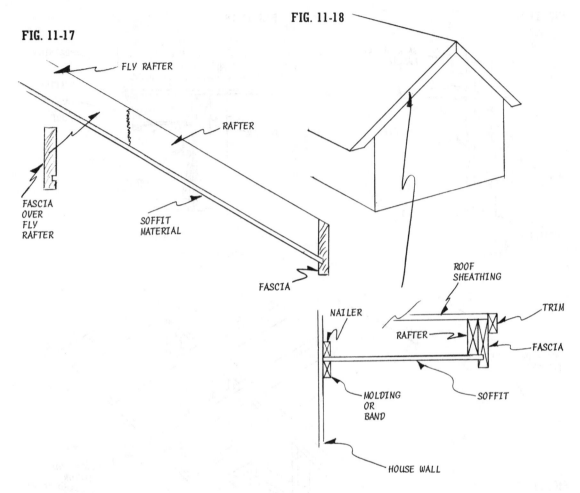

FIG. 11-17

FIG. 11-18

FLY RAFTER

RAFTER

FASCIA OVER FLY RAFTER

SOFFIT MATERIAL

FASCIA

ROOF SHEATHING

TRIM

NAILER

RAFTER

FASCIA

MOLDING OR BAND

SOFFIT

HOUSE WALL

and the fly rafter, and a fascia is added to cover the rafter and the edge of the soffit (**Figure 11-17**).

For narrower overhangs where lookout rafters are not used, a nailer, which runs like a rafter and has its bottom edge on the same plane as the bottom edge of the fly rafter, is attached to the wall so the inboard edge of the soffit may be secured (**Figure 11-18**).

When the side cornices are done the same way — that is, without lookouts so the soffit attaches directly to the rafters — the fascias join at the bottom end and the construction is complete except for the addition of trim.

Unlike the hip roof, where boxed cornices are continuous and similar around the entire structure, the gable roof requires what is called a *return*, where cornices along the side of the house meet those of the end wall (**Figure 11-19**). While there are many ways to design the return, a simple and practical one consists of widen-

FIG. 11-19

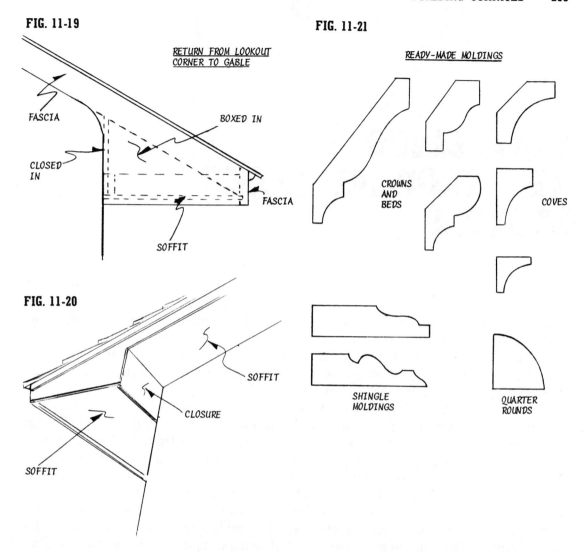

RETURN FROM LOOKOUT
CORNER TO GABLE

FASCIA

BOXED IN

CLOSED
IN

FASCIA

SOFFIT

FIG. 11-20

SOFFIT

CLOSURE

SOFFIT

FIG. 11-21

READY-MADE MOLDINGS

CROWNS
AND
BEDS

COVES

SHINGLE
MOLDINGS

QUARTER
ROUNDS

ing the fascia at the end of the side cornices. It's not likely that this can be done with a single piece of material that is long enough to run from the ridge to the eaves or wide enough to cover the cornice. But do plan to use as few pieces as possible and to work so you get tight joints, preferably with a T&G arrangement at the mating edges.

The back of the side cornice is sealed with a piece of soffit material used as a closer (**Figure 11-20**). It will probably be easier to add this and the corner will be more weathertight if you do it before working on the fascia.

Some of the many varieties of ready-made moldings that are usable to trim cornice and roof edges are shown in **Figure 11-21**.

12

INSTALLING THE ROOF-DRAINAGE SYSTEM

All drainage systems on sloped roofs, regardless of the materials used, consist basically of gutters and downspouts that work together to collect runoff water from the roof and direct it so it can't cause foundation seepage. The discharge can flow into drainpipes that run to storm sewers, or it can be directed to a natural runoff area, perhaps the street, so that it can't form a swamp on adjacent ground. Once it was common to collect the water in barrels placed under the downspouts, or in cisterns, to be stored for future use.

Disposal methods depend on the amount of rainfall and snow in the area. When it is high, you must plan to avoid problems. When it is low or moderate, you can often make out merely by directing the water away from the foundation.

The size of the gutters depends on the amount of water they will receive, and this is affected by the area of the roof. Workable, standardized guides say that a 4-inch gutter is the minimum and is usable for a roof that is not larger than 750 square feet. Use 5-inch gutters when the roof area approaches 1,400 square feet, and 6 inches thereafter. Downspouts that are 3 inches in diameter will do if the roof is under 1,000 square feet; use 4-inch downspouts if the area is greater. The size of the downspouts is of less importance than their number and location. One outlet for each 600 square feet of roof area is a minimum. A gutter run that is less than 20 feet can do with a single outlet at one end. Use an outlet at each end when the run is greater.

Don't figure the square footage of a roof on a horizontal plane. Figure its real area, including dormer roofs, overhangs, etc. All planes will collect water during any rainfall, especially if there is wind. The higher the ridge, the more true roof area there is and the faster water will be carried into the gutters. So if you have a very steep roof, you might plan to be more generous with gutter sizes regardless of the roof area.

PLAN THE INSTALLATION

Gutters will carry water more easily to downspouts if they are sloped a minimum of ½ inch for every 20 feet of run. The greater the slope, the faster the water travel, so 1 inch per 20 feet is not out of line, except that it is visually disturbing. This off-horizontal line annoys some people, and so they set the gutters dead level. Water will still escape, but the lack of slope is bound to create water pockets that will eventually cause problems.

A solution is to install a level wooden trough that is deep enough to contain metal gutters with good slope, but this is custom work and requires extra time and materials. It's also possible to design special waterways, usually called eave troughs, which are attached to roof edges and look like no more than trim. They can be preassembled so the bottom slopes but there is still a horizontal face, which really becomes a fascia. Such constructions should be done with waterproof glue and a decay-resistant wood like heart grades of cedar or redwood. Inside joints should be sealed with caulking and all surfaces treated with a preservative.

Usually, however, standard ready-to-install metal gutters and downspouts are used and the necessary slope is accepted.

A typical installation is shown in **Figure 12-1**. There is no law that tells you where you must place a downspout, but appearance is a factor. Usually it's a good idea to locate them at the corners of a building and away from doorways. They can drop vertically from the gutters, but if the roof has an overhang they will be

FIG. 12-1

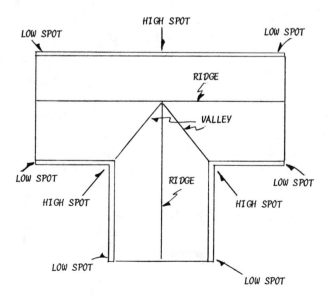

eyesores. It's better to use standard elbows so the downspout can be directed back to the house wall and tied there as inconspicuously as possible.

Ready-made gutter systems are available in various sizes, and in shapes that include half-round, K-style, and ogee. Materials include galvanized steel, copper, aluminum, and even vinyl. Galvanized-steel gutters are used most often, probably because they are the least expensive. Now they are available with a baked-on acrylic coat so you don't have to apply the initial prime coat that is usually required before galvanized steel can be painted.

Metals used in gutters do vary in thickness; 26-gauge is quite strong but 28-gauge is more commonly used because it reduces cost. The most frequently used thicknesses in aluminum gutter material are .032 and .027. Always work with fasteners and other parts that match the gutter material. Don't, for example, use aluminum screws on steel or the opposite. An electrolytic action can cause quick corrosion when contacting metals are dissimilar.

Though the designs and the method of hanging may differ, all gutter systems include the components that are necessary to make runs, turn corners, connect to downspouts, and so on (**Figure 12-2**).

FIG. 12-2

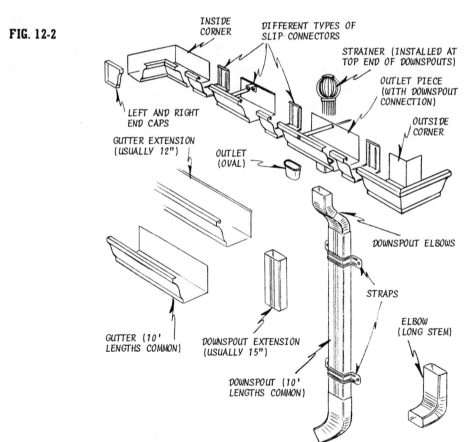

Some gutters are attached against the fascia and can be installed when the roof is complete. The easiest of these uses gutter spikes that pass through both sides of the gutter and are driven into the end of every other rafter (**Figure 12-3**). A sleeve which fits inside the gutter and through which the spike passes keeps the gutter from buckling.

FIG. 12-3

SPIKE WITH SLEEVE

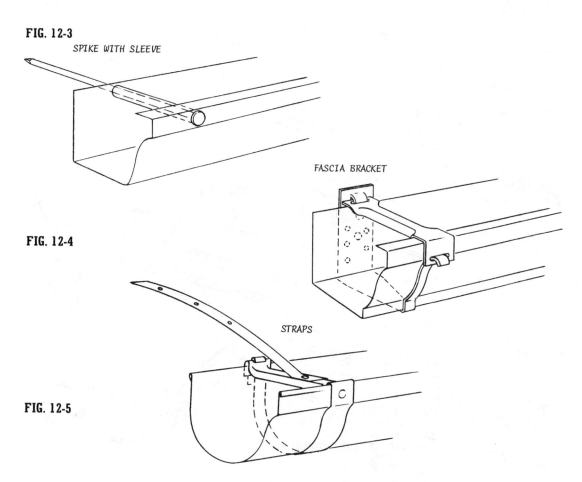

FASCIA BRACKET

FIG. 12-4

STRAPS

FIG. 12-5

Another type works with brackets that are installed against the fascia at alternate rafters (**Figure 12-4**). These may vary in design and the manufacturer may even suggest a different spacing, but basically, the idea is to attach the brackets and then lock the gutters in place with a strap that is part of the assembly.

Strap types differ, but all work like the one shown in **Figure 12-5** and are installed after the roof sheathing is in place and before the finish roofing goes on. Spacing of the straps (or hangers) is about 48 inches when they are galvanized steel. If the system is copper or aluminum, the spacing is reduced to 30 inches.

A more sophisticated system includes a flange, or roof apron, which is attached to the edge of the roof over sheathing to act as a flashing (**Figure 12-6**). The designs can be one-piece or two-piece, with a lock connection between the flange and the gutter. The flange is attached with nails about every 12 inches. The hangers hook into a lip at the front edge of the gutter and are nailed over every other rafter.

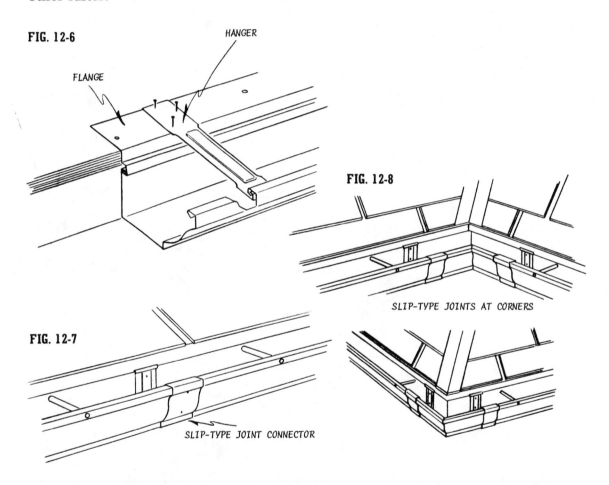

FIG. 12-6

HANGER

FLANGE

FIG. 12-8

SLIP-TYPE JOINTS AT CORNERS

FIG. 12-7

SLIP-TYPE JOINT CONNECTOR

Most systems include slip connectors for joining sections of gutters (**Figure 12-7**). They are designed to grip without further attention except the addition of a caulking compound to waterproof the joint, but a wise extra step is to drill holes and lock the joint with sheet-metal screws or Pop-Rivets.

The slip connectors are also used wherever the gutter turns a corner (**Figure 12-8**). Here, as with all connections, make the joints as smooth as possible so you won't create dams.

FIG. 12-9

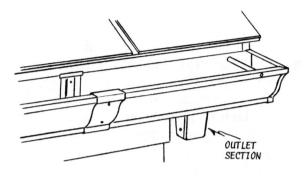

OUTLET
SECTION

FIG. 12-11

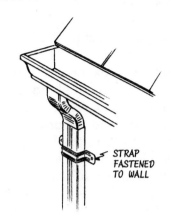

STRAP
FASTENED
TO WALL

FIG. 12-10

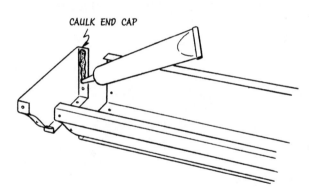

CAULK END CAP

FIG. 12-12

HOW TO CONNECT GROUNDING WIRE

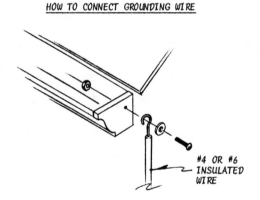

#4 OR #6
INSULATED
WIRE

Install outlet sections at all downspout positions (**Figure 12-9**). End all runs with an end cap secured with mechanical fasteners plus a good amount of caulking (**Figure 12-10**).

Downspout connections are made with elbows and, if necessary, short lengths of downspout extensions (**Figure 12-11**). Special straps are nailed in place to hold the downspout snugly against the house.

If it is necessary in your area, you can ground gutters by connecting them to a pipe in the ground with a length of #4 or #6 insulated solid conductor wire connected to the gutter as shown in **Figure 12-12**. Check this system out against local codes before you do it.

DOING THE JOB

A good place to start the installation is at the point which is farthest away from the downspout area. This can be at the end of a roofline so there will be a one-directional slope, or at a midpoint so the gutters will slope to both left and right. Drive a nail at the high point and then stretch a line to the other end, dropping it in relation to the slope you are using. Start placing gutters at the high point and run from there to the opposite end, or ends, attaching material so the top edge of the gutter follows the line. Downspouts will have to be installed after the house siding is up, but since you are involved with metal work at this point you may wish to preassemble parts so they will be ready when needed.

The metals involved can be cut easily with a hacksaw and a pair of metal-cutting snips. Be careful, though; maybe even wear gloves, since sharp edges result. Finish all cuts with a file or emery paper to remove burrs.

Vinyl materials may also be cut with a fine-tooth saw, but, as with metal materials, support the work solidly and keep the blade at a low angle. Vinyl gutter systems are assembled with a special cement provided by the manufacturer.

13

WATERPROOFING ROOF JOINTS

There are few roof designs that are not interrupted by intersections of various roof planes or by house components such as chimneys, soil stacks, vents, or even skylights. The joints that occur require special attention so water can't pass through them, or from them to enter under regular roofing material. The areas are treated by flashing, which is a procedure involving special materials such as galvanized metals, aluminum, or copper or more common materials like roll roofing or even asphalt shingles.

Some of the constructions may be done before the finish roof is applied; others are done together with the final steps. What and where depends a good deal on how you are finishing the roof, but all the procedures are based on the water-flow principle already described in Chapter 12. Water must drain *over* roofing material, not *under* it.

The most critical areas are at valleys, around chimneys, and anywhere a roof surface abuts a vertical wall. Of the three, the valley areas can cause the most problems. This is not hard to understand when you consider that valleys are depressions that collect water from various directions. Snow can lodge there and may become ice dams to interrupt the natural flow of water, which will then seek any weak point in the cover to escape. The slope of the roof is another factor. The less there is, the greater the danger of leakage.

There are some situations, especially under asphalt shingles, where codes (and practicality) permit the use of 90-pound mineral-surfaced asphalt roll roofing as the flashing material. This is applied as a double coverage, with the first piece about 18 inches wide and placed with the mineral side down, and the second piece about 36 inches wide and placed with the mineral side up (**Figure 13-1**). It's okay to use joints, if the job can't be done with single pieces, but the top piece should overlap the bottom one by at least 12 inches and mating surfaces should be bonded with asphalt cement. A good application procedure is to nail down one edge of the flashing and then do the other edge while the material is pressed firmly into the valley. Coat the edges of the first ply with a 3- or 4-inch-

FIG. 13-1

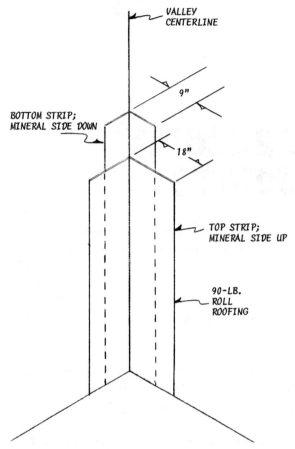

wide band of asphalt cement before you add the second ply. Use regular asphalt-shingle nails, but place them so they will be amply covered by the final roof material. Careful workers will coat those nail areas with cement, doing it before other materials hide them.

After the flashing is installed, it's wise to mark lines that will tell where courses of shingles should end. Start by using a straight board or by snapping a chalk line to indicate the centerline of the valley. Then do another one on each side so you will know the width of the waterway. The width at the bottom should be more than what you start with at the top; it should increase by ⅛ inch per foot. Minimum widths can be 4 to 6 inches; 5 inches is a happy compromise. Thus if a valley is 12 feet long, the waterway would be 5 inches wide at the top and 6½ inches wide at the eaves.

When you get to placing the shingles, the valley-end one in each course will be trimmed to follow the line you have established. Usually the top outer corner of the shingle is cut off at a 45 degree angle and the edge of the shingle is cemented to the flashing (**Figure 13-2**).

FIG. 13-2

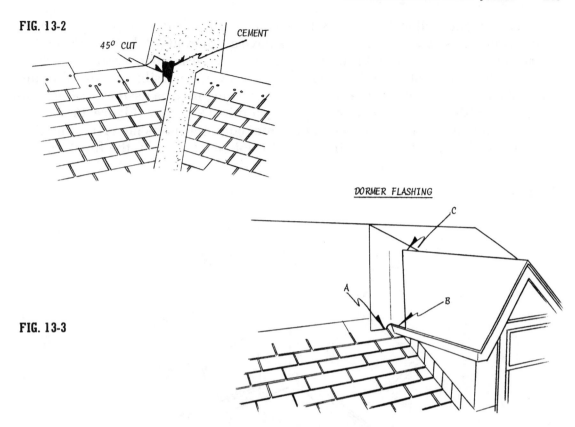

FIG. 13-3

Valleys at gable dormers usually have short runs, so one-piece flashing is feasible. The best procedure is to make a paper pattern for one side, which when flipped, can be used for the opposite valley as well. The cut at C in **Figure 13-3** is a notch to permit flashing pieces to overlap at the ridge while the remainder lies flat on the roof. Edge B should extend beyond the eave of the dormer, while edge A should be long enough so at least 2 inches of it can lie flat on the roof. The corner between A and B should be cut as a curve so the flashing won't crack when it is pressed into the valley.

Valley flashing can be done with metal, and usually is when the roof cover is wood shingles or shakes. The steeper the slope, the narrower the flashing must be. For example, 12 inches is okay when the slope is 7 in 12 or over; 18 inches will do for slopes that run from 4 in 12 to 7 in 12; 24 inches should be used for slopes under 4 in 12.

Check the valley area carefully for any projecting nailheads before you place the flashing. Premium construction calls for a layer of 15-pound or 30-pound roofing felt to be placed under the metal and for both surfaces to be painted with a good grade of metal paint. If joints are necessary, the top piece should overlap the bottom one by at least 6 inches.

Ready-made valley flashing with a splash rib or ridge running down its center is often used when adjacent slopes have a different pitch. The purpose of the rib, or standing seam as it is often called, is to serve as a barrier so water flowing from the steeper slope will not cross the valley and possibly be forced under the roof cover on the other side. The special flashing may be used as a safety factor, if you wish, even when adjoining roof slopes are similar (**Figure 13-4**).

Let valley flashing extend beyond the roof edge at least as far as the shingles. Do not nail through it at any point that will not be covered by the finish roof.

FIG. 13-4

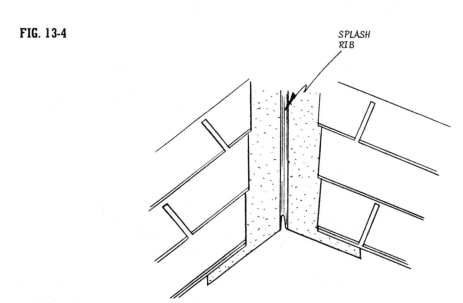

SPLASH RIB

AROUND CHIMNEYS

The joint between a chimney and the roof is waterproofed by using base flashing, step flashing, and counter flashing (**Figure 13-5**).

Since a chimney is established on its own foundation and the wood members around it are subject to expansion and contraction, the flashing is done so movement can occur without breaking the water seal. The precaution is taken where the counter flashing overlaps the others, No bond is used here unless it is an elastic cement. Actually, if the installation is done correctly with sufficient laps, water will not be able to enter behind the flashings, even without cement.

The base flashing is installed after the roof cover has reached the base of the chimney. Follow the pattern shown in **Figure 13-6** so the material can be folded to fit snugly around the chimney base. A similar piece, which is applied last, can be cut for the back of the chimney.

FIG. 13-5

FIG. 13-6

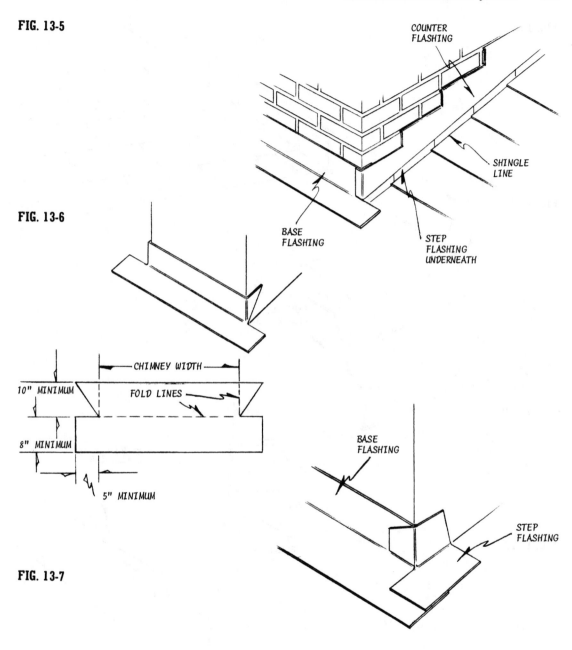

FIG. 13-7

The first piece of step flashing is placed as shown in **Figure 13-7**, while other individual pieces are set in position at the end of each course of shingles as shown in **Figure 13-8**. The pieces of step flashing are bent to form a right angle, with each leg 5 to 6 inches wide. The length should be 2 to 3 inches more than the

FIG. 13-8

FIG. 13-9

FIG. 13-10

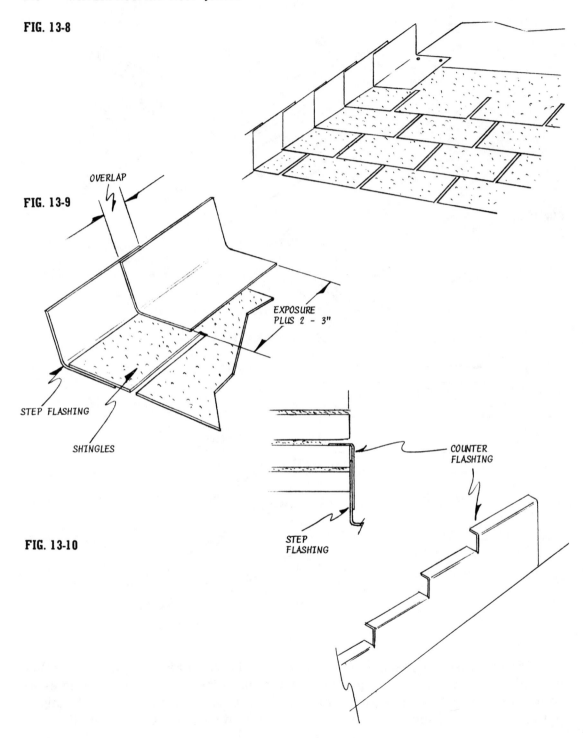

OVERLAP

EXPOSURE
PLUS 2 - 3"

STEP FLASHING

SHINGLES

COUNTER
FLASHING

STEP
FLASHING

FIG. 13-11

exposure used on the shingles. Since the flashing is placed along with the shingles, you automatically get a good overlap at all joints (**Figure 13-9**). Secure each piece with one or two nails along the top edge.

Counter flashing is done with metal sheets or a single preformed piece. In each case a flange at the top, as shown in **Figure 13-10**, must penetrate the mortar joint in the masonry at least 1 inch. Side flashing, which is stepped to conform to the slope of the roof, is often done simply by embedding separate sheets in the masonry joints and then bending them down over the step flashing when the mortar has set. This can be accomplished while the chimney is being erected.

If the chimney is up, the mortar is raked out of the joints by careful work with a chisel and hammer, leaving a groove as shown in **Figure 13-11**, and the joints are packed with fresh mortar or a bituminous mastic after the flashing is in place. The counter flashing at the front and the back of the chimney is done with straight pieces.

Flashing around a chimney, except for the counter flashing, is often done with 90-pound mineral-surfaced roofing material when the roof finish is asphalt shingles. In such cases, the various pieces of flashing may be bonded to the roof and to each other with mastic. We'll show more about flashing a shake roof in Chapter 14.

A SADDLE FOR THE CHIMNEY

A saddle, often called a cricket, is a gablelike construction done on the high side of the chimney to break down the area so water can flow off more freely and so snow will have less chance of accumulating (**Figure 13-12**). This is a nuisance to make and should be done only if the chimney is exceptionally wide or the climate is exceptionally wet.

FIG. 13-12

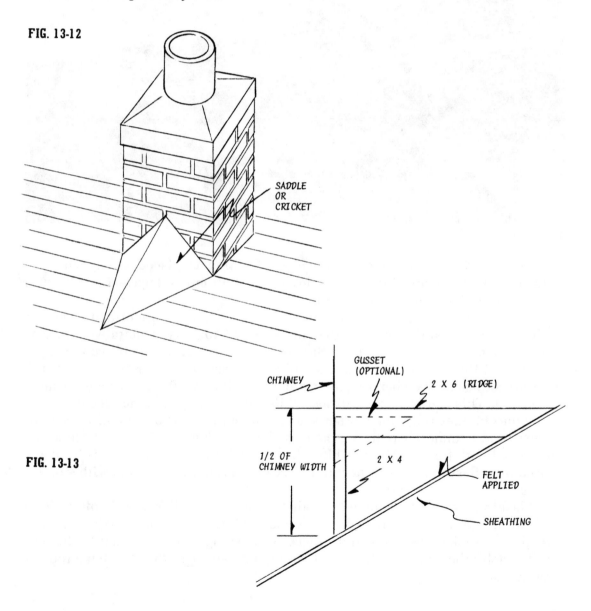

SADDLE
OR
CRICKET

GUSSET
(OPTIONAL)

CHIMNEY

2 X 6 (RIDGE)

1/2 OF
CHIMNEY WIDTH

2 X 4

FELT
APPLIED

SHEATHING

FIG. 13-13

The construction is always done after the roof sheathing is on and consists of a ridge that is nailed to the sheathing at one end and is supported by a 2×4 at the chimney (**Figure 13-13**).

The cover, preferably, is done with two pieces of exterior-grade plywood, sawed so joints are as tight as possible. The structure will be strong enough after the cover is nailed along the ridge and across bottom edges into the roof. Saddles are seldom covered with roofing material, so full coverage with flashing is in order. The best way to do this is to work with sheets of paper and masking tape to create a pattern you can use to cut the flashing in one piece. Allow ample material against the chimney so the flashing can be folded up to serve as step flashing and be covered by the counter flashing. The other edges should also be generous so they can be folded flat on the roof and sealed with a mastic. Use as few nails as possible to attach the flashing and place them so they will be amply covered by the finish roof. The shingles that must be cut to follow the line of the saddle should be embedded in cement.

FLASHING AT A VERTICAL WALL

This is done following the procedure outlined for the installation of step flashing along the sides of a chimney, and is required wherever a vertical wall intersects with a roof plane (**Figure 13-14**). It can occur, for example, at the wall-base of a partial second story or where dormers intrude. The roofing, to that point, and the step flashing are placed before the vertical wall is covered with siding. The bottom edge of the siding overlaps the flashing, but allow sufficient clearance so those bottom edges can be painted.

FIG. 13-14

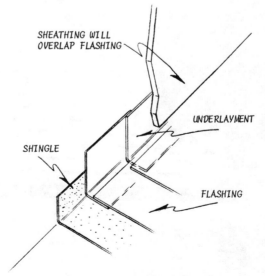

SHEATHING WILL
OVERLAP FLASHING

UNDERLAYMENT

SHINGLE

FLASHING

14

INSTALLING THE FINISH ROOFING

AN ASPHALT SHINGLE ROOF

The improvements in asphalt roofing products and methods of installation are so impressive they are being used more and more in all types of modern constructions. Innovations in textures to suit various architectural styles, more colors, heavier weights—all contribute visual appeal to a product that is practical to begin with. The old-fashioned objection to the "flatness" of an asphalt roof is no longer justified, because of newer shingles that create deep shadow lines and provide a thick roof with a rustic appearance. An example is the "Architect 70" shingle (made by Bird & Son) which is textured and has random edges, and is available in various colors (**Figure 14-1**).

FIG. 14-1

There isn't enough room in this book to discuss all such products, but information may be obtained by writing to the companies listed in the back of the book.

Asphalt roofing products are called saturated felts, shingles, and roll roofing. Saturated felt, which is used as an underlayment over solid sheathing, is made by impregnating dry felt with coal tar or asphalt. It is purchased by weight—for example, 15-pound or 30-pound, the weight of the amount of the material that will cover 100 square feet when placed as a single ply. In other words, one will weigh 15 pounds per 100 square feet of roof, the other 30 pounds. The weight of 100 square feet of uninstalled felt will be somewhat less, of course, since it is installed with overlaps.

When a felt has already been impregnated with asphalt is further strengthened with a coating of weather-resistant material (such as asphalt), it becomes roll roofing (**Figure 14-2**). Mineral granules are often bonded to the surface of such materials to provide protection from the sun and to add fireproofing qualities.

FIG. 14-2

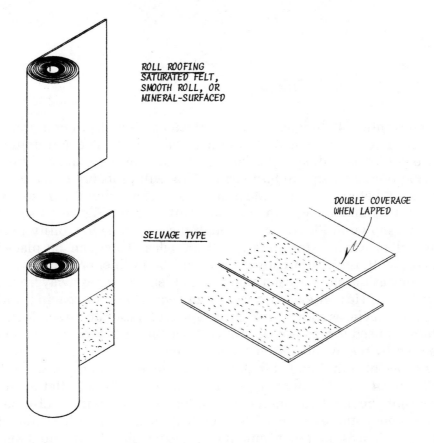

ROLL ROOFING
SATURATED FELT,
SMOOTH ROLL, OR
MINERAL-SURFACED

SELVAGE TYPE

DOUBLE COVERAGE
WHEN LAPPED

The granules also make it possible to add various colors, and they contribute to durability. It's possible to do an entire roof using such material, although this isn't commonly done on modern residences. You might want to consider it, though, for outbuildings such as garden sheds. Especially when lapped as shown in **Figure 14-3**, it makes a durable roof, and it is easy to install.

FIG. 14-3

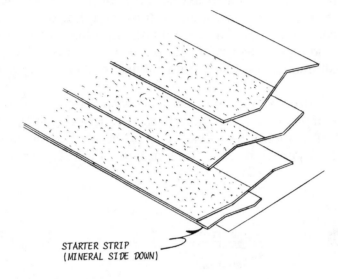

STARTER STRIP
(MINERAL SIDE DOWN)

Most conventional shingles have a surface coating of mineral granules and are available in various weights, patterns, and colors. The minimum recommended weight for asphalt shingles has increased over the years and now stands at about 235 pounds for square-butt strips. This will probably be subject to local codes, so check before you buy. Most applications are done with strips having two or three tabs and being either square-butt or hex (**Figure 14-4**). They are overlapped as shown in **Figure 14-5**, though there is some flexibility in how you line up the slots when using square-butt shingles. They can be placed at the center of tabs so those in alternate courses will be in line, or the spacing can be random as long as the slots in each course are at least 3 inches away from those in the course below. Hex shingles do not permit variations in spacing. Individual shingles are available, as shown in **Figure 14-6**, and these can be set down in standard patterns or used to create special effects, as long as the exposure and the laps recommended by the manufacturer are followed.

Other types of individual shingles are designed to interlock when they are placed (**Figure 14-7**). The intent is to make the shingles stay flat even in high wind. Some of the modern nonlocking shingles come with seal-type tabs to provide the same wind resistance. It's not a bad idea, even with regular shingles, to put a small dab of asphalt roof cement under each tab to help hold them down.

FIG. 14-4

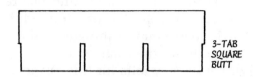

3-TAB
SQUARE
BUTT

FIG. 14-5

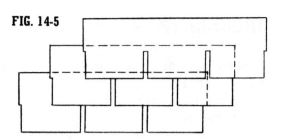

2-TAB
HEXAGONAL

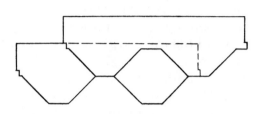

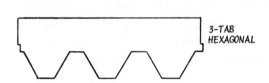

3-TAB
HEXAGONAL

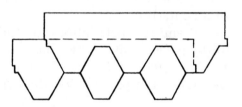

FIG. 14-7

FIG. 14-6

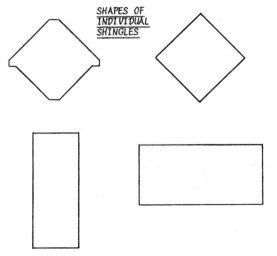

SHAPES OF
INDIVIDUAL
SHINGLES

LOCK-DOWN
SHINGLES

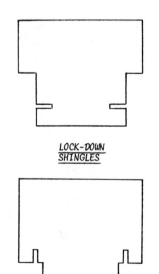

UNDERLAYMENT

A cover of roofing felt, usually 15-pound, is placed over the sheathing before shingles are nailed down. The purpose of the cover is threefold. It provides additional weather protection for the roof, it is a barrier between the shingles and the wood, and it keeps the sheathing dry until you get to do the final cover.

It's important not to use coated material or very heavy felts for this purpose, since they will become vapor barriers, which can lead to the accumulation of moisture, and maybe frost, under the final cover. The 15-pound felt usually does a good job, but the requirements can be affected by the type of shingles and the slope of the roof, so give some consideration to the recommendations of the manufacturer.

All underlayments should have a minimum 4-inch side lap and 2-inch top lap. They should be secured with the minimum number of staples or nails required to keep wind from blowing them off the roof until the final cover is in place. The laps at hips and across valleys should be at least 6 inches on each side of a centerline.

Premium installations include metal drip edges along perimeters (**Figure 14-8**). These may be purchased ready-made, usually formed from 26-gauge galvanized steel and with the top flange 3 to 4 inches wide. Note that the underlayment is placed over the drip edge at eaves but under it at a slope.

FIG. 14-8

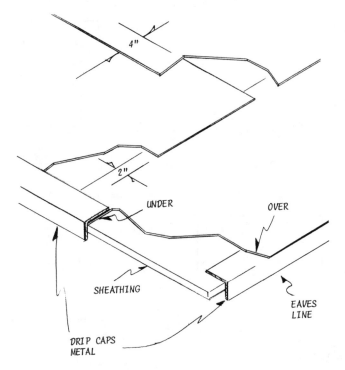

NAILING HOW-TO

Asphalt roof nails have very sharp points and head diameters that can run from ³⁄₈ to ⁷⁄₁₆ inch. Most installations today are done with modern 11- or 12-gauge galvanized steel nails that do not have smooth shanks (**Figure 14-9**). The length of the nail must be enough to allow almost full penetration of the sheathing.

FIG. 14-9

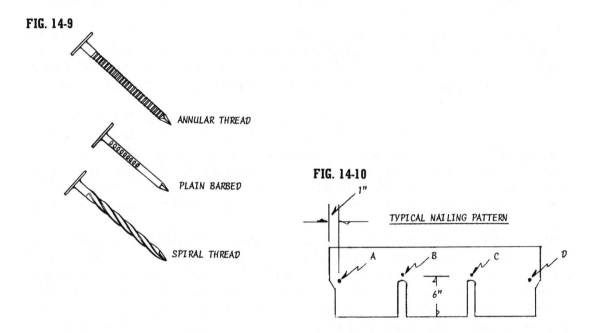

FIG. 14-10

The number of nails you use, where you place them, and the sequence in which you drive them are critical factors for a premium-quality roof. A typical pattern using a minimum of four nails is shown in **Figure 14-10**, on a three-tab square-butt shingle. Nail each strip at one end after it has been aligned with its neighbor, and then continue nailing toward the opposite end. Don't nail down the ends and then the center, since this can cause the strip to buckle. Keep the nail shank perpendicular to the slope as you drive so the nailhead won't tilt and maybe cut the corner. Don't hammer so much that you sink the nailhead into the surface of the material. If the nail doesn't hit solid sheathing, remove it, plug the hole with asphalt cement, and drive another nail close by.

A lot of nailing is required on jobs like this, so you might check out the possibility of renting power equipment. Tools designed specifically for the work, driving special roofing staples, are acceptable. Manufacturers of roofing covers can advise what type of staples can be used.

DOING THE SHINGLING

The job can be started from one end of the roof or from a centerline. In either case it's a good idea to place a line of shingles "dry" to see how they will fall at roof ends or at valleys if they are involved. Then you adjust one way or the other for uniform placement and end cuts that do not fall, for example, in a slot area. You can still work from a line that you snap from the ridge to the eaves, even though it may not indicate the center of the roof.

Use a starter strip under the first course of shingles to add bulk and to back up openings such as slots. The starter strip should be 10 to 12 inches wide and can be cut from mineral-surfaced roll roofing that matches the shingles both in weight and color, or it can be done with strips of the shingling material placed so the slots point toward the ridge.

Place the starter strip so it overhangs the roof edge a bit and secure it with nails spaced to be covered by the first course of shingles and so they are in from the edge of the roof about 4 inches. Snap a chalk line across the starter strip so you will have a guide for the top edge of the first course of shingles. The first course and the following courses are placed as shown in **Figure 14-11**. Note that the layout aligns the slots in alternate courses. To do it differently, you must reduce the length of the strips that fall at the side of the roof. This can be planned for when you do the dry run we talked about so you can still work from a centerline mark. For example, to get a random pattern as in **Figure 14-12**, you would remove a different amount from the end strip in each of four or five successive courses, repeating the procedure on the following sets of courses until you reach the ridge.

Hexagonal shingles do not permit a random pattern. Begin with the same kind of starter strip but place each course so the tabs align with the cutouts in the bottom course (**Figure 14-13**).

AT RIDGES AND HIPS

You can buy special shingles for use at ridges and hips, but it's just as easy to make them by cutting rectangular pieces, measuring about 9 by 12 inches, from mineral-surfaced roll roofing. These are bent on the centerline and applied as shown in **Figure 14-14,** with no more than a 5-inch exposure. Maintain a 1-inch edge distance on the nails and place them so they will be well covered by succeeding pieces. Place those on the ridge so the "open" side will not face the direction of prevailing winds. Pieces formed as hip and ridge covers should not be bent when they are cold or they may acquire surface cracks or even break. Warm them in the sun or, if necessary, near a fire or with a torch.

FIG. 14-11

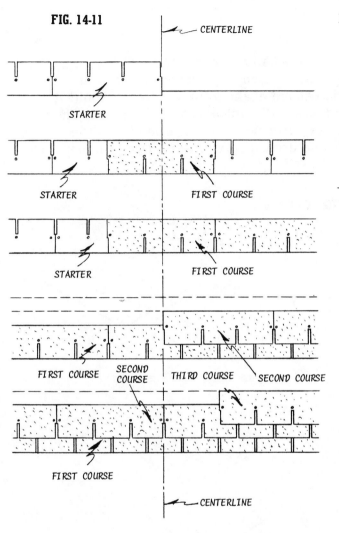

STARTER

STARTER   FIRST COURSE

STARTER   FIRST COURSE

FIRST COURSE   SECOND   THIRD COURSE   SECOND COURSE
                  COURSE

FIRST COURSE

CENTERLINE

FIG. 14-12

TABS DO NOT ALIGN

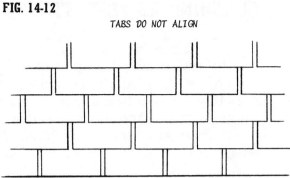

RIDGE SHINGLING

FIG. 14-14

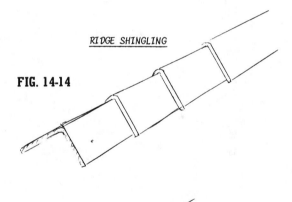

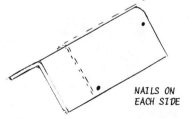

NAILS ON
EACH SIDE

FIG. 14-13

TABS MUST ALIGN

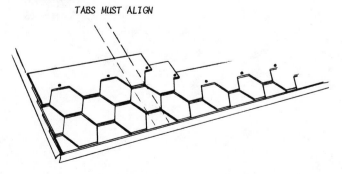

FLASHING AT VENT STACKS

This can be done with asphalt material as shown in **Figure 14-15**. Cut the shingle so it fits snugly over the pipe and then cement down a special cover that you cut from roll roofing. Apply enough plastic cement at the joint so you can mold it smoothly up a section of the pipe and over adjacent edges of the cover. The shingle in the next course is placed over the cover and cut to fit around the pipe.

A ready-made metal collar can also be used (**Figure 14-16**). Use plumber's tape to seal the joint between the collar and the pipe.

FIG. 14-15   **FIG. 14-16**

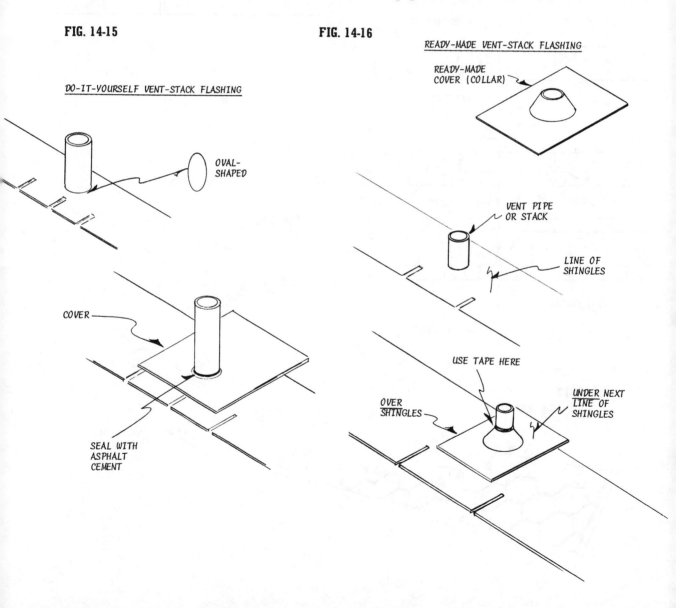

WOOD SHINGLES

Wood shingles have the appeal that all wood has. They are a natural material, are nice to work with, weather to pleasant colors, and are durable. They are taper-sawed from decay-resistant species like cypress, redwood, and cedar, with cedar having the edge because of characteristics that include a low expansion and contraction ratio in relation to moisture content, a nice even grain, and a high impermeability to liquid. Unfortunately, wood does not resist fire too well, so wood shingles may be frowned on in some areas unless thay have been specially treated with a fire retardant. It's a point to check.

The shingles may be put down over closed or open sheathing (spaced boards); the latter is often used because it reduces sheathing costs and contributes a ventilation factor that helps the cover to dry out more quickly. Consider solid sheathing when you wish to gain insulation and additional protection from the weather. An underlayment is not usually required regardless of the sheathing design, but if you wish to add it to guard against air infiltration, use a light, unsaturated felt or a building paper that has been sized with rosin.

Shingle lengths are 16, 18, and 24 inches, and each can be used with different exposures, depending on the thickness of the cover you want. You can see in **Figure 14-17** how the thickness increases as the exposure decreases.

FIG. 14-17

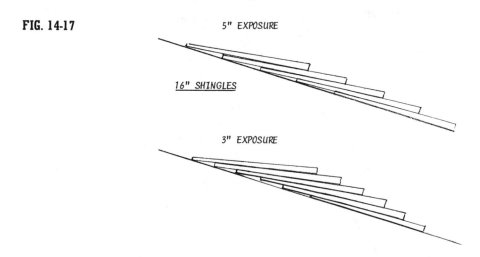

The correct amount of exposure depends on the slope of the roof. In most situations, the recommended exposure is as shown in the table in **Figure 14-18**. The 4-in-12 or steeper roof will have a 3-ply cover, while those with lesser slopes will have a 4-ply cover. The 3-ply cover is considered the minimum.

FIG. 14-18.   EXPOSURE RECOMMENDATIONS FOR SHINGLES

ROOF PITCH	SHINGLE LENGTH	EXPOSURE
4 in 12	16″	5″
or	18″	5½″
steeper	24″	7½″
less than 4 in 12	16″	3¾″
to	18″	4¼″
3 in 12	24″	5¾″
less than 3 in 12	NOT RECOMMENDED	

NOTE: Above exposures assume the use of No. 1 grade shingles

Shingles do have to be cut, and while you can do this with a saw the job will go faster and be easier if you rent or buy a special shingler's hatchet (**Figure 14-19**). This is a lightweight tool with two sharp edges and an adjustable gauge you can set to check exposure as you work. The corrugated head makes it less likely that the tool will slide off nailheads when you strike.

FIG. 14-19

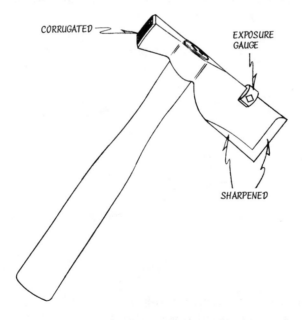

Work only with rust-resistant nails, choosing sizes as recommended in **Figure 14-20**. The nails may be aluminum or hot-dipped, zinc-coated steel.

FIG. 14-20   RECOMMENDED NAIL SIZES

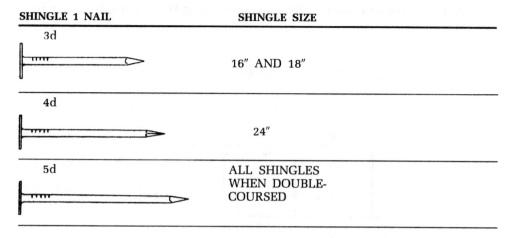

SHINGLE 1 NAIL	SHINGLE SIZE
3d	16″ AND 18″
4d	24″
5d	ALL SHINGLES WHEN DOUBLE-COURSED

The shingles you get in a bundle will vary in width but, regardless, never use more than two nails per shingle, placed ¾ inch from the edge and so they will be covered by that same distance by the next course (**Figure 14-21**).

FIG. 14-21

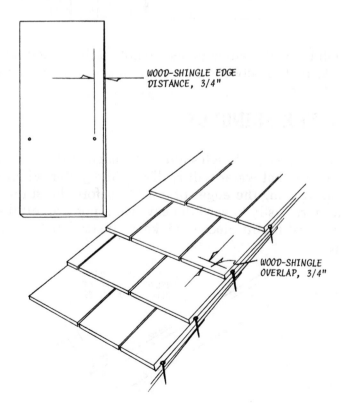

WOOD-SHINGLE EDGE
DISTANCE, 3/4″

WOOD-SHINGLE
OVERLAP, 3/4″

Overlap shingles as shown in **Figure 14-22**. Note the ¼-inch space between shingles to allow for expansion that might cause buckling if the shingles were butted.

FIG. 14-22

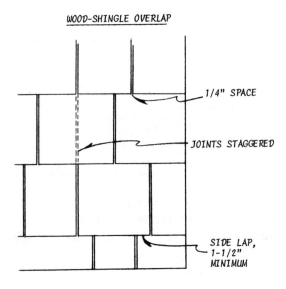

WOOD-SHINGLE OVERLAP

1/4" SPACE

JOINTS STAGGERED

SIDE LAP,
1-1/2"
MINIMUM

Note that joints in adjacent courses are not on the same line. Maintain a side lap of at least 1½ inches between the joints in successive courses.

INSTALLING THE SHINGLES

Start with a double layer of shingles along the edge of the roof, letting them project enough to assure that water will spill into the gutter (**Figure 14-23**). Often a cant strip is placed along the edges of a gable before the shingles are placed so water will flow more easily toward the eaves and not drip off the gable end (**Figure 14-24**). This is a difficult piece to make, so instead just nail down a length of beveled siding.

FIG. 14-23

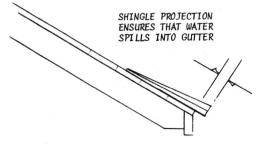

SHINGLE PROJECTION
ENSURES THAT WATER
SPILLS INTO GUTTER

FIG. 14-24

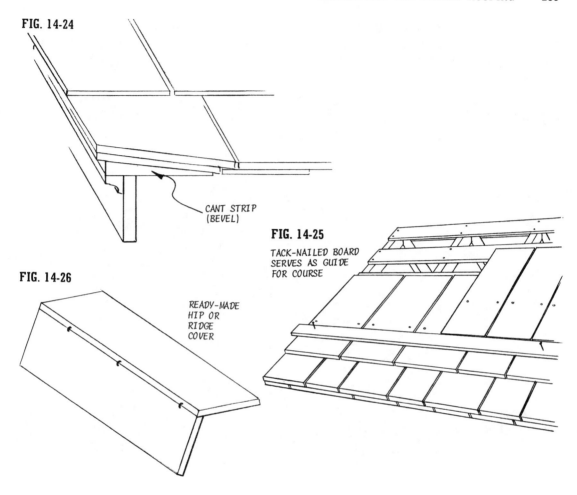

CANT STRIP
(BEVEL)

FIG. 14-25

TACK-NAILED BOARD
SERVES AS GUIDE
FOR COURSE

FIG. 14-26

READY-MADE
HIP OR
RIDGE
COVER

Succeeding courses of shingles should be placed by following a chalk line that you snap or by tack-nailing a straight piece of wood against which you can butt the courses of shingles you are nailing (**Figure 14-25**). It won't hurt to measure occasionally from the ridge to the last course of shingles you have placed to be sure you are keeping the courses parallel. As you will discover quickly, you can't align the shingles along the tops, because their legths vary somewhat.

Courses that abut a valley should be started there, and with wide pieces of shingling. Make the miter cuts you will need very carefully so course ends will match neatly. Never end a course with a very narrow shingle. It's better to reduce the width of others in the same course so you can end with a wide piece.

Ready-made units you can use to cover hip and ridge joints are available (**Figure 14-26**). Or you can make your own by beveling the mating edges of selected pieces of regular shingling. These are nailed together and placed with regular exposure but with the overlap moving left and right in alternate courses (**Figure 14-27**). Those on the ridge are placed so the open ends will not face prevailing winds (**Figure 14-28**). The first course in both hip and ridge covers is doubled and placed as shown in **Figure 14-29**.

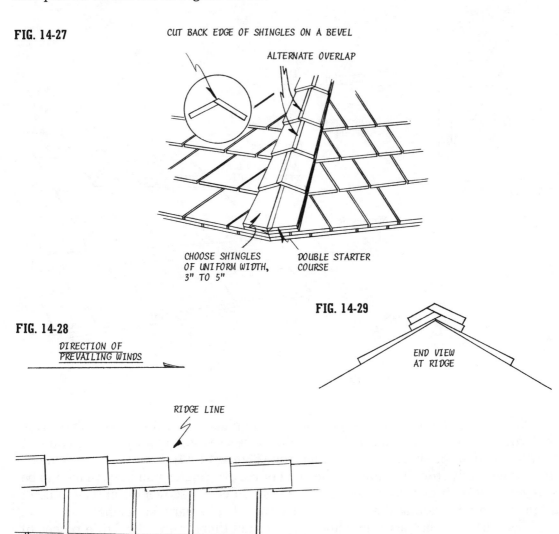

FIG. 14-27

CUT BACK EDGE OF SHINGLES ON A BEVEL

ALTERNATE OVERLAP

CHOOSE SHINGLES OF UNIFORM WIDTH, 3" TO 5"

DOUBLE STARTER COURSE

FIG. 14-29

END VIEW AT RIDGE

FIG. 14-28

DIRECTION OF PREVAILING WINDS

RIDGE LINE

Be sure to attach the cover pieces with nails that are long enough to penetrate the sheathing, and place them so they won't be exposed to the weather.

Two types of roof junctures are shown in **Figure 14-30** to demonstrate how to handle breaks in a roof line. The flashing should be galvanized metal and at least 8 inches wide so 4-inch flanges will cover the roof slope and the area that follows. Be sure the flashing covers the nails in the last course of shingles. Note, in each example, that the new area begins with doubled starter courses.

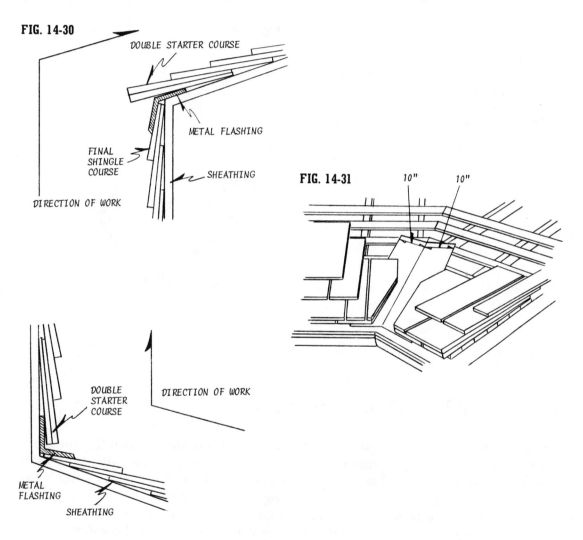

FIG. 14-30

DOUBLE STARTER COURSE

METAL FLASHING

FINAL SHINGLE COURSE

SHEATHING

DIRECTION OF WORK

FIG. 14-31

10"   10"

DOUBLE STARTER COURSE

DIRECTION OF WORK

METAL FLASHING

SHEATHING

Do valley flashing very carefully, since these areas can be serious trouble spots. Use sheets that will extend a minimum of 10 inches on each side of the centerline whenever the roof pitch is less than 12 in 12 (**Figure 14-31**). The extension may be reduced to 7 inches on steeper roofs.

The chimney-flashing procedure already described in Chapter 13 applies here, but check **Figure 14-32** and you will see how generous you can be with base and counter flashing to assure that water will run off and not in.

Use ready-made collars around all pipes that come through the roof, placing them as shown in **Figure 14-33** so that the waterflow principle is followed.

FIG. 14-32   **FIG. 14-33**

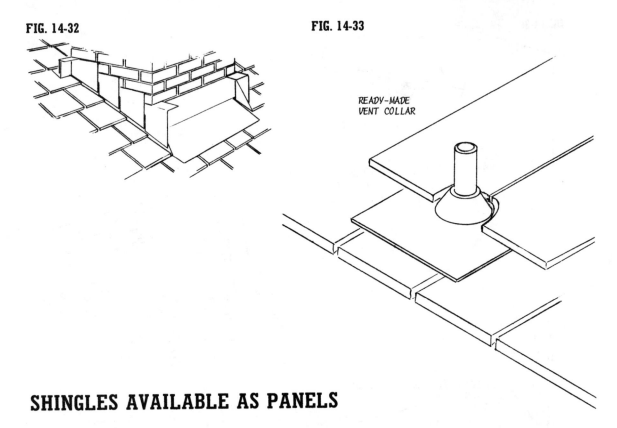

READY-MADE
VENT COLLAR

SHINGLES AVAILABLE AS PANELS

One of the newer roofing products on the market, from Shakertown Corporation, is an 8-foot panel which is composed of 18-inch cedar shingles bonded to ½-inch sheathing-grade plywood (**Figure 14-34**). The system lets you apply 16 shingles at a time together with sheathing, with nailing required only at the rafters.

The installation begins by nailing down special starter panels that are correctly positioned by placing a temporary starter strip across the rafters or by snapping a chalk line (**Figure 14-35**). All starter panels must end on the centerline of a rafter, but those that follow can join between rafters as long as such joints are well staggered. Plywood clips can be used at all end joints in order to get a stronger underlayment (**Figure 14-36**).

FIG. 14-34

FIG. 14-35

RIDGE

RAFTER

STARTER
STRIP

STARTER
PANEL

FIG. 14-36

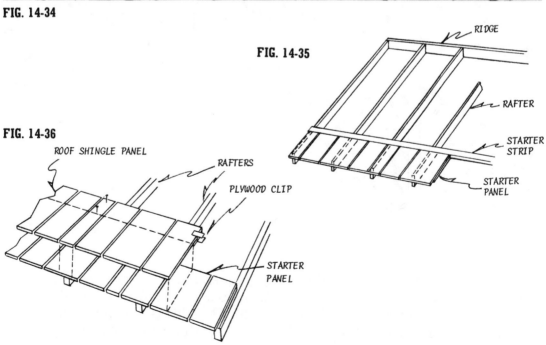

ROOF SHINGLE PANEL

RAFTERS

PLYWOOD CLIP

STARTER
PANEL

269

The nailing schedule calls for two 6d box-head rust-resistant nails at each rafter, with one placed about 1 inch down from the top edge of the plywood and the other about 2 inches up from the lower edge of the plywood. The placement of the nails is important, since it permits the shingles to lie flat but does not interfere with proper ventilation of the cover (**Figure 14-37**).

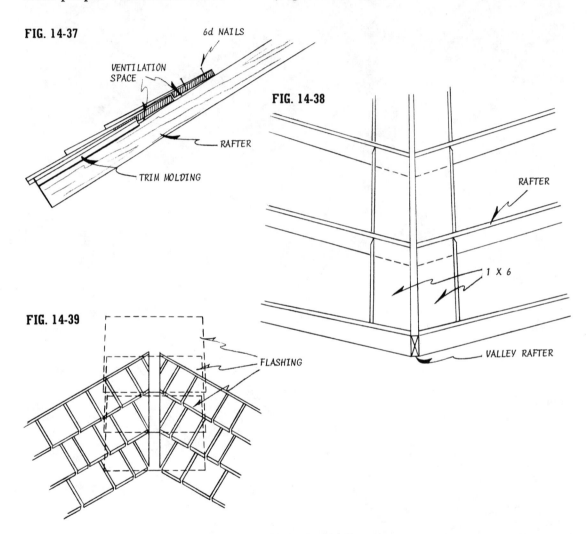

FIG. 14-37

FIG. 14-38

FIG. 14-39

Conventional flashing procedures apply generally, but a special design is recommended for valleys. Here, 1×6s or 2×6s are placed parallel to the valley rafter and so that top edges are flush (**Figure 14-38**). The idea is to supply a solid support for the flashing, which is installed in step fashion to protect the plywood from the weather (**Figure 14-39**).

Hip and ridge units can be made by utilizing cutoffs from regular panels. Application is done conventionally, but with a narrow piece of roofing felt used as an underlayment (**Figure 14-40**).

The shingle-panel system is made for roofs with a 4-in-12 rise or steeper. A similar system that is done with hand-split shakes instead of shingles is also available.

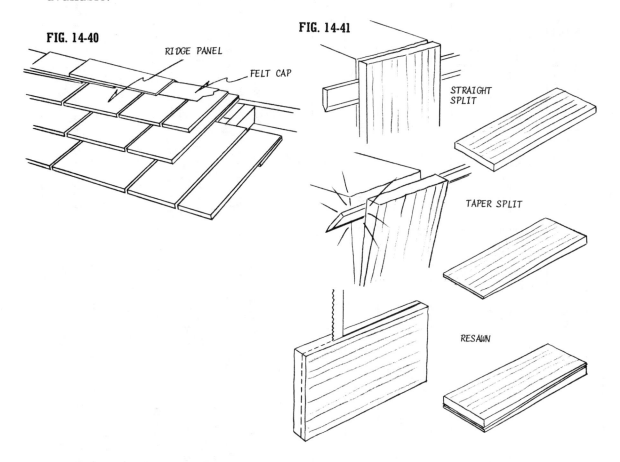

FIG. 14-40

RIDGE PANEL

FELT CAP

FIG. 14-41

STRAIGHT SPLIT

TAPER SPLIT

RESAWN

THE SHAKE ROOF

Roof covers done with heavy shinglelike pieces of wood go far back in the history of this country. Atlantic Coast colonists made them from various species of wood, including white pine, but today the emphasis is on red cedar, because of its fine grain, light weight, absence of pitch or resin, and natural resistance to decay.

Shakes are made in different ways, as shown in **Figure 14-41**. Resawn shakes are actually straight splits which are then cut on a bandsaw to get two shakes,

FIG. 14-42.   SHAKE SPECIFICATIONS

Shake Type, Length and Thickness	No. of courses per bundle	No. of bundles per square	Approximate coverage (in sq. ft.) of one square, when shakes are applied with ½" spacing, at following weather exposures (in inches):								
			5½	6½	7	7½	8½	10	11½	14	16
18" × ½" to ¾" Resawn	9/9 (a)	5 (b)	55 (c)	65	70	75 (d)	85 (e)	100 (f)			
18" × ¾" to 1¼" Resawn	9/9 (a)	5 (b)	55 (c)	65	70	75 (d)	85 (e)	100 (f)			
24" × ⅜" Handsplit	9/9 (a)	5		65	70	75 (g)	85	100 (h)	115 (i)		
24" × ½" to ¾" Resawn	9/9 (a)	5		65	70	75 (c)	85	100 (j)	115 (i)		
24" × ⅜" to 1¼" Resawn	9/9 (a)	5		65	70	75 (c)	85	100 (j)	115 (i)		
24" × ½" to ⅝" Tapersplit	9/9 (a)	5		65	70	75 (c)	85	100 (j)	115 (i)		
18" × ⅜" True-Edge Straight-Split	14 (k) Straight	4								100	112 (l)
18" × ⅜" Straight-Split	19 (k) Straight	5	65 (c)	75	80	90 (j)	100 (i)				
24" × ⅜" Straight-Split	16 (k) Straight	5		65	70	75 (c)	85	100 (j)	115 (i)		
15" Starter-Finish Course	9/9 (a)	5	Use supplementary with shakes applied not over 10" weather exposure.								

(a) Packed in 18"-wide frames.

(b) 5 bundles will cover 100 sq. ft. roof area when used as starter-finish course at 10" weather exposure; 6 bundles will cover 100 sq. ft. wall area when used at 8½" weather exposure; 7 bundles will cover 100 sq. ft. roof area when used at 7½" weather exposure; see footnote (m).

(c) Maximum recommended weather exposure for three-ply roof construction.

(d) Maximum recommended weather exposure for two-ply roof construction; 7 bundles will cover 100 sq. ft. roof area when applied at 7½" weather exposure; see footnote (m).

(e) Maximum recommended weather exposure for sidewall construction; 6 bundles will cover 100 sq. ft. when applied at 8½" weather exposure; see footnote (m).

(f) Maximum recommended weather exposure for starter-finish course application; 5 bundles will cover 100 sq. ft. when applied at 10" weather exposure; see footnote (m).

(g) Maximum recommended weather exposure for application on roof pitches between 4-in-12 and 8-in-12.

(h) Maximum recommended weather exposure for application on roof pitches of 8-in-12 and steeper.

(i) Maximum recommended weather exposure for single-coursed wall construction.

(j) Maximum recommended weather exposure for two-ply roof construction.

(k) Packed in 20" wide frames.

(l) Maximum recommended weather exposure for double-coursed wall construction.

(m) All coverage based on ½" spacing between shakes.

each with a thin end and a butt end and each with one sawn and one hand-split face. The multiple grooves that result from splitting act as channels that lead water off the roof. Also, splitting separates the wood fibers and leaves cell walls intact, so the shake absorbs little of the runoff. All quality shakes are made from the heartwood of the tree, since this is the decay-resistant area. The amount of sapwood in a shake is strictly controlled, usually limited to no more than $1/8$ inch along an edge.

The width of the shakes you get in a bundle will vary considerably, but lengths and thicknesses — the latter within tolerances — are standardized. Much information about the various types of shakes and the amounts required for different applications is contained in the table in **Figure 14-42**, which was made by the Red Cedar Shingle & Handsplit Shake Bureau. Few roofing materials can equal handsplit shakes for a naturally rustic, deeply textured finish. Since they are also adaptable to various environments and have good insulating qualities, they are justifiably popular. Variations in application techniques are also possible, like the random pattern in **Figure 14-43**, which puts even more emphasis on the hand-hewn effect.

FIG. 14-43

BASIC APPLICATION OF SHAKES

Sheathing can be solid, or spaced 1×6 boards as in **Figure 14-44**, the latter being practical in snow-free areas. The slope of the roof should be 4 in 12 or more with a maximum exposure of 10 inches for 24-inch shakes and 7½ inches for 18-inchers. Premium installations are often done by reducing the exposures to get a superior 3-ply coverage.

FIG. 14-44

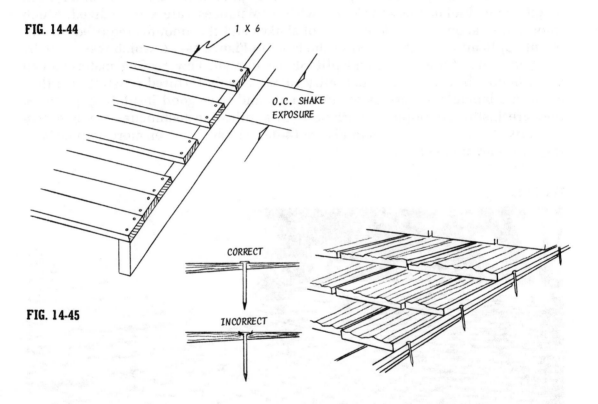

FIG. 14-45

Work with hot dipped zinc-coated 6d nails unless the combination of shake thickness and weather exposure make a longer nail advisable. The nail must be long enough to penetrate almost the full thickness of the sheathing. Use only two nails to a shake, placing them about 1 inch from edges and so they will be covered by 1 to 2 inches of the following course. Drive the nails only until the head meets the surface of the shake—don't sink them (**Figure 14-45**).

Leave a gap of about ½ inch between shakes and stagger the joints. All joints in one course should be offset at least 1½ inches from those in the following course.

DOING THE JOB

Start by placing a 36-inch-wide strip of 15-pound roofing felt over the sheathing along the eaves line. The starter course is always doubled, as in **Figure 14-46**, but may be tripled if you wish to bulk the edge. A cantstrip, like the one we described for shingles, may be placed on gable slopes (**Figure 14-47**). Nail on the cant strip before starting to nail shakes.

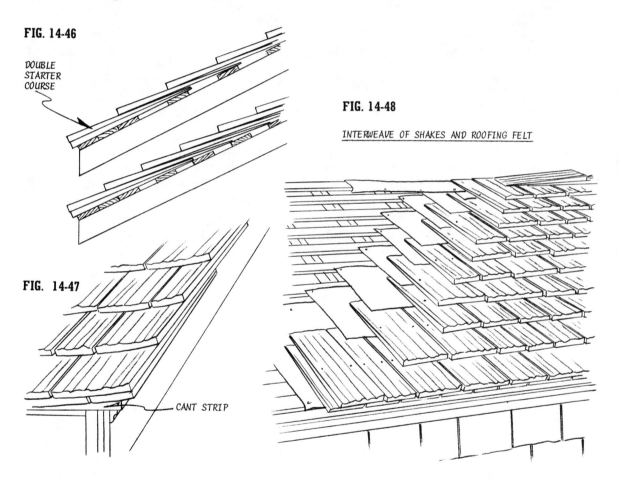

FIG. 14-46

DOUBLE
STARTER
COURSE

FIG. 14-48

INTERWEAVE OF SHAKES AND ROOFING FELT

FIG. 14-47

CANT STRIP

Courses of shakes are interwoven with 18-inch-wide strips of 15-pound roofing felt. Place the felt so its bottom edge will be away from the butt end of the course already in place by about twice the weather exposure. This would be 20 inches if you are working with 24-inch shakes and a 10-inch exposure, so the felt would cover the top 4 inches of the shakes and also 14 inches of the sheathing (**Figure 14-48**).

FIG. 14-49

Select uniform shakes when you get to the final course at ridge and hip lines. Place an 8-inch-wide strip of felt over the crown before nailing ridge or hip units (**Figure 14-49**). These units, which you can buy ready-made or assemble yourself using shakes about 6 inches wide, get the same exposure given the roof shakes. As with shingles, the starter course for both hips and ridges is doubled and the crown joint moves left and right as the courses are placed (**Figure 14-50**).

FIG. 14-50

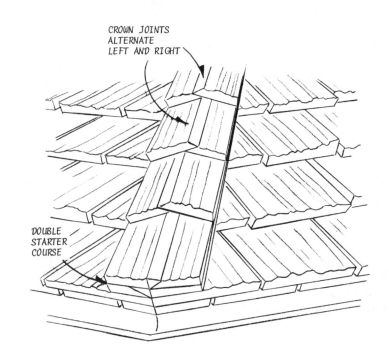

CROWN JOINTS
ALTERNATE
LEFT AND RIGHT

DOUBLE
STARTER
COURSE

Conventional base, step, and counter flashing is used around chimneys, but be sure that any flashing positioned under the shakes extends for at least 6 inches (**figures 14-51** and **14-52**).

FIG. 14-51

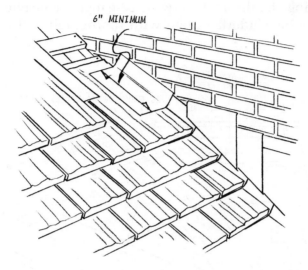

6" MINIMUM

FIG. 14-52

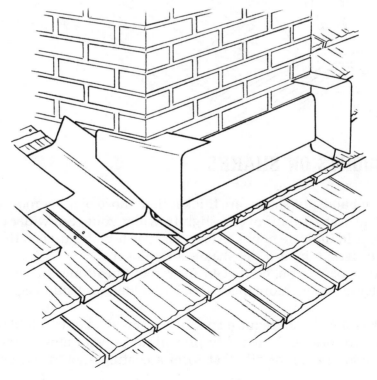

Paint both surfaces of all flashing metal with a heavy-bodied lead-and-oil paint or a bituminous paint and allow it to dry before placement. It's recommended that you don't do the painting until you have bent the metal to correct angles.

Valley flashing should be done with 26-gauge, or heavier, sheets that are at least 20 inches wide. Underlay them with roofing felt and use 6-inch headlaps (**Figure 14-53**).

FIG. 14-53

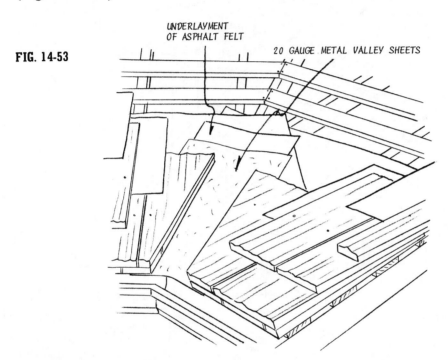

UNDERLAYMENT OF ASPHALT FELT

20 GAUGE METAL VALLEY SHEETS

MORE THOUGHTS ON SHAKES

The butt lines of handsplit shakes are thicker than those of other roofing materials and tend to shed water runoff over a slightly wider zone. So, wider gutters than normal rules would indicate might be in order. Use 5-inch gutters if the shakes are ¾ inch thick or less; 6-inch if the shakes are thicker.

Construction paper can be placed over roof sheathing, in addition to the roofing-felt interlay, to get more insulation and added protection against air infiltration.

Most shake roofs do not require special treatments, but fungicidal chemicals can be used where heat and humidity prevail, to prevent moss, fungus, or mildew. Heavy-bodied paints or oils that form a coating should not be used on a shake roof.

THE BUILT-UP ROOF

The name describes this type of roofing. It's literally "made" on the job by placing strips of felt individually and bonding them to each other and to the deck with hot asphalt or pitch (**Figure 14-54**).The final step is to mop the entire surface of the roof and then cover it with, usually, a layer of gravel or crushed stone or chips of marble. The design is very common on flat roofs, but may be used on slopes that are not more than 2 or 3 in 12. The installation is quite durable when the job is done correctly. Often a built-up roof is called a 10-, 15-, or 20-year roof depending on the coverage, and this, of course, relates to the exposure of the felt plies.

FIG. 14-54

This is the kind of job you might pass on to professional people. The procedure is not very complicated, but there are problems involved in heating and maintaining the temperature of the hot materials (**Figure 14-55**). It's also a messy job but one that can be done rather quickly by people who have done it often.

FIG. 14-55

The job is usually started by placing a base layer of 30-pound saturated roofer's felt. This goes down dry with edges lapped about 4 inches and is secured with special large-headed roofing nails. The special purpose of the dry layer is to prevent the bonding material from penetrating joints in the deck. Other layers of saturated felt are mopped on with the hot tar (**Figure 14-56**).

FIG. 14-56

FIG. 14-57

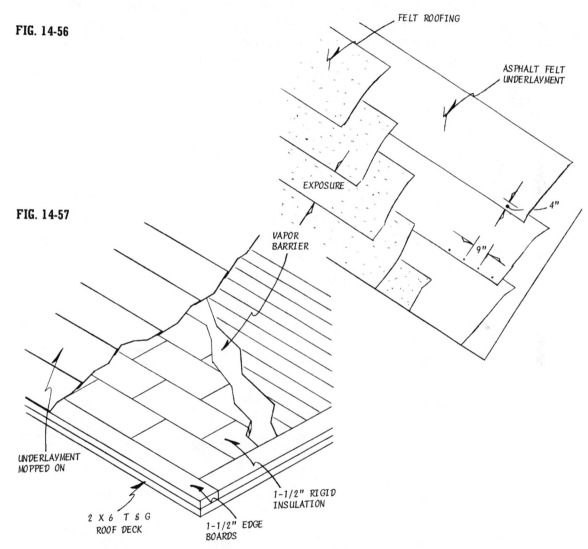

FELT ROOFING

ASPHALT FELT UNDERLAYMENT

EXPOSURE

VAPOR BARRIER

4"

9"

UNDERLAYMENT MOPPED ON

2 X 6 T & G ROOF DECK

1-1/2" EDGE BOARDS

1-1/2" RIGID INSULATION

The slight-slope roof over my work area was done with a built-up roof. I placed a layer of rigid insulation first and then mopped on the underlayment material (**Figure 14-57**). I also placed a vapor barrier under the insulation. This is a good plan when the underside of the roof decking is the ceiling.

Eave lines should be finished with a metal edging or flashing which serves as a water drip and keeps gravel from dropping off the roof. This can be done simply, as shown in **Figure 14-58**, with the stop nailed in place and the roof flange covered with first an 8-inch and then a 10-inch strip of felt mopped in place. A more sophisticated system, like the one shown in **Figure 14-59**, might be in order since this is the area that water will flow toward. The special webbing shown in the sketch is a reinforcement and will be available from the roofing-supply dealer. Water runoff is accomplished by forming holes through the roof for downspout outlets which are sealed with felt and the bonding material.

FIG. 14-58

FIG. 14-59

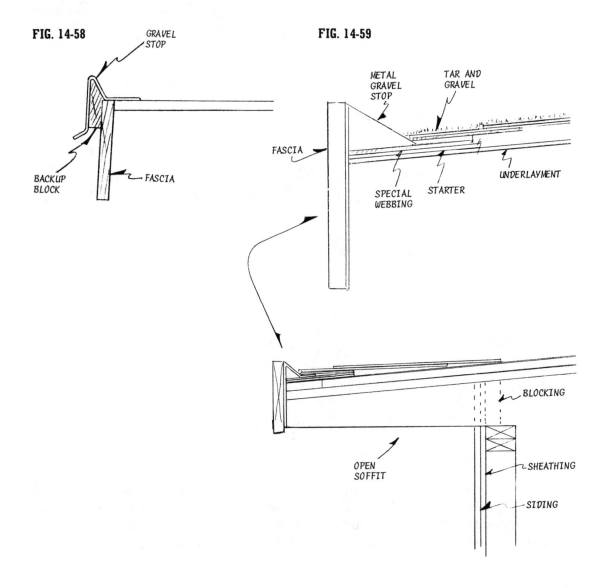

A built-up roof, flat or sloped, that abuts a vertical wall should make a turn up onto the wall siding after passing over a cant strip that you make for the purpose (**Figure 14-60**). Keep the flashing away from the top of the cant strip, but let it extend 5 or 6 inches over the siding.

FIG. 14-60

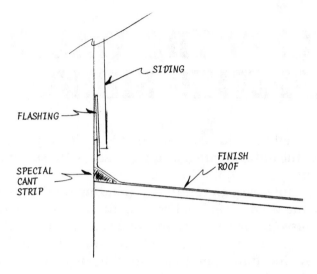

You have some control over the appearance of the roof because you can choose the color of the stone used as a final cover. White marble chips are good reflectors of the sun's heat, but check to see if neighbors on a higher slope might be bothered by the glare. Use gravel that will meet a screen size of $\frac{1}{4}$ inch to $\frac{5}{8}$ inch. If it's smaller it will be easily washed or blown away.

15

SHEATHING THE WALLS AND APPLYING SIDING

Wall sheathing is a tight outside covering that is nailed directly to the framework. After it is applied, the only openings you will see are those left for doors and windows. The sheathing is a smooth, flat base on which you apply the final cover — siding — and can supply enough rigidity, depending on the material and installation, to eliminate the need for special bracing in the frame itself. Sheathing is not included in all house constructions; some codes do not make it mandatory. Yet it should be considered, if not for its structural value, for the fact that it can just about eliminate air infiltration and can provide insulation. The most common materials include sheathing grades of plywood, boards, and special structural insulating boards. The insulating boards, shown in **Figure 15-1**, are basically fiberboards that are coated or impregnated with asphalt or are treated in other ways so the product is water-resistant. It is available in various densities and in a nail-base type that permits direct nailing of siding. Usually, though, special nails are required.

FIG. 15-1

Since the products vary in composition and come in different sizes, the type and length of nails and the nail spacing should follow the directions of the manufacturer. Often, local building regulations are written to cover the application. Products that come in 4×8 sheets are usually placed vertically, since this permits complete perimeter nailing. Products that come in 2-foot widths are installed horizontally (**Figure 15-2**). All joints are staggered and meet at the midpoint of a framing member. Check to see if the sheathing can be used without corner bracing in the wall frame.

FIG. 15-2

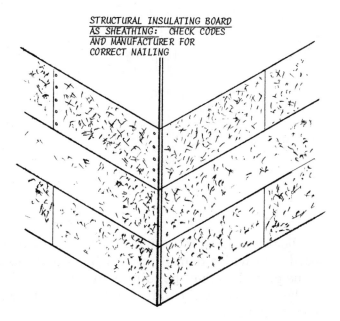

STRUCTURAL INSULATING BOARD AS SHEATHING: CHECK CODES AND MANUFACTURER FOR CORRECT NAILING

SHEATHING WITH BOARDS

Boards used for sheathing should have a nominal 1-inch thickness and, preferably, should not be more than 6 or 8 inches wide. Square-edge boards are commonly used, shiplap or tongue-and-groove material will provide a much tighter cover. Random-length boards that are both side- and end-matched are available and are convenient to use since they permit end joints to occur anywhere. End joints on other types should occur on a framing member. Actually, wall-sheathing material does not differ radically from what is used as roof sheathing or subflooring. Generally, a #2 or #3 common grade of any of the softwoods is suitable.

Board sheathing is often applied horizontally, as in **Figure 15-3**, but only because it is easier to do and usually involves less waste than a diagonal pattern. However, a wall frame with horizontal board sheathing should be reinforced with corner bracing, while one with diagonal board sheathing doesn't require it.

FIG. 15-3

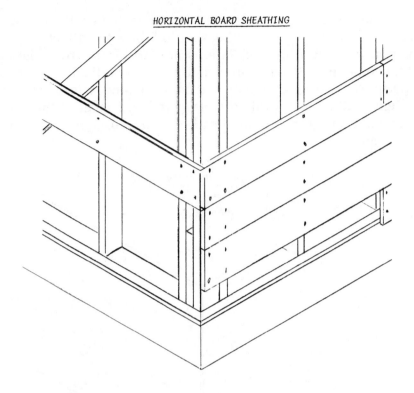

HORIZONTAL BOARD SHEATHING

Nail horizontal boards to each framing member with two 8d nails. Use three 8d nails if the boards are wider than 8 inches. It's a good idea for the bottom sheathing board to be placed so that the joint between sill and joist and between joist and studs will be spanned by the nails used in a single board. The idea is to tie framing members to the sill.

Attach diagonal sheathing by using three 8d nails at the corner post and two 8d nails at each stud (**Figure 15-4**). Add a nail at each place if the boards are wider than 8 inches.

End joints, regardless of whether the boards are placed horizontally or diagonally, should occur at the midpoint of a stud unless the boards are end-matched. Stagger all joints, preferably by at least two stud spaces.

SHEATHING WITH PLYWOOD

Plywood is justifiably popular as sheathing material because the sheets cover large areas quickly and add strength and rigidity to the structure. It may be applied either vertically or horizontally (**Figure 15-5**). Horizontal application is recommended if a siding, like shingles, must be nailed directly into the sheathing.

FIG. 15-4

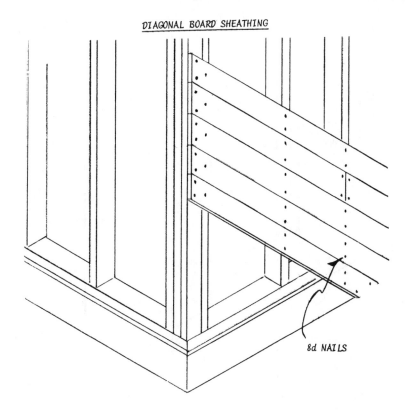

DIAGONAL BOARD SHEATHING

8d NAILS

In general, the horizontal application provides more stiffness if applied loads are perpendicular to the surface, while a vertical application counters racking forces to a greater degree. This is why panels placed vertically, at least at all corners, will usually eliminate the need for diagonal bracing in the wall frame.

When a horizontal application is advisable, you can include 2×4 nailing blocks in the wall frame as shown in **Figure 15-5** to allow complete perimeter nailing, and this can contribute enough anti-racking strength to make frame corner-bracing unneccesary.

The sheathing can be attached with common smooth, annular, spiral-thread, or galvanized box 6d nails if it is less than ½ inch thick. Use 7d or 8d nails if the plywood is ½ inch thick or thicker. Space panel edges about ⅛ inch; ends, about 1/16 inch. The spacing should be doubled if wet or humid conditions prevail. The thicker plywood panels are usually recommended when the stud spacing is greater than 16 inches.

FIG. 15-5

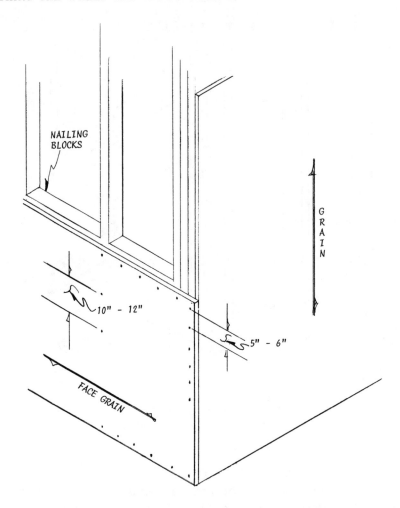

Plan the installation so no piece of plywood will be shorter than the span across three studs. All vertical joints must be over a framing member and should be staggered.

SHEATHING AT THE SILL

Sheathing at the sill can be done in several ways (**Figure 15-6**). It can end on the subfloor if the wall framing is set back to accommodate it (A). This does not provide a tie between the wall and floor frames, and should be reinforced with anchor straps running vertically from the sill at, at least, every other stud position. Use steel straps that are long enough to extend above the subfloor about 18 inches.

FIG. 15-6

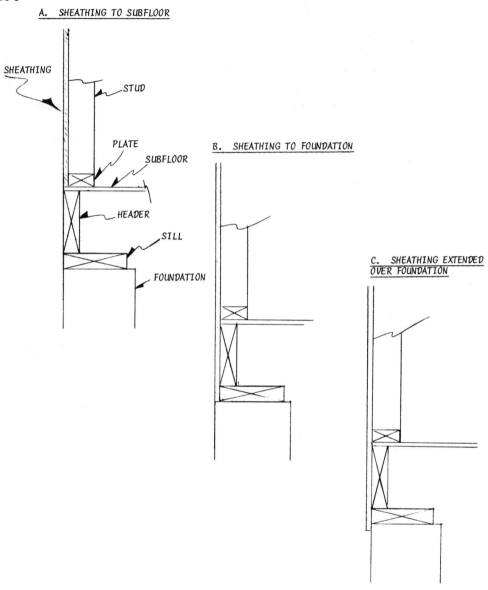

A. *SHEATHING TO SUBFLOOR*

SHEATHING

STUD

PLATE

SUBFLOOR

HEADER

SILL

FOUNDATION

B. *SHEATHING TO FOUNDATION*

C. *SHEATHING EXTENDED OVER FOUNDATION*

A second method (B) has the sill and the floor and wall frames set back from the edge of the foundation the thickness of the sheathing. This does make a strong connection between framing components, especially when the sheathing is plywood or diagonal boards, but it's a bigger nuisance to do than setting the frames flush with the foundation and placing the sheathing so it extends a bit below the sill (C).

SHEATHING PAPER

Sheathing paper is applied directly to the wall frame if sheathing is not used, and over sheathing that is done with boards. It is often called building paper and sometimes "breathing paper," since its purpose is to allow the movement of vapor but oppose other moisture. It can be 15-pound asphalt or rosin paper, but not a coated felt or a waterproof paper. The material also acts to prevent air infiltration.

Apply the paper horizontally with the first course at the bottom of the wall and with succeeding plies having a minimum 4-inch headlap (**Figure 15-7**).

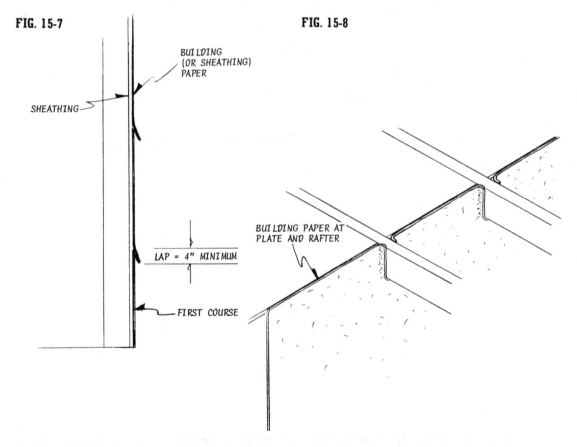

FIG. 15-7

SHEATHING

BUILDING (OR SHEATHING) PAPER

LAP = 4" MINIMUM

FIRST COURSE

FIG. 15-8

BUILDING PAPER AT PLATE AND RAFTER

You can use roofing nails to attach the paper, but use only enough to keep the cover from being blown off until you get to attaching the siding. Use some care where the paper must be cut to accommodate rafters and such, forming slits so narrow flanges can turn corners to seal joints (**Figure 15-8**).

How you end at the top plate will depend on the structure, but the point to remember is that part of the paper's job is to guard against air infiltration. Two methods are shown in **Figure 15-9**.

FIG. 15-9

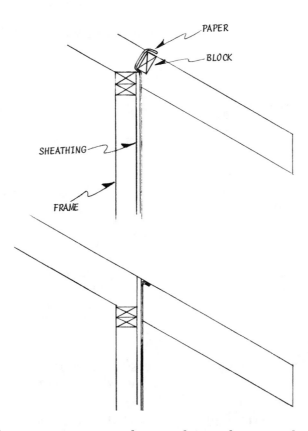

Ordinarily, the paper is not used over plywood or any sheet material that is water-resistant, but it's a good idea anyway to cut strips of it about 10 inches wide and use them to seal around all door and window openings.

Sheathing paper *is* used over plywood if the walls will be finished with stucco, as in **Figure 15-10**, but it is not required behind brick veneer if the 1-inch air space shown in **Figure 15-11** is provided.

PLYWOOD SIDING

All panels in the three basic siding categories are exterior grades made with a fully waterproof glue. The common size is 4×8 feet, but 9-foot and 10-foot lengths are also available. A special plywood lap siding with a maximum width of 2 feet

FIG. 15-10

PLYWOOD SHEATHING BEHIND STUCCO

FIG. 15-11

PLYWOOD SHEATHING BEHIND BRICK VENEER

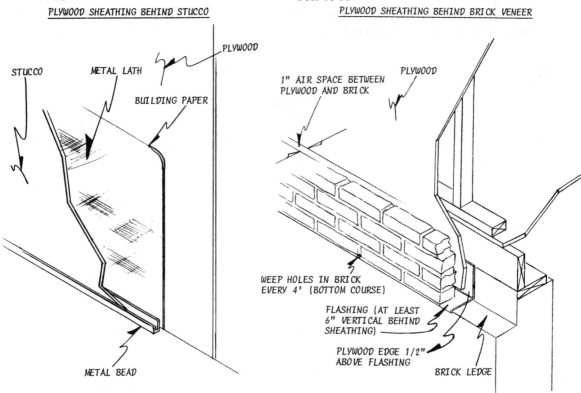

STUCCO

METAL LATH

PLYWOOD

BUILDING PAPER

METAL BEAD

1" AIR SPACE BETWEEN PLYWOOD AND BRICK

PLYWOOD

WEEP HOLES IN BRICK EVERY 4' (BOTTOM COURSE)

FLASHING (AT LEAST 6" VERTICAL BEHIND SHEATHING)

PLYWOOD EDGE 1/2" ABOVE FLASHING

BRICK LEDGE

comes in 16-foot lengths. The panels are surface-veneered with various wood species, but those most available are fir, redwood, and cedar. Surface patterns and textures vary, as shown in **Figure 15-12**.

Medium-density overlaid plywood (MDO) is especially recommended for paint finishes. Its surface is a resin-treated sheet that is hot-bonded to the panel to provide a uniformly smooth base that is excellent for paint. MDO comes as lap siding as well as in panel form.

Another product, 303 siding, includes many surface textures and patterns, most of them developed for a stain finish. The grade of surface veneer allows tight knots and even knotholes that have been repaired with a synthetic, weatherproof filler that results in an acceptable, rustic appearance. If this is objectionable, there are 303 panels available manufactured with a minimum of the imperfections. The material is back-stamped to indicate whether it should be used with studs on 16-inch centers, or may be used with 24-inch stud spacing. Single-wall construction —no sheathing underlay required—is possible and may be permitted by local codes in some areas.

Texture 1-11 plywood is a 303 panel done with shiplapped edges and paral-

FIG. 15-12   <u>FACE</u>   <u>SECTION</u>

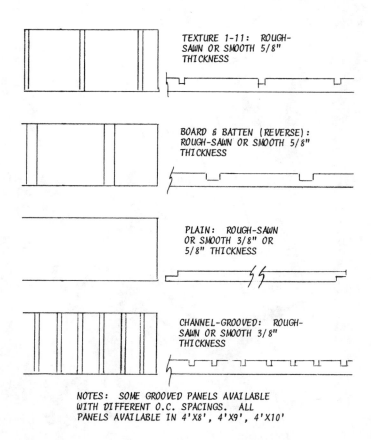

TEXTURE 1-11: ROUGH-
SAWN OR SMOOTH 5/8"
THICKNESS

BOARD & BATTEN (REVERSE):
ROUGH-SAWN OR SMOOTH 5/8"
THICKNESS

PLAIN: ROUGH-SAWN
OR SMOOTH 3/8" OR
5/8" THICKNESS

CHANNEL-GROOVED: ROUGH-
SAWN OR SMOOTH 3/8"
THICKNESS

NOTES: SOME GROOVED PANELS AVAILABLE
WITH DIFFERENT O.C. SPACINGS. ALL
PANELS AVAILABLE IN 4'X8', 4'X9', 4'X10'

lel grooves (**Figure 15-13**). It comes in a ⅝-inch thickness only but with a variety of groove spacings and many surface textures, including overlaid.

Whether or not you include sheathing and building paper in a plywood-siding installation depends on the thickness of the material you select, local codes, economy, and your personal preferences. The two-layer system, even though it's more expensive and time-consuming, does provide broad options in siding patterns and frame spacing plus easier nailing schedules. The extra weathertightness is obvious.

INSTALLATION OF PLYWOOD SIDING

Plywood panels are usually installed vertically, since this permits complete perimeter nailing without having to add nailing blocks to the frame (**Figure**

FIG. 15-13

FIG. 15-14

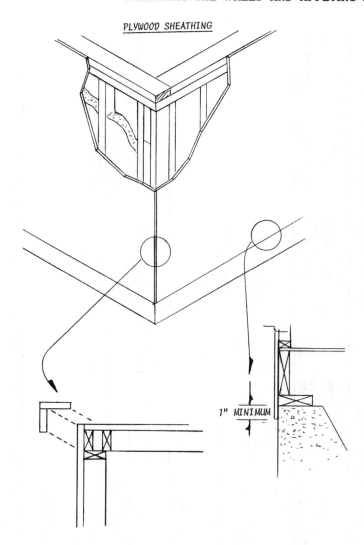

PLYWOOD SHEATHING

1" MINIMUM

15-14). Installation details relating to various plywood types and thicknesses are given in the table in **Figure 15-15**. These are recommendations made by the American Plywood Association, and even though they have been successfully field-tested and are accepted by major regional building codes, any move away from the norm, such as 24-inch stud spacing, must be checked against *local* codes.

Let all vertical joints occur over a stud; leave a $1/16$-inch gap between all panel edges. Over long runs, the accumulation of gaps between panels may require that you trim an occasional panel to keep its edge close to the centerline of a framing member. Caulk the joints or treat the plywood edges with a water repellent. This

FIG. 15-15.   RECOMMENDATIONS FOR EXTERIOR PLYWOOD SIDING OVER SHEATHING (ALL SPECIES GROUPS)

Type	PLYWOOD SIDING		MAX. STUD SPACING (INCHES)		NAIL SIZE (USE NONSTAINING BOX, SIDING OR CASING NAILS)	NAIL SPACING (IN.)	
	Description	Nominal Thickness (in.)	Face Grain Vertical	Face Grain Horizontal		Panel Edges	Intermediate
Panel Siding	MDO EXT-APA	5/16	16[d]	24	6d for panels 1/2" thick or less; 8d for thicker panels	6	12
		3/8	16[d]	24			
		1/2 & thicker	24	24			
	303-16 o.c. Siding EXT-APA	5/16 & thicker	16[d]	24			
	303-24 o.c. Siding EXT-APA	7/16 & thicker	24	24	(a)		
Lap Siding	MDO EXT-APA	5/16	—	16[b]	6d for siding 3/8" thick or less; 8d for thicker siding	4" @ vertical butt joints; one nail per stud along bottom edge	8" @ each stud, if siding wider than 12"
		3/8	—	16[b]			
		1/2 & thicker	—	24			
	303-16 o.c. Siding EXT-APA	5/16 or 3/8	—	16[b]			
	303-16 o.c. Siding EXT-APA 303-24 o.c. Siding EXT-APA	7/16 & thicker	—	24	(c)		

(a) Use next regular nail size when sheathing (other than plywood or lumber) is thicker than 1/2".
(b) May be 24" with plywood or lumber sheathing.
(c) Use next larger nail size when sheathing is other than plywood or lumber, and nail only into framing.
(d) May be 24" with approved nailable sheathing, if panel is also nailed 12" o.c. between studs.

FIG. 15-16

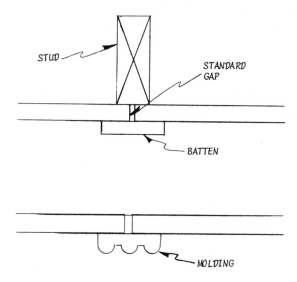

isn't necessary if the edges are shiplapped or if the joint will be covered with a batten, but it's not a bad thing to do anyway. Batten material can be plain strips of wood or molding (**Figure 15-16**). Attachment nails should pass between the edges of the plywood and penetrate the studs at least 1 inch.

If you wish to add some decorative detail to plain plywood siding you can do so by using batten material to form patterns like those shown in **Figure 15-17**. This is easier to do if the walls include sheathing. All nails should be long enough to penetrate the sheathing.

Shiplapped edges should also occur over a stud and join as shown in **Figure 15-18**. Some shiplapped edges are designed so the joint is a wide groove; others come together more tightly.

Outside corners deserve special attention to make them weathertight. A neat, unobtrusive design can be done by using a rabbet joint, but one- or two-piece corner boards are more practical and easier to do (**Figure 15-19**). You can do the one-piece affair by sawing material that matches the siding or using a readymade molding called a corner guard. If you do the former, the piece you cut out can be used to seal an inside corner.

It's a good idea to caulk the edges of panels that meet at an inside corner. A neat appearance results if you do the job carefully, but adding a corner, which can be a plain square strip of wood or something like quarter-round molding will complete the seal and make the panel joint less critical (**Figure 15-20**).

Horizontal joints should be backed up by nailing blocks placed between studs whether codes call for them or not, unless the frame has been covered with

FIG. 15-17

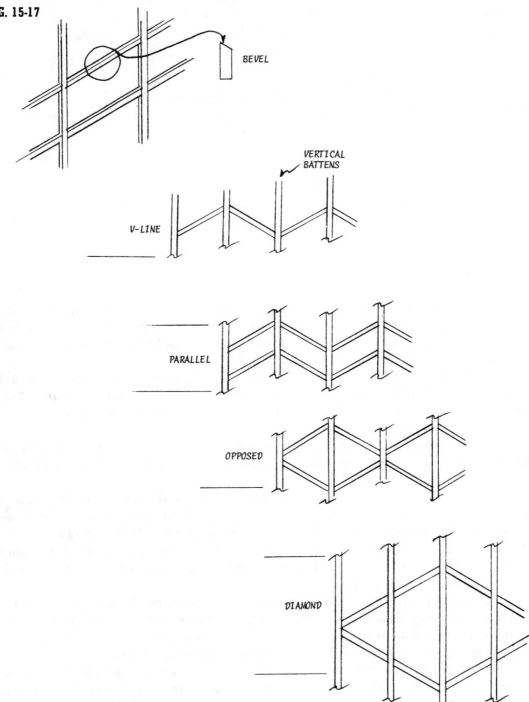

BEVEL

VERTICAL
BATTENS

V-LINE

PARALLEL

OPPOSED

DIAMOND

FIG. 15-18

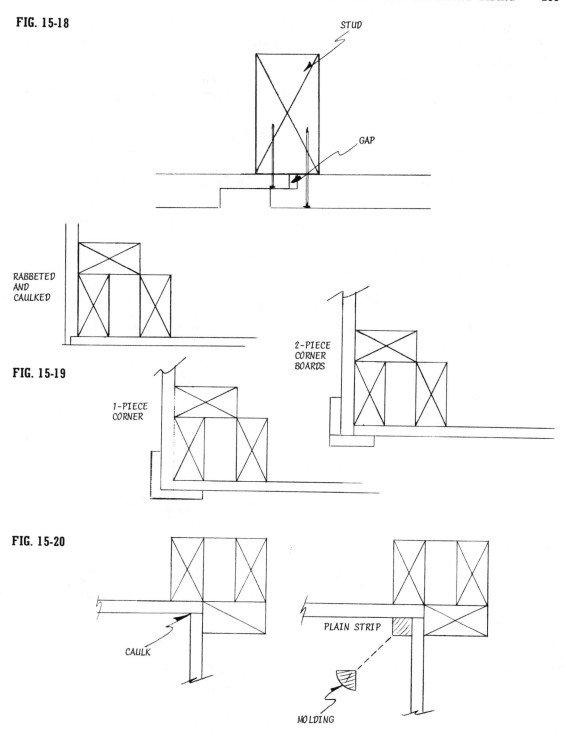

STUD

GAP

RABBETED
AND
CAULKED

FIG. 15-19

1-PIECE
CORNER

2-PIECE
CORNER
BOARDS

FIG. 15-20

CAULK

PLAIN STRIP

MOLDING

sheathing. The joints should be well caulked and may be covered with battens. A beveled top edge that will shed water, as shown in **Figure 15-21**, makes more sense than a square edge.

Special flashing, installed as shown in **Figure 15-22**, is available, or you can work with moldinglike *water tables*, which are more decorative (**Figure 15-23**).

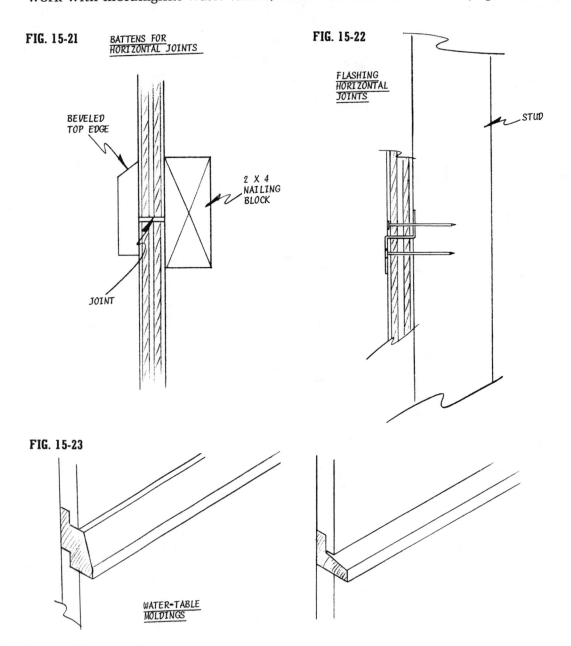

FIG. 15-21   BATTENS FOR HORIZONTAL JOINTS

BEVELED TOP EDGE

2 X 4 NAILING BLOCK

JOINT

FIG. 15-22

FLASHING HORIZONTAL JOINTS

STUD

FIG. 15-23

WATER-TABLE MOLDINGS

Plywood lap siding is installed with pieces cut from conventional shingles placed behind all end joints (**Figure 15-24**). Caulk all the vertical joints or treat the plywood edges with a water repellent. Joints must occur over a stud unless sheathing is used, and they should be staggered. Check the table shown previously (**Figure 15-15**) for the approved nailing pattern.

Inside and outside corners should be mitered and caulked. They can stay as is if tight enough, or you can cover them with the same kind of corner boards suggested for flat siding. Another system is to install special metal corners over each course of siding as it is installed (**Figure 15-25**).

FIG. 15-24

FIG. 15-25

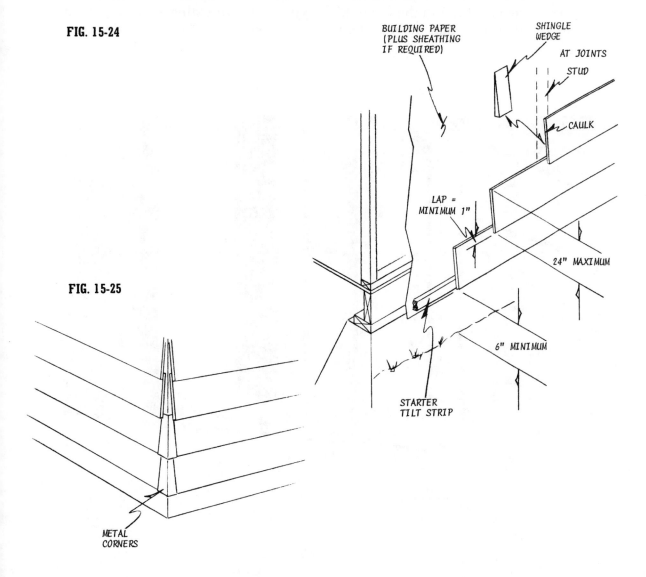

LUMBER SIDING

Lumber siding is solid wood made in the form of boards that can be plain or shaped in various ways to create special effects. It is usually applied over sheathing, but in mild climates it may be permissible to nail it directly to studs. If the latter procedure is followed, the wall frame should be reinforced with corner bracing.

Most types are placed horizontally as in **Figure 15-26**, because the shadow lines, which are especially prominent in bevel siding, help to lower the house profile so it seems closer to the earth. Most types of board siding are available in various wood species that include the western pines and cedars and redwood.

FIG. 15-26

There are numerous patterns, but those that have proved most popular are shown here together with some installation details. When the specifications say that a board has a smooth side and a rough side, it means that either side may be exposed to the weather, the deciding factor being whether you wish to have a natural or stained finish or a painted one. Common patterns are *bevel* and *bungalow*, **Figure 15-27**; Dolly Varden, **Figure 15-28**; *drop*, **Figure 15-29**; *channel rustic*, **Figure 15-30**; *tongue-and-groove*, **Figure 15-31**; and *log cabin*, **Figure 15-32**.

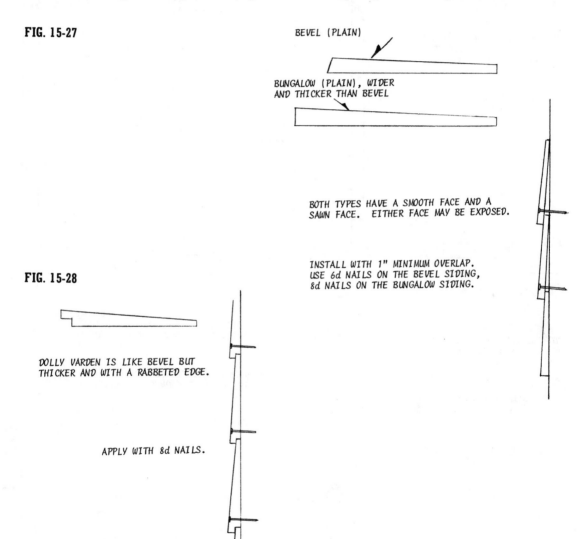

FIG. 15-27

BEVEL (PLAIN)

BUNGALOW (PLAIN), WIDER
AND THICKER THAN BEVEL

BOTH TYPES HAVE A SMOOTH FACE AND A
SAWN FACE. EITHER FACE MAY BE EXPOSED.

INSTALL WITH 1" MINIMUM OVERLAP.
USE 6d NAILS ON THE BEVEL SIDING,
8d NAILS ON THE BUNGALOW SIDING.

FIG. 15-28

DOLLY VARDEN IS LIKE BEVEL BUT
THICKER AND WITH A RABBETED EDGE.

APPLY WITH 8d NAILS.

FIG. 15-29

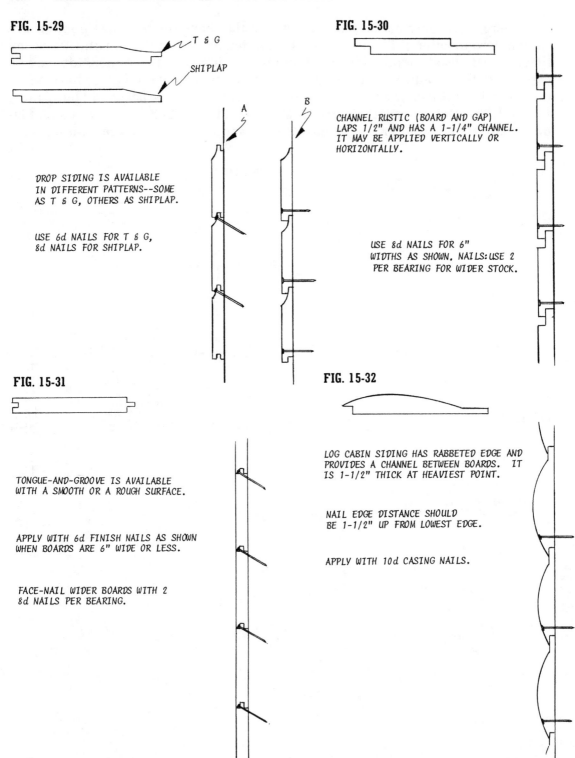

T & G

SHIPLAP

DROP SIDING IS AVAILABLE
IN DIFFERENT PATTERNS--SOME
AS T & G, OTHERS AS SHIPLAP.

USE 6d NAILS FOR T & G,
8d NAILS FOR SHIPLAP.

A

B

FIG. 15-30

CHANNEL RUSTIC (BOARD AND GAP)
LAPS 1/2" AND HAS A 1-1/4" CHANNEL.
IT MAY BE APPLIED VERTICALLY OR
HORIZONTALLY.

USE 8d NAILS FOR 6"
WIDTHS AS SHOWN. NAILS: USE 2
PER BEARING FOR WIDER STOCK.

FIG. 15-31

TONGUE-AND-GROOVE IS AVAILABLE
WITH A SMOOTH OR A ROUGH SURFACE.

APPLY WITH 6d FINISH NAILS AS SHOWN
WHEN BOARDS ARE 6" WIDE OR LESS.

FACE-NAIL WIDER BOARDS WITH 2
8d NAILS PER BEARING.

FIG. 15-32

LOG CABIN SIDING HAS RABBETED EDGE AND
PROVIDES A CHANNEL BETWEEN BOARDS. IT
IS 1-1/2" THICK AT HEAVIEST POINT.

NAIL EDGE DISTANCE SHOULD
BE 1-1/2" UP FROM LOWEST EDGE.

APPLY WITH 10d CASING NAILS.

Board siding that is applied vertically is usually a shiplap or a tongue-and-groove type, or plain boards that are applied in one of the ways shown in **Figure 15-33**, with two horizontal lines of 2×4 blocks placed between studs to provide nailing surfaces.

FIG. 15-33

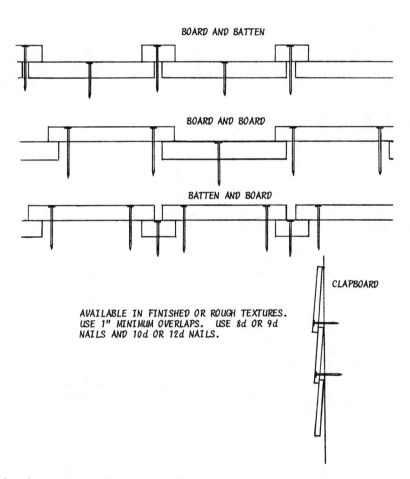

BOARDS

BOARD AND BATTEN

BOARD AND BOARD

BATTEN AND BOARD

CLAPBOARD

AVAILABLE IN FINISHED OR ROUGH TEXTURES. USE 1" MINIMUM OVERLAPS. USE 8d OR 9d NAILS AND 10d OR 12d NAILS.

Wood siding varies from ½ inch to 1½ inches in thickness, and widths run from 4 inches to 12 inches. The Western Wood Products Association table in **Figure 15-34** lists the surfacing requirements and tells the nominal as well as the dressed sizes.

The table in **Figure 15-35** provides factors so you can determine the exact amount of material you need when working with the five basic types of wood paneling. To use it, multiply the square footage you must cover by the area factor.

FIG. 15-34.   WESTERN WOOD SIDING SIZES

	NOMINAL SIZE		DRESSED DIMENSIONS		
PRODUCT	Thickness In.	Width In.	Thickness In.	Width In.	
BEVEL SIDING	½	4	$^{15}/_{32}$ butt, $^3/_{16}$ tip	3½	
For western red cedar sizes	¾	5	¾ butt, $^3/_{16}$ tip	4½	
(See footnote*)		6		5½	
WIDE BEVEL SIDING	¾	8	¾ butt, $^3/_{16}$ tip	7¼	
(Colonial or Bungalow)		10		9¼	
		12		11¼	
				Face	Overall
RABBETED BEVEL SIDING	¾	6	$^5/_8$ by $^5/_{16}$	5	5½
(Dolly Varden)	1	8	$^{13}/_{16}$ by $^{13}/_{32}$	6¾	7¼
		10		8¾	9¼
		12		10¾	11¼
RUSTIC AND DROP SIDING	1	6	$^{23}/_{32}$	5⅛	5⅜
(Dressed and matched)		8		6⅞	7⅛
		10		8⅞	9⅛
		12		10⅞	11⅛
RUSTIC AND DROP SIDING	1	6	$^{23}/_{32}$	5	5⅜
(Shiplapped, ⅜-in. lap)		8		6¾	7⅛
		10		8¾	9⅛
		12		10¾	11⅛
RUSTIC AND DROP SIDING	1	6	$^{23}/_{32}$	4$^{15}/_{16}$	5$^7/_{16}$
(Shiplapped, ½-in. lap)		8		6⅝	7⅛
		10		8⅝	9⅛
		12		10⅝	11⅛
LOG CABIN SIDING	1½ (6/4)	6	1½″ at	4$^{15}/_{16}$	5$^7/_{16}$
		8	thickest	6⅝	7⅛
		10	point	8⅝	9⅛
TONGUE & GROOVE (T&G)	1 (4/4)	4	¾	3⅛	3⅜
SANDED 2 SIDES AND		6		5⅛	5⅜
CENTER-MATCHED JOINT		8		6⅞	7⅛
CREATES FLUSH SURFACE		10		8⅞	9⅛
		12		10⅞	11⅛

* Western Red Cedar Bevel Siding available in ½″, ⅝″, ¾″ nominal thickness. Corresponding surfaced thick edge is $^{15}/_{32}$″, $^9/_{16}$″ and ¾″. Widths 8″ and wider ½″ off.

NAILS

Work with high-tensile-strength aluminum nails or galvanized nails that are mechanically plated. Hot-dipped galvanizing is okay, but the degree of coating protection varies. Nail shanks should be ring-threaded or spiral-threaded and, preferably, have blunt points, since these are less likely to split the wood. If you find that a board does have a tendency to split—and this can happen especially near an end—drill small pilot holes before you drive the nails.

FIG. 15-35. **COVERAGE ESTIMATOR**

	NOMINAL SIZE	WIDTH		AREA FACTOR*
		Dress	Face	
SHIPLAP	1 × 6	5½	5⅛	1.17
	1 × 8	7¼	6⅞	1.16
	1 × 10	9¼	8⅞	1.13
	1 × 12	11¼	10⅞	1.10
TONGUE AND GROOVE	1 × 4	3⅜	3⅛	1.28
	1 × 6	5⅜	5⅛	1.17
	1 × 8	7⅛	6⅞	1.16
	1 × 10	9⅛	8⅞	1.13
	1 × 12	11⅛	10⅞	1.10
SANDED 4 SIDES	1 × 4	3½	3½	1.14
	1 × 6	5½	5½	1.09
	1 × 8	7¼	7¼	1.10
	1 × 10	9¼	9¼	1.08
	1 × 12	11¼	11¼	1.07
PANELING PATTERNS	1 × 6	5⁷/₁₆	5¹/₁₆	1.19
	1 × 8	7⅛	6¾	1.19
	1 × 10	9⅛	8¾	1.14
	1 × 12	11⅛	10¾	1.12
BEVEL SIDING (1″ lap)	1 × 4	3½	3½	1.60
	1 × 6	5½	5½	1.33
	1 × 8	7¼	7¼	1.28
	1 × 10	9¼	9¼	1.21
	1 × 12	11¼	11¼	1.17

* Allowance for trim and waste should be added.

INSTALLATION OF LUMBER SIDING

Study the wall you must cover and plan the placement of courses to coordinate with openings. Ideally, boards should fit against the top and the bottom of a window, for example, without notching. If notching must be done, do it carefully for the sake of appearance and a tight fit. Often you can make adjustments where boards lap as long as you stay within the minimum requirements; you can even choose a siding width that is compatible with the layout.

Horizontal siding should be placed dead level, so snapping a chalk line, at least for the first course you place, is essential. Professionals make a special story pole that they use at each end of the wall and at openings to mark the line of courses. To make one, use a strip of wood that will reach from the soffit to 1 inch below the top of the foundation and mark it off in equal board-width spaces but allowing for overlaps. Another way is to use a scrap piece of the siding as a gauge to mark course heights at the ends or any intermediate point along a run.

Most one-story installations are started at the bottom of the wall, although it may be more convenient to do gable ends first, as in **Figure 15-36**, especially when scaffolding is secured to the house frame. Note that here the siding was applied before the finish roof. This is often done with gables—frequently complete sheathing is done before or while the roof is being finished. This adds rigidity and a good deal of weather protection to the structure right off.

FIG. 15-36

Bevel siding is begun with a starter strip that has a thickness equal to the thin edge of the board, placed as shown in **Figure 15-37**, so that the first course will have the correct slant. The extension of the first course below the starter strip provides a drip edge so water can fall free.

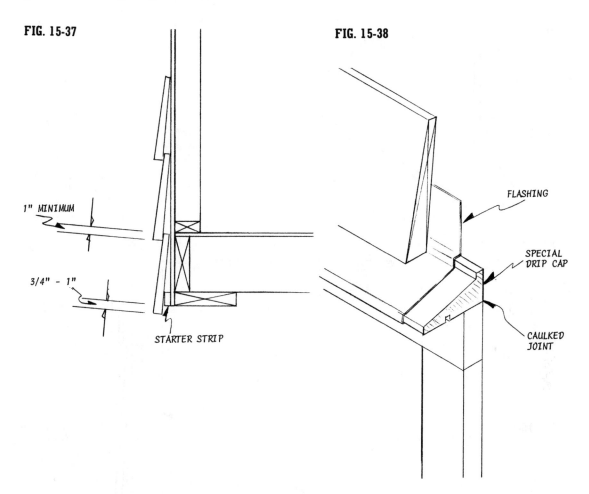

FIG. 15-37

1" MINIMUM

3/4" - 1"

STARTER STRIP

FIG. 15-38

FLASHING

SPECIAL DRIP CAP

CAULKED JOINT

Let all end joints occur over a stud, and stagger those in adjacent courses as far apart as possible. Make all end cuts very carefully so boards will butt tightly. Work with a fine-tooth crosscut saw or a crosscut blade if you are using power equipment. Remember these will be visible joints and deserve special attention. Using a caulking compound in the joint to provide weathertightness makes sense.

Joints around openings are detailed in the later chapters on doors and windows, but, briefly, siding should fit snugly against casings and over special drip caps, that are put in place as shown in **Figure 15-38**.

To add a design feature, gable ends are often covered with a siding that runs perpendicular to what is used over the wall. The break between the two can be made weatherproof and attractive by using the same type of drip caps that are shown in **Figure 15-39**.

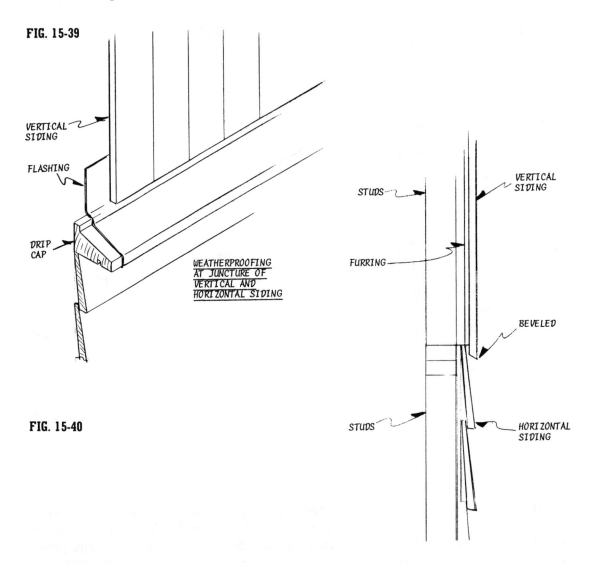

FIG. 15-39

VERTICAL SIDING

FLASHING

DRIP CAP

WEATHERPROOFING AT JUNCTURE OF VERTICAL AND HORIZONTAL SIDING

STUDS

FURRING

VERTICAL SIDING

BEVELED

FIG. 15-40

STUDS

HORIZONTAL SIDING

Another way is to place 1×3 furring strips over the studs or the sheathing of the gable so the siding placed there will project beyond what is used below (**Figure 15-40**). Note that the bottom edge of the upper siding is beveled to serve as a drip edge.

CORNERS

There are many options on outside corners, depending on how you want the job to appear and the time you care to spend doing it (**Figure 15-41**). All siding, whether flat or beveled, can be mitered or butted against corner boards. Ready-made metal corners, because they go on quickly and are neat and weatherproof, are used extensively by commercial people (**Figure 15-42**). For practical reasons, the job should be done with a minumum number of joint lines, but there is no reason why you can't do it your way as long as you remember that weather-tightness is as important as appearance.

FIG. 15-41

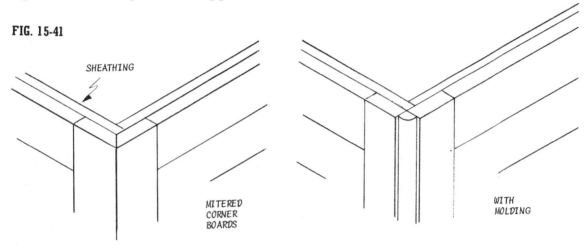

SHEATHING

MITERED
CORNER
BOARDS

WITH
MOLDING

FIG. 15-42

For inside corners, the most practical and the neatest-looking joint is done with a square corner strip against which the siding is butted (**Figure 15-43**). Basically the same idea applies when the siding ends against a vertical, such as a casing around a door or window (**Figure 15-44**).

FIG. 15-43

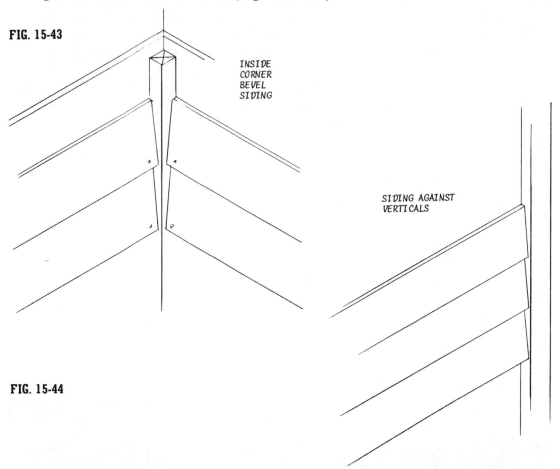

INSIDE
CORNER
BEVEL
SIDING

SIDING AGAINST
VERTICALS

FIG. 15-44

In all situations it pays to apply a caulking compound between all mating surfaces. This should be done neatly, especially if the siding will be left natural or stained.

Premium construction calls for applying a prime coat of paint on the back surfaces of the siding before it is applied. If the finish will be natural or stained, then use a water repellent instead of the prime coat. A prime coat or a water repellent should also be used on outside surfaces as soon as possible after installation. Some sidings, especially in the plywoods, are available with basic protective coatings already applied.

HARDBOARD SIDING

Hardboards have entered the siding field with considerable impact. As you probably know, hardboards are a manufactured product but are made from timber and engineered to retain wood's natural fibers and binding agents. As a siding, it is specially formulated to withstand exposure.

There are many choices in styles. You can do vertical or horizontal treatments with smooth or textured surfaces and in contemporary or traditional patterns—even a skip-trowel texture (Masonite's "Stuccato") which says "stucco" but which appears simply by nailing panels into place. Most examples can be stained or painted or can be obtained prefinished in lap or panel form (Colorlok) so that the job requires no attention after installation. Typical colors are white, brown, yellow, green, and bronze.

When prefinished siding is used, all marking and cutting should be done from the back surface to avoid marring the finish. Exposed nails are coated with a touch-up paint, or you can use special nails that are color-matched to the siding. When driving the latter, it's a good idea to work with a plastic-covered hammer head.

Apply the siding to studs or sheathing, but always over a building paper that will form a continuous vapor barrier.

Some of the lap sidings are actually systems that incorporate preshaped metal forms as starter strips and as seals for lap and vertical butt joints (**Figure 15-45**). Similar pieces are supplied so you can end inside and outside corners

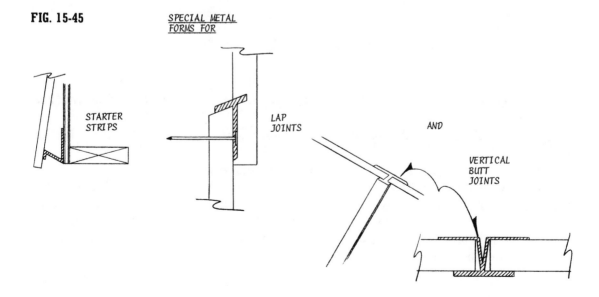

FIG. 15-45

SPECIAL METAL FORMS FOR

STARTER STRIPS

LAP JOINTS

AND

VERTICAL BUTT JOINTS

FIG. 15-46

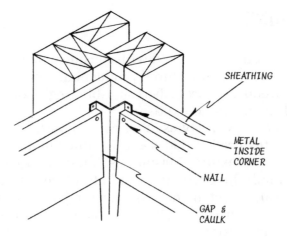

SHEATHING

METAL
INSIDE
CORNER

NAIL

GAP &
CAULK

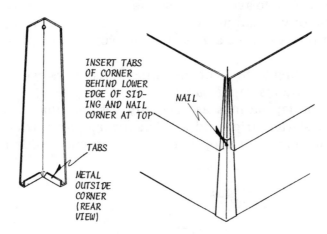

INSERT TABS
OF CORNER
BEHIND LOWER
EDGE OF SID-
ING AND NAIL
CORNER AT TOP

NAIL

TABS

METAL
OUTSIDE
CORNER
(REAR
VIEW)

neatly (**Figure 15-46**). Other types are applied in more conventional fashion (**Figure 15-47**).

Like lap siding, panels can be applied over sheathing or directly to studs (**Figure 15-48**).

Joints and corners are done as shown in **Figure 15-49**.

Always leave a slight gap where siding butts against trim; never force pieces together when you are doing a butt joint in a course. Fill all gaps with caulking. Color-matched caulking is available for use with prefinished siding materials.

Butt joints should always occur over a stud unless the wall is sheathed *and* you are using a system that includes metal butt-joint moldings.

FIG. 15-47

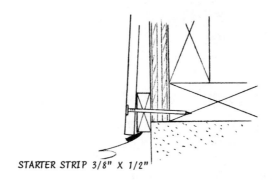

STARTER STRIP 3/8" X 1/2"

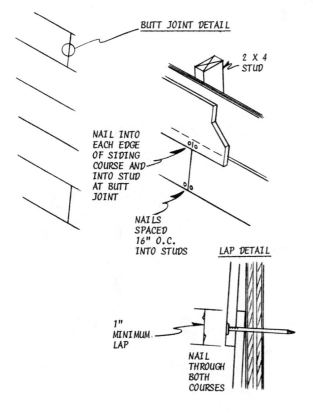

BUTT JOINT DETAIL

2 X 4
STUD

NAIL INTO
EACH EDGE
OF SIDING
COURSE AND
INTO STUD
AT BUTT
JOINT

NAILS
SPACED
16" O.C.
INTO STUDS

LAP DETAIL

1"
MINIMUM
LAP

NAIL
THROUGH
BOTH
COURSES

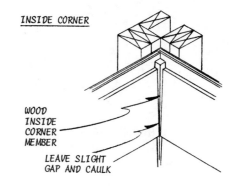

INSIDE CORNER

WOOD
INSIDE
CORNER
MEMBER

LEAVE SLIGHT
GAP AND CAULK

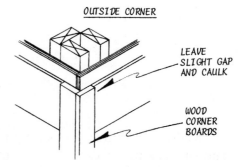

OUTSIDE CORNER

LEAVE
SLIGHT GAP
AND CAULK

WOOD
CORNER
BOARDS

FIG. 15-48

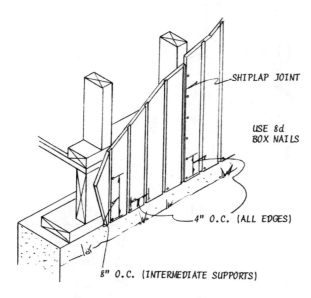

HARDBOARD PANEL OVER SHEATHING

SHIPLAP JOINT

USE 8d
BOX NAILS

4" O.C. (ALL EDGES)

8" O.C. (INTERMEDIATE SUPPORTS)

MAINTAIN 3/8" EDGE DISTANCE

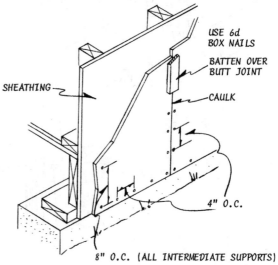

HARDBOARD OVER STUDS

USE 6d
BOX NAILS

BATTEN OVER
BUTT JOINT

SHEATHING

CAULK

4" O.C.

8" O.C. (ALL INTERMEDIATE SUPPORTS)

MAINTAIN 3/8" EDGE DISTANCE

Matching soffit systems might be available to go along with the siding. This is true of Masonite's Colorlok, which, in addition to the parts shown in **Figure 15-50**, includes aluminum fascia caps and roof edging.

FIG. 15-49

HARDBOARD PANEL JOINTS

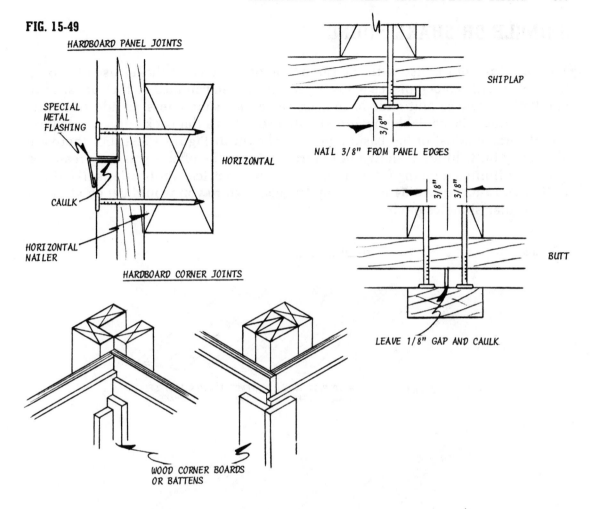

SPECIAL
METAL
FLASHING

CAULK

HORIZONTAL
NAILER

HORIZONTAL

SHIPLAP

NAIL 3/8" FROM PANEL EDGES

3/8"

BUTT

3/8" 3/8"

LEAVE 1/8" GAP AND CAULK

HARDBOARD CORNER JOINTS

WOOD CORNER BOARDS
OR BATTENS

FIG. 15-50

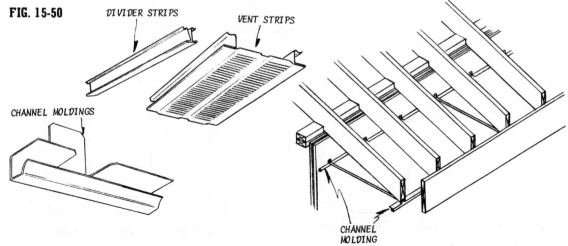

DIVIDER STRIPS

VENT STRIPS

CHANNEL MOLDINGS

CHANNEL
MOLDING

SHINGLE OR SHAKE SIDING

You can side your house with the same type of shingles or shakes used to cover the roof or with special pieces whose edges have been trimmed so they are strictly parallel and whose butt ends have been sawed for squareness. In addition, this type of shingle is available machine-treated so it has a flat back but a striated exposed surface. Most of the basic types are shown in **Figure 15-51**, but there are also fancy-butt shingles, shown in **Figure 15-52**, so the effects you can create are almost unlimited, ranging from more formal, symmetrical applications like those in **figures 15-53** and **15-54** to the deep, rugged texture you would get by working with regular shingles or shakes.

FIG. 15-51   TYPES OF SHINGLES AND SHAKES

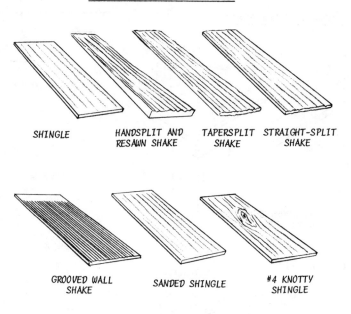

SHINGLE   HANDSPLIT AND   TAPERSPLIT   STRAIGHT-SPLIT
             RESAWN SHAKE   SHAKE   SHAKE

GROOVED WALL   SANDED SHINGLE   #4 KNOTTY
SHAKE   SHINGLE

FIG. 15-52   FANCY BUTT SHINGLES

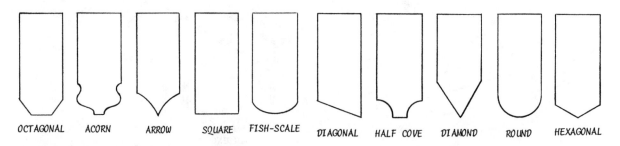

OCTAGONAL   ACORN   ARROW   SQUARE   FISH-SCALE   DIAGONAL   HALF COVE   DIAMOND   ROUND   HEXAGONAL

FIG. 15-53

FIG. 15-54

Many shingles especially designed for sidewall application are available prepainted or stained.

Premium-quality fancy-butt shingles are packed in cartons which contain 160 pieces to cover about a third of a square. Detailed application instructions are included in each carton, but here is a quick explanation.

The shingles are 16 inches long and approximately 5 inches wide and are intended for concealed nailing with a 6-inch exposure to the weather. The nails are hidden by placing them a maximum of 1 inch above the butt line of succeeding courses. Widths do vary a bit, so spacing may have to be adjusted to maintain a desired pattern precisely. Exterior walls must be done over solid sheathing. If used indoors, as they often are, they may be applied to spaced nailing strips.

Regular shingles or shakes may be applied to solid sheathing with an underlay of building paper. Use only as many nails or staples as you need to keep the paper in place until the shingles are applied. When the cover is against open studs or over non-wood sheathing, nailing strips are installed (open sheathing) with spacing determined by the exposure of the shingles (**Figure 15-55**). The

FIG. 15-55

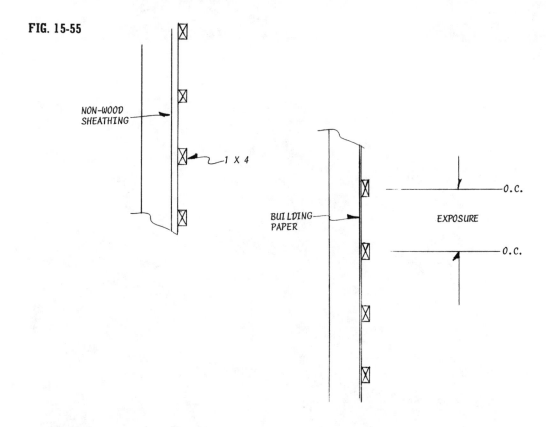

FIG. 15-56

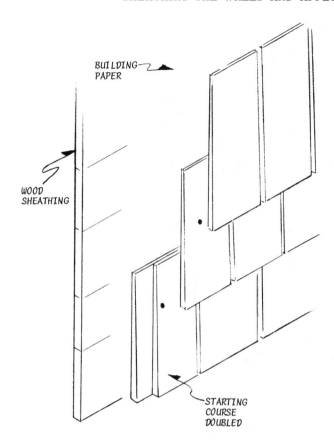

BUILDING PAPER

WOOD SHEATHING

STARTING COURSE DOUBLED

cover may be applied in a single course, as shown in **Figure 15-56**, in which case the exposure should be a bit less than half of the shingle length. For example, use 7½-inch exposure for 16-inch shingles, a 8½-inch for 18-inch shingles, and 11½-inch for 24-inch shingles. These suggestions apply for top-grade shingles. It's a good idea to reduce the exposure if you work with anything but a No. 1 grade.

Double-coursing, which is actually two layers of shingles — one placed directly over the other, as shown in **Figure 15-57** — provides deeper shadow lines, a generally more solid appearance, and vastly increased protection from the weather. The design also allows you to increase the exposure: 12 inches is okay for 16-inch shingles, 14 inches for 18-inch shingles, and 16 inches for 24-inch shingles. With double-coursing you can work with two grades of shingles — a quality material for the exposed course but a No. 4 grade (often called undercoursing) for the bottom one. The top cover is placed so the butt end extends below that on the under course by about ½ inch (**Figure 15-58**).

FIG. 15-57

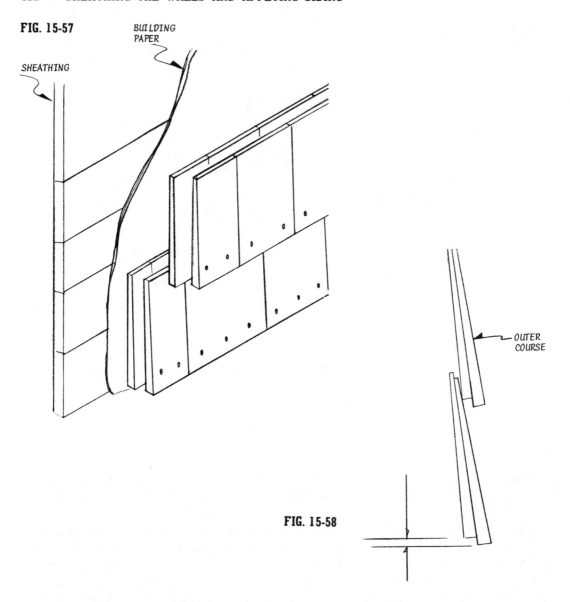

SHEATHING

BUILDING PAPER

OUTER COURSE

FIG. 15-58

Plan installations as if you were applying horizontal board siding. You might be able to make slight adjustments in exposures so the butt ends of shingles will even out with the top and bottom lines of openings. This will look neater and will minimize cutting.

You can place the shingles guided by a chalk line that you snap across the wall, but a more convenient method is to tack-nail a guide strip on which you can rest the shingles as you nail them (**Figure 15-59**).

FIG. 15-59

GUIDE STRIP

Single-coursing doesn't differ from a roofing job except in the area of exposure, which we have already discussed. All nails holding one course should be concealed by the shingles in the following course. Use two nails to a shingle, using a ¾-inch edge distance and placing them about 1 inch above the butt line of the next course. Use three nails if the shingle is wider than 8 inches, centering the third one between the first two.

Nails for all applications should be rustproof and long enough to penetrate the sheathing, or nailing strips if they are used; 3d nails are usually okay for single-coursing and 5d for a double-coursing.

Always use two layers of shingles at the bottom of the wall unless you are double-coursing. Here, use a double layer of under-course shingles so the first line will actually be triple-thick (**Figure 15-60**).

FIG. 15-60

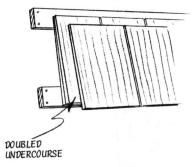

DOUBLED
UNDERCOURSE

Space the bottom layer of shingles in all courses about ⅛ inch apart, but do not nail them in standard fashion. A single nail or even a staple placed somewhere near the top will do to hold the part in place until the top pieces are added. The final course can be set tight—that is, without spacing—and secured with a minimum of two nails having a ¾-inch edge distance and placed about 2 inches above the *butt* line (**Figure 15-61**). This, of course, results in exposed nails, so it's essential here to use a type that is rustproof and, preferably, has a small head. Be sure the nail is long enough to penetrate the sheathing.

FIG. 15-61

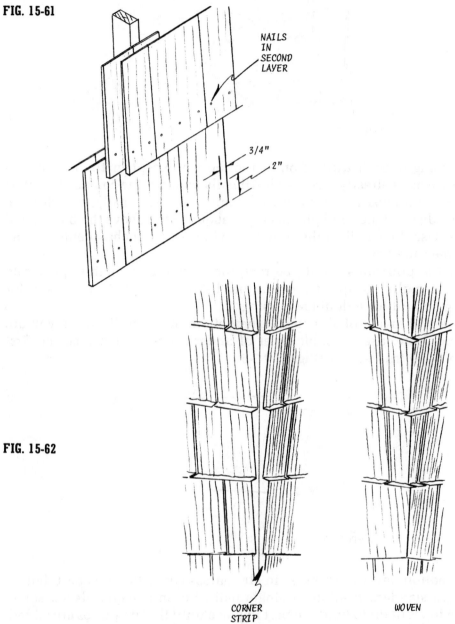

NAILS
IN
SECOND
LAYER

3/4"

2"

FIG. 15-62

CORNER
STRIP

WOVEN

Inside corners can be butted against a corner strip or they can be woven by alternating how the course ends butt against each other (**Figure 15-62**).

Or as shown in **Figure 15-63**, outside corners may also be woven or they can end in miter joints.

FIG. 15-63

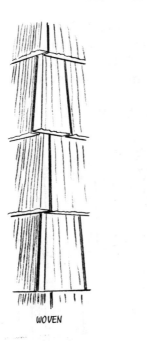

WOVEN

MITERED

Shingles that end against a vertical trim piece — around doors and windows — are just butted. A bead of caulking to cover the joint between trim and sheathing, under the shingle, won't hurt.

One of the newer products on the market lets you apply shingles or shakes in panel form. Like the roofing panels described in Chapter 14, these are authentic cover materials in various styles and colors that are bonded to backer boards. Various constructions are used in the manufacture of the panels (**Figure 15-64**). The

FIG. 15-64

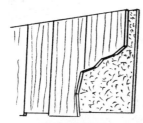

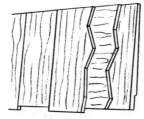

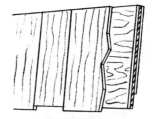

2-PLY PANELS

INDIVIDUAL SHAKES OR SHINGLES ARE ELECTRONICALLY BONDED TO A TAPERED INSULATION BACKER-BOARD TO FORM A PANEL 46-3/4" LONG

3-PLY PANELS

A COMBINATION OF TEXTURED SHAKES OR SHINGLES AND UNDER-COURSING SHINGLE BACKING WITH A CROSS-BIND CORE OF PLYWOOD VENEER WHICH ARE ELECTRON-ICALLY BONDED TOGETHER TO FORM A RIGID PANEL 8' LONG

4-PLY PANELS

THE AUTHENTIC TEXTURE OF BARN SHAKES BACKED WITH 5/16" PLYWOOD TO FORM A RUGGED 8' PANEL IN THE POPULAR 7' COLONIAL EXPOSURE

intent is to make them suitable for use directly on studs or over sheathing or nailing strips (**Figure 15-65**). The three-ply and four-ply panels may even be applied directly to studs that are 24-inch O.C. Matching corners, as shown in **Figure 15-66**, and color-matched nails are part of the systems so the job looks neat

FIG. 15-65

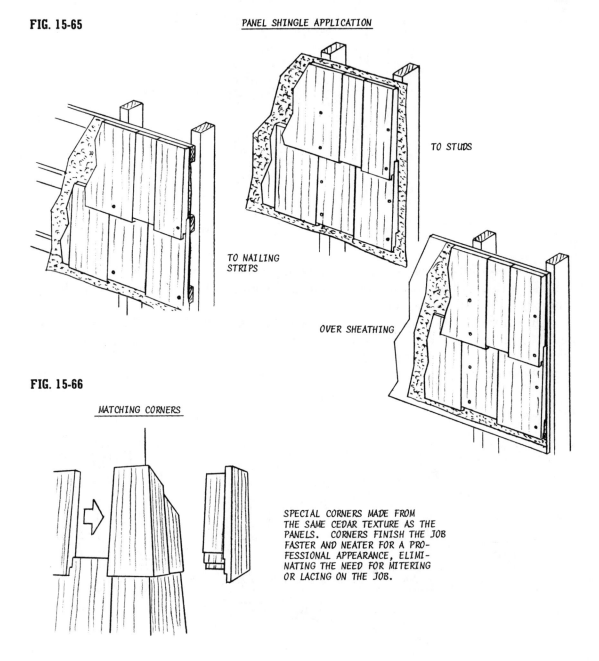

PANEL SHINGLE APPLICATION

TO STUDS

TO NAILING STRIPS

OVER SHEATHING

FIG. 15-66

MATCHING CORNERS

SPECIAL CORNERS MADE FROM THE SAME CEDAR TEXTURE AS THE PANELS. CORNERS FINISH THE JOB FASTER AND NEATER FOR A PRO- FESSIONAL APPEARANCE, ELIMI- NATING THE NEED FOR MITERING OR LACING ON THE JOB.

and the need to miter or weave around turns is eliminated. The products are packed in bundles or cartons that include detailed installation instructions, so there is no point in repeating them here. Check our address list at the back of the book for where to get more information. Be sure, if the instructions suggest a 24-inch stud spacing, that you check out this spacing against local codes.

Also check the addresses for locations of manufacturers who produce sidings of materials like aluminum or vinyl. Most such systems are special enough to include particular devices that make the installation simpler and waterproof. These can include metal starter strips, joint moldings, backer boards that may or may not add insulation, special fasteners and clips, and so on.

Vinyl siding is a recent development. Actually, the product has developed to the point where it can be used to do an entire house cover. In the example in **Figure 15-67**, the fascias, soffits, porch ceiling, siding, and gutter system are all vinyl.

The gutter system, shown more closely in **Figure 15-68**, is virtually maintenance free and can be used on any house design.

FIG. 15-67

FIG. 15-68

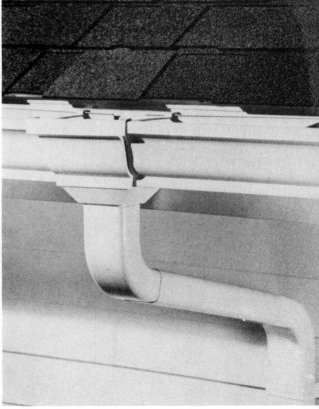

16

INSTALLING DOORS

After walls are sheathed, exterior doors (and windows) should be installed, so that the house can be secured quickly and so the final cover can coordinate with the frames and trim in and around openings. Interior doors are placed after walls are covered and, often, after the finish floor is down.

All doors and windows today are factory-assembled units that come to the site ready for placement in the rough openings. Some door units are completely assembled—the door is hinged, hung in the frame, and ready for other hardware —but others have frames in knock-down form and you assemble them on the job. It's not unreasonable, however, to do the complete job yourself if you have a special design in mind or want to work with a particular material unavailable in ready-made form. I, and others, have used reject doors with considerable savings in out-of-pocket costs. An example was a heavy, paneled door costing better than $100 but purchased at one third the price because one of the panels was not perfectly square. Installed as is, the door's defect has gone unnoticed.

Another tested procedure is to buy inexpensive flush doors and work on them, with moldings for example, so they become custom units (**Figure 16-1**).

FIG. 16-1

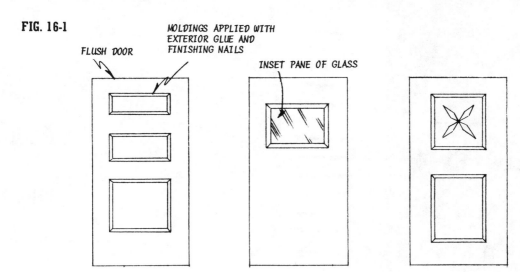

FLUSH DOOR

MOLDINGS APPLIED WITH
EXTERIOR GLUE AND
FINISHING NAILS

INSET PANE OF GLASS

Most exterior doors are flush or paneled, often including "lights" which are merely inset panes of glass (**Figure 16-2**). The basic parts of a paneled door are shown in **Figure 16-3**. There are also special decorator doors, like the Simpson

FIG. 16-2

FLUSH

PANELED

PANELS AND "LIGHTS"

FIG. 16-3

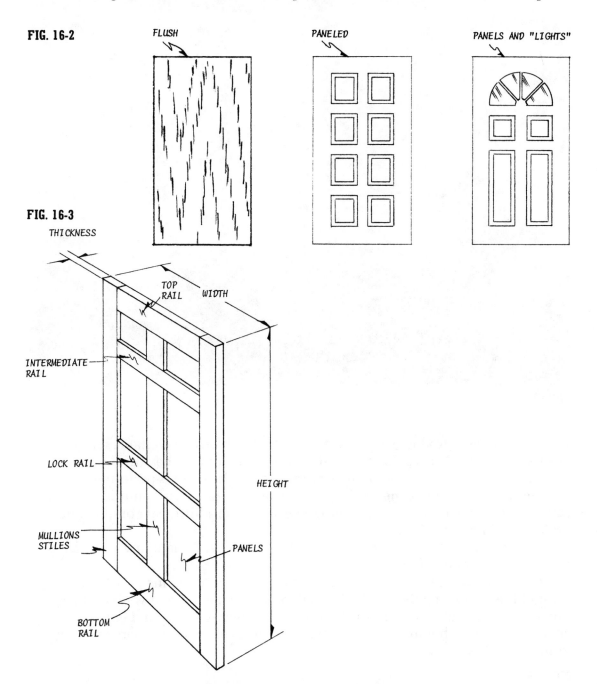

THICKNESS

TOP RAIL

WIDTH

INTERMEDIATE RAIL

LOCK RAIL

HEIGHT

MULLIONS
STILES

PANELS

BOTTOM RAIL

FIG. 16-4

Timber Company product shown in **Figure 16-4**, which are available ready-made but have a custom-designed look that sets off the entrance beautifully.

Flush doors can be either hollow or solid-core; the latter is preferred for exterior applications. Many times, utility entrances are done with simple doors and trim, while the main door and its setting will include additional components that become architectural features.

The detail doesn't have to be more than a cap with matching pilasters that you can make yourself (**Figure 16-5**).

Commercial caps in various styles are available ready to install (**Figure 16-6**). Most of them can be purchased with matching pilasters. If you wish to do something more elaborate yourself, a construction scheme is shown in **Figure 16-7**. All such assemblies should be weathertight in themselves and wherever they are joined with the wall.

FIG. 16-5

FIG. 16-6   COMMERCIAL CAPS

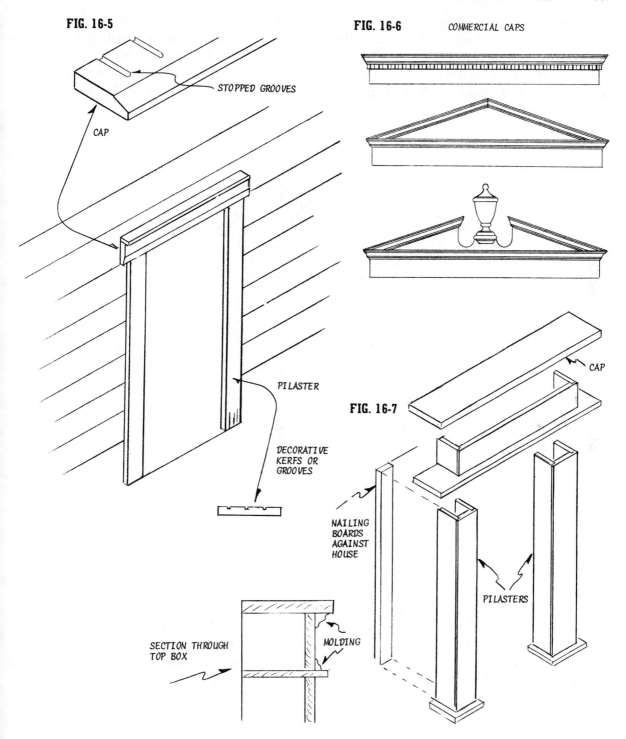

STOPPED GROOVES

CAP

PILASTER

DECORATIVE
KERFS OR
GROOVES

FIG. 16-7

CAP

NAILING
BOARDS
AGAINST
HOUSE

PILASTERS

SECTION THROUGH
TOP BOX

MOLDING

Utility entrances, especially those not protected by overhangs, can be finished as shown in **Figure 16-8**. View the top as a frame that will be finished to match the roof. The structure can be installed after the siding is up, but it will look less like an afterthought if you do it against sheathing. The latter procedure also permits adding adequate flashing along the top edge before siding is placed.

FIG. 16-8

UTILITY ENTRANCE

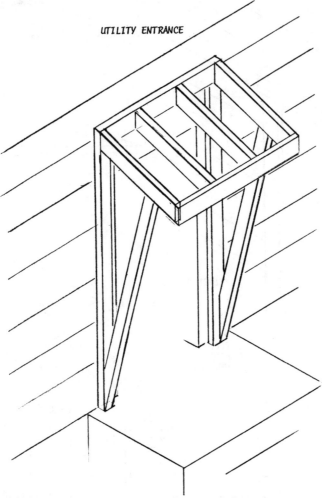

The main entrance door should be 36 inches wide and can be as much as 7 feet high, although 6 feet 8 inches is fairly standard. Other doors can be narrower, but don't go under 30 inches. Since all doors are passageways for furnishings as well as people there is nothing wrong with using the 36-inch width everywhere. All frame installations must be done very carefully so components will be plumb and square. A correctly installed door will remain open in any position, which will not be the case if the frame has a tilt.

THE FRAME

Door frames consist of jambs and stops, as shown in **Figure 16-9**, and, on exterior doors, a sill. Jambs and stops can be made from softwoods, but the sill, which must withstand considerable traffic, should be a hardwood like oak.

FIG. 16-9

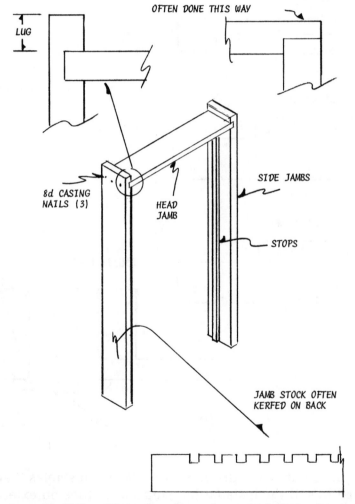

Jamb designs do differ. Plain flat ones, done with ¾ inch stock, are mostly for interior doors, while the other designs shown in **Figure 16-10** are for exterior applications. Jambs must be as wide as the total wall thickness so that the gap that will exist after the frame is installed in the rough opening can be neatly covered with trim (**Figure 16-11**).

FIG. 16-10

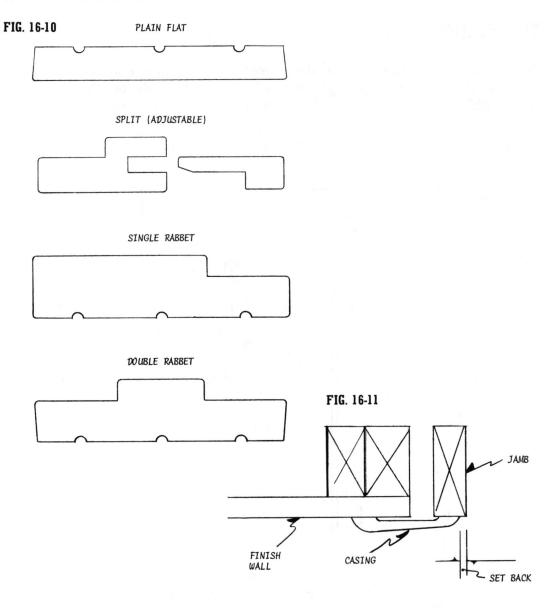

PLAIN FLAT

SPLIT (ADJUSTABLE)

SINGLE RABBET

DOUBLE RABBET

FIG. 16-11

JAMB

FINISH
WALL

CASING

SET BACK

Jambs come in standard widths or as two-piece adjustables that can be narrowed or widened to suit the installation. Jambs and stops on exterior doors are one-piece affairs, since this adds to weathertightness. Also, the material used is thicker—running from $1\frac{1}{4}$ inches to as much as $1\frac{5}{8}$ inches—since the frames must support heavy doors and possibly screens or storm doors. Homemade types should resemble **Figure 16-12**. Rabbet cuts are made along both edges when it is necessary to accommodate more than the main door. Rabbet depth is $\frac{1}{2}$ inch. The

FIG. 16-12   HOMEMADE, ONE-PIECE JAMB MEMBER

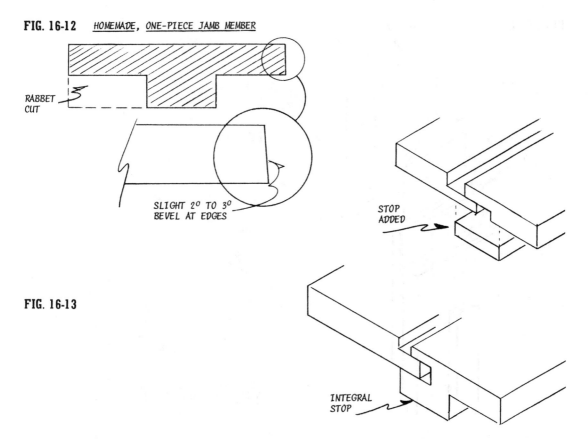

RABBET
CUT

SLIGHT 2° TO 3°
BEVEL AT EDGES

STOP
ADDED

FIG. 16-13

INTEGRAL
STOP

width of the inside rabbet equals the door thickness. The 2- or 3-degree bevel along the edges assures a good fit for the trim.

If you wish to make your own adjustable jambs, you can do the job in either of two ways, as shown in **Figure 16-13**. The added stop should be used on inside doors only to ensure weathertightness.

INSTALLING THE JAMBS AND STOPS

Assemble the frame and then slip it into the rough opening. Use height blocks under the side jambs of interior installations if you are doing the job before the finish floor is down. This is to provide room under the jambs for the thickness of the final floor cover.

Work with a level to be sure the top jamb is horizontal and the side ones are plumb. The frame is secured and adjustments are made by using tapered shims

FIG. 16-14

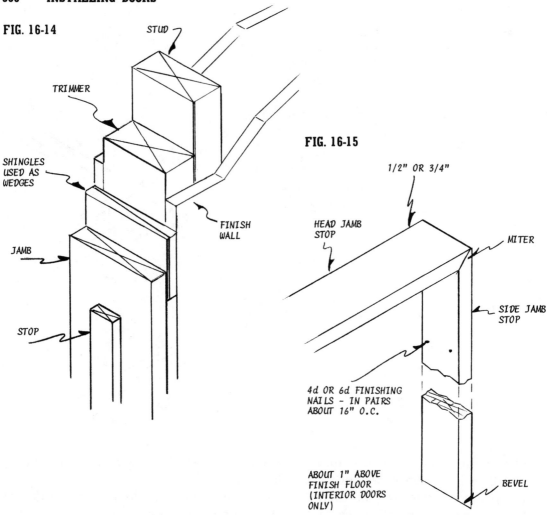

FIG. 16-15

(regular shingles) between the jambs and the trimmers (**Figure 16-14**). To maintain the correct distance between the side jambs, use a length of wood as a spacer between the jambs at the floor line. You should have four or five shims on each side of the frame, placing some of them so they will be approximately in hinge and lock areas. The head jamb, or the top edges of the side jambs, should not bear against the header.

After wedges are adjusted so the frame is rigid and straight, drive finishing or casing nails, long enough to penetrate the trimmer about 1 inch, through jambs and shims. Place the nails so they will be hidden by the stops.

Stops are located to accommodate the thickness of the door and are installed as shown in **Figure 16-15**. Do the hammering carefully so you don't mar the wood.

Sink the nailheads with a nail set so you can hide them with wood dough. Stops are supplied as part of a prehung door assembly. You can make your own by working with plain strips of wood or by using ready-made stop molding in various shapes (**Figure 16-16**).

FIG. 16-16

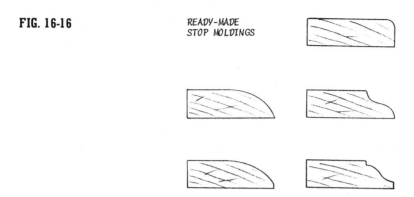

READY-MADE
STOP MOLDINGS

INSTALLING THE SILL

There are various ways to install the sill for an exterior door. Much depends on whether its surface will be flush with the finish floor. Often the sill is rabbeted and it is secured directly to the subfloor so part of its width extends outside the house frame (**Figure 16-17**).

FIG. 16-17

FIG. 16-18

FIG. 16-19

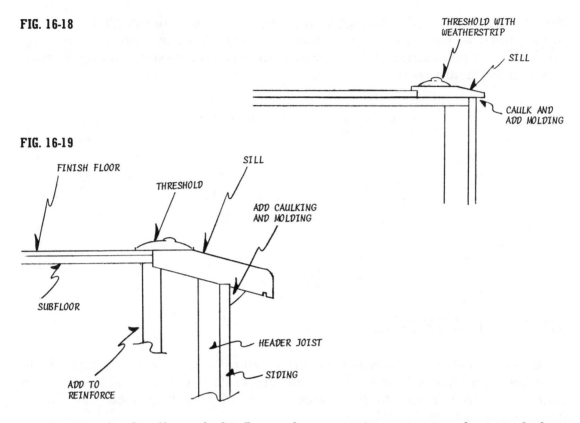

Custom-made sills might lie flat as shown in **Figure 16-18** and can include a small rabbet to accommodate the finish floor. This does leave a slight ledge but it can be minimized, even beveled, so it will not be a tripping hazard.

More complicated installations call for trimming the joist header (**Figure 16-19**). Add whatever reinforcement is necessary to support and supply perimeter nailing for the flooring. In such cases, the threshold, often a metal-and-vinyl one as shown in **Figure 16-20** is placed to cover the joint between sill and floor.

FIG. 16-20

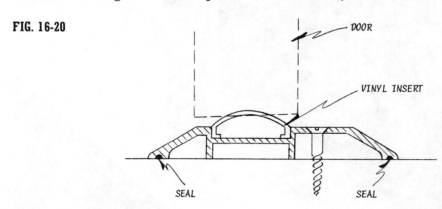

The sill installation is critical, since it must bear much traffic and must be weathertight and waterproof. Ready-to-install exterior door assemblies are available with complete weatherstripping, often an interlocking type, in place around jambs and at the sill. In such cases, the installation must be done in line with the design of the unit.

INSTALLING HINGES AND OTHER HARDWARE

Since interior doors are usually hollow-core and thinner and narrower than exterior units, $1\frac{3}{8}$ inches thick by 32 inches wide being fairly standard, they are much lighter and can be placed with two hinges instead of three. However, hinges do more than just let the door swing. They guard against distortion in addition to supplying support, and so a third hinge on all doors is a good safety factor.

FIG. 16-21

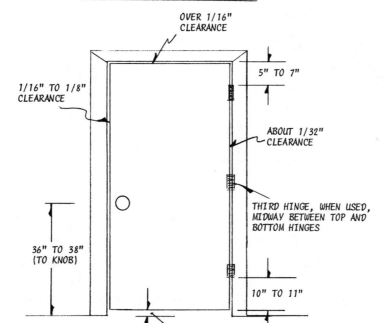

HINGE PLACEMENT AND DOOR CLEARANCE

OVER 1/16"
CLEARANCE

1/16" TO 1/8"
CLEARANCE

5" TO 7"

ABOUT 1/32"
CLEARANCE

THIRD HINGE, WHEN USED,
MIDWAY BETWEEN TOP AND
BOTTOM HINGES

36" TO 38"
(TO KNOB)

10" TO 11"

ENOUGH TO CLEAR FINISH FLOOR INSIDE;
ENOUGH FOR WEATHERSTRIP DETAILS ON
OUTSIDE

Hinge locations and correct clearances are shown in **Figure 16-21**. Don't hang the door any tighter than this or you will surely have sticking problems later.

FIG. 16-22

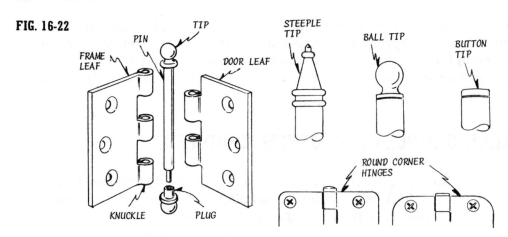

Work with loose-pin butt hinges as shown in **Figure 16-22**, since they make it possible to remove the door after installation without having to take out screws. The size of the hinge goes along with the size of the door; see the table of suggested sizes in **Figure 16-23**.

FIG. 16-23.   HOW TO CHOOSE HINGES FOR DOORS

THICKNESS OF DOOR	WIDTH OF DOOR	SUGGESTED HINGE HEIGHT
¾ up to 1⅛ cabinet doors	up to 24	2½
⅞, 1⅛ screen or combination door	up to 36	3
1⅜	up to 32	3½
	over 32	4
1¾	up to 36	4½
	over 36 to 48	5
2	up to 42	5 or 6

Some butt hinges are reversible, some are not, so check how the door will operate (often called the "hand" of the door) and then decide, if necessary, whether you need right-hand or left-hand hinges (**Figure 16-24**). Remember too that some door locks must be chosen according to whether the door swings left or right.

If you are installing the hinges, place the door in the frame and use shims to hold it there while you mark hinge locations on both the door and the jambs. Use one leaf of the hinge as a pattern to mark its outline on the door edge, following the backset recommendations in **Figure 16-25**. Incise along the lines you have made by working with a sharp knife and then remove the waste with a chisel (**Fig-**

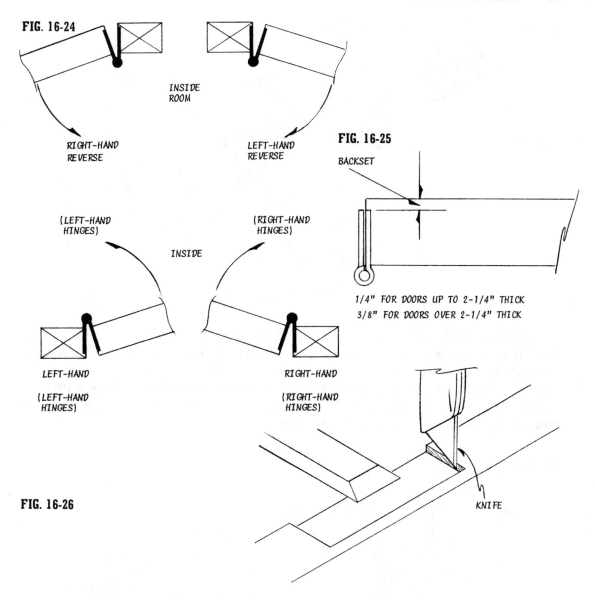

FIG. 16-24

INSIDE
ROOM

RIGHT-HAND
REVERSE

LEFT-HAND
REVERSE

FIG. 16-25

BACKSET

1/4" FOR DOORS UP TO 2-1/4" THICK
3/8" FOR DOORS OVER 2-1/4" THICK

(LEFT-HAND
HINGES)

(RIGHT-HAND
HINGES)

INSIDE

LEFT-HAND

(LEFT-HAND
HINGES)

RIGHT-HAND

(RIGHT-HAND
HINGES)

FIG. 16-26

KNIFE

ure 16-26). The depth of the mortise or "gain" should be just enough for the hinge to be flush with the wood. Follow the same procedure on the jamb, or, as some do, install the hinges on the door first and then prop up the door in correct open position so you can trace the hinge outline on the jamb. In any case, the job should be done with some precision so the hinges will seat correctly. The mortising can be done with a router driving a square-end bit. Special templates that almost eliminate layout work are available for use with a router.

FIG. 16-27

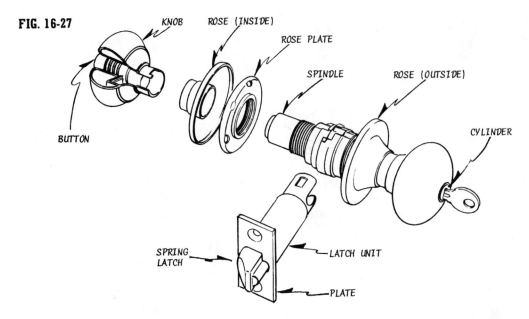

Cylinder locks of the type shown in **Figure 16-27** are very common in residential constructions. They are simple to install, requiring only a large hole through the surface of the door, a smaller one drilled through from the edge, and a mortise for the latch plate (**Figure 16-28**). The small hole can be done with a bit and brace, or you can use a portable drill. The large one requires a special bit, or you can use a hole saw and portable drill.

FIG. 16-28

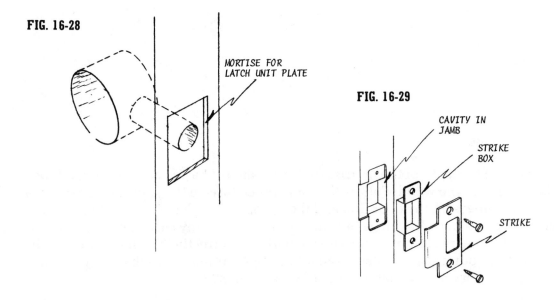

FIG. 16-29

It's best to install the lock first so you can use it to mark the correct position of the strike plate on the jamb. The strike plate requires a cavity, which you can form by drilling overlapping holes, for the strike box and a mortise for the strike plate (**Figure 16-29**). You can leave the lock hardware installed, but it's a good idea to remove it until after finish coats have been applied to the woodwork.

All lock sets come complete with installation instructions, and usually with cardboard templates you use to mark hole locations—and this applies to any type of security lock you may wish to include. Read all such instructions carefully before starting installations.

You will probably find it easier to install stops for interior doors after the doors are hung and the lock is in place. Work with the door closed and place the stops so there will be about 1/32-inch clearance between them and the door. Often the stop on the lock side is placed snugly against the door, since this will eliminate rattle.

INSTALLING DOOR TRIM

Trim, called *casing*, is used around doors to cover the gap that exists between the jambs and the walls. Two basic installations are shown in **Figure 16-30**. The miter joint is used more often; the butt joint can't be used on all types of ready-made

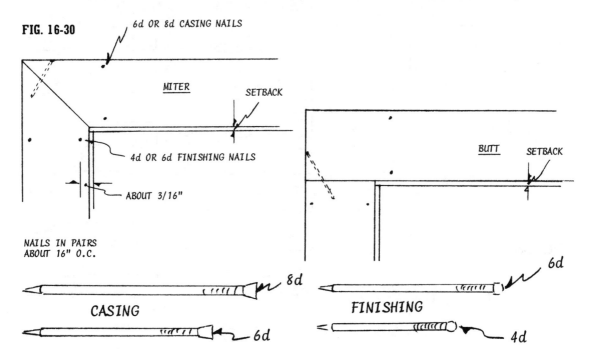

FIG. 16-30

6d OR 8d CASING NAILS

MITER SETBACK

4d OR 6d FINISHING NAILS

ABOUT 3/16"

NAILS IN PAIRS
ABOUT 16" O.C.

BUTT SETBACK

CASING 8d 6d

FINISHING 6d 4d

casing but is a simple way to go if you decide to do your own trim. Homemade pieces can be simple, square-edged strips, or you can work on them to add the kind of detail shown in **Figure 16-31**.

FIG. 16-31

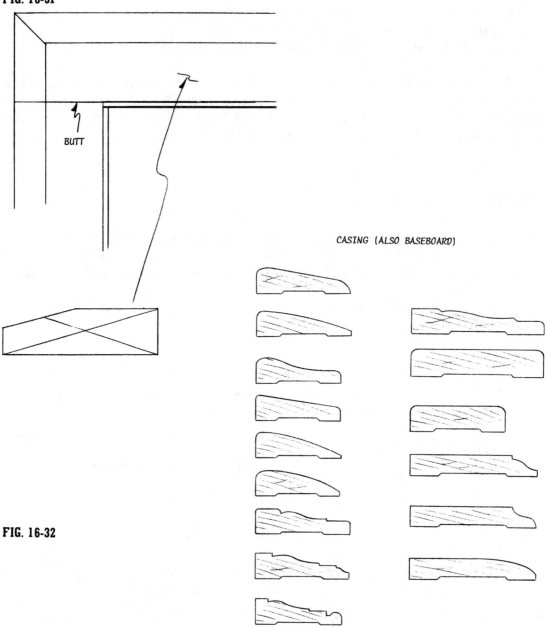

BUTT

CASING (ALSO BASEBOARD)

FIG. 16-32

There is much variety in the kind of casing material that is available ready to install. Some of the pieces shown in **Figure 16-32** are also called baseboards. How you go depends on the appearance you want. Simpler casing can blend in with the walls, while the fancier ones become decorative details.

DOORS THAT FOLD

Bi-fold doors come in styles that include solid or louvered panels (**Figure 16-33**). They are usually used on closets but often are installed between open rooms to serve as dividers when needed. They look good, and they don't require as much swing room as regular doors do.

 FIG. 16-33

BI-FOLD DOORS
WITH LOUVERED PANELS

Most are available in two- or four-door designs with installation possibilities as shown in **Figure 16-34**. They are installed after the rough opening has been finished with jambs and trim, and all require special pieces of hardware that are placed following instructions that come with the kits (**Figure 16-35**).

In essence, they are supported at the jambs by hinges or pivots and at the outer edge of each panel by a hanger which rides a track that is secured to the head jamb.

Folding doors include the accordion types, which are made by assembling many narrow panels of wood or other materials. They work, and are installed,

FIG. 16-34   TWO- AND FOUR-DOOR BI-FOLDS

PIVOT
HERE OR

HINGE

PIVOT   HINGE   PIVOT

PULL   PULL

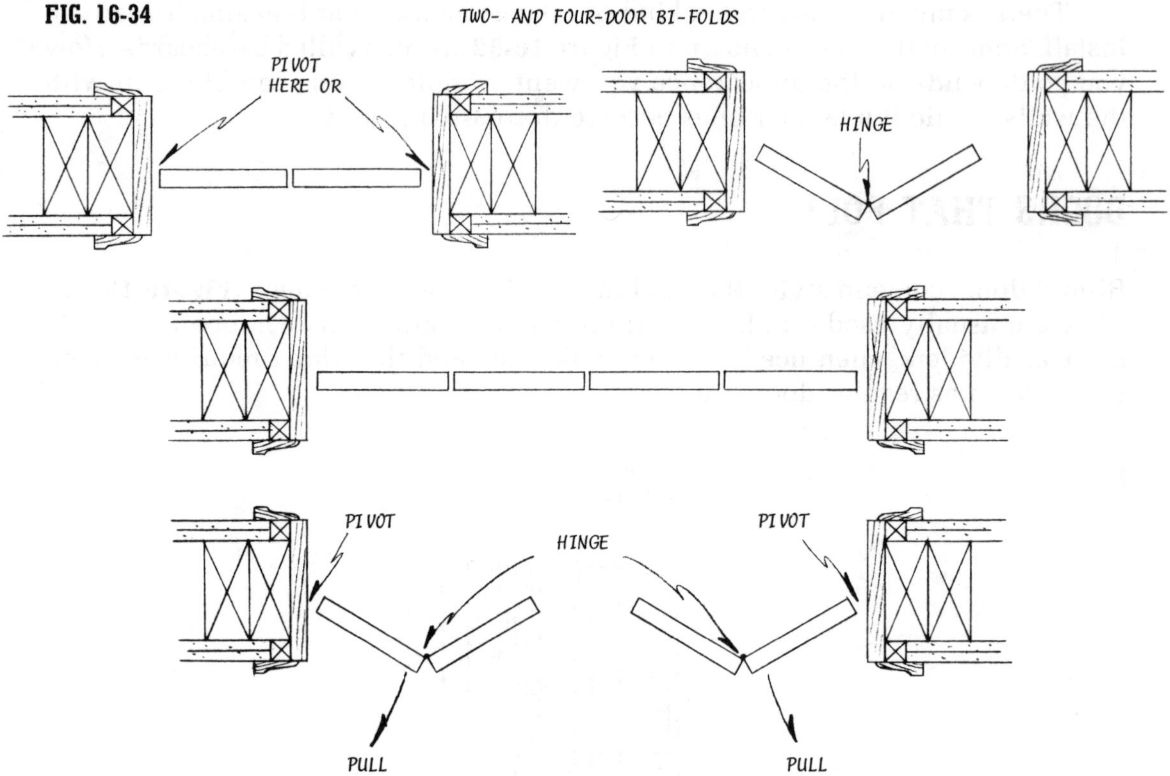

FIG. 16-35

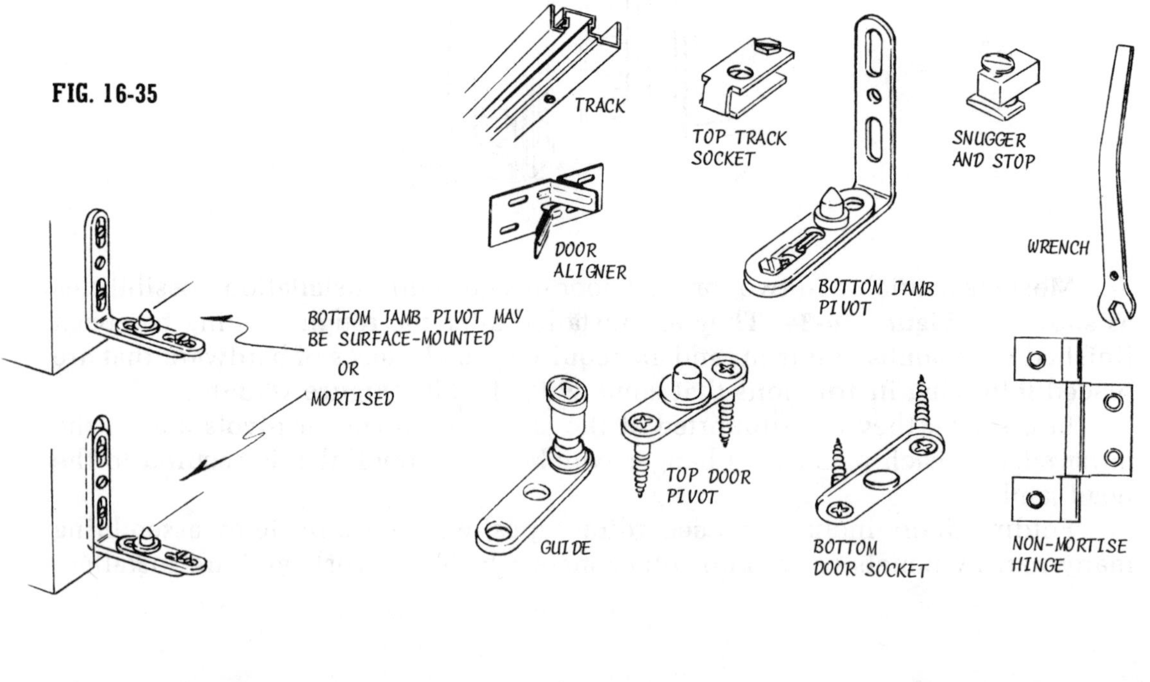

TRACK

TOP TRACK
SOCKET

SNUGGER
AND STOP

WRENCH

DOOR
ALIGNER

BOTTOM JAMB
PIVOT

BOTTOM JAMB PIVOT MAY
BE SURFACE-MOUNTED

OR

MORTISED

GUIDE

TOP DOOR
PIVOT

BOTTOM
DOOR SOCKET

NON-MORTISE
HINGE

pretty much like the bi-folds and are available in sizes to suit various openings. Because the panels are narrow and fold back on themselves, they require very little floor space when they are opened.

SLIDING DOORS THAT BYPASS

These units hang from and move on double tracks which are usually installed against the underside of a conventional head jamb. Some tracks, like the one shown in **Figure 16-36**, have a built-in fascia; others are hidden by adding a trim

FIG. 16-36

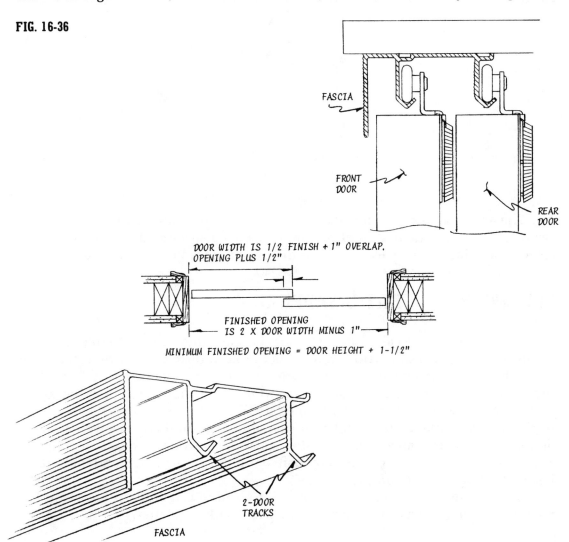

FASCIA

FRONT DOOR

REAR DOOR

DOOR WIDTH IS 1/2 FINISH + 1" OVERLAP. OPENING PLUS 1/2"

FINISHED OPENING IS 2 X DOOR WIDTH MINUS 1"

MINIMUM FINISHED OPENING = DOOR HEIGHT + 1-1/2"

2-DOOR TRACKS

FASCIA

strip after the installation is complete. It's also possible to get special head jambs in which the tracks are, or may be, recessed so the top edge of the doors will be flush with the underside of the jamb. Like the bi-folds, hardware for sliding-door installations is available in kit form to suit various door thicknesses. A floor guide is included so the doors will remain perpendicular regardless of the hanging method (**Figure 16-37**). Good units will include an adjuster built into the hangers so the doors may be adjusted for correct height after they have been installed.

FIG. 16-37

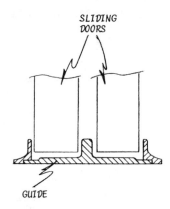

SLIDING
DOORS

GUIDE

The rough opening is framed conventionally and completed before the doors are hung. Bypass sliders do not require any swing space and are often used on closets and similar areas even though the design does not permit total access.

SLIDING DOORS THAT DISAPPEAR

These space savers are called pocket doors because they slide into openings that are designed as part of partitions. This is the kind of thing you plan for in advance, buying preassembled units that you install as you erect the wall frame (**Figure 16-38**). You can view the job as you would a regular rough opening except that it must be wide enough to accommodate the pocket as well as the door opening. Units are available in a number of standard sizes for double as well as single doors with pocket frames made of steel as well as wood. Track and roller assemblies are much like those used on conventional sliding doors and include adjusters for door height and plumbness.

Good units will have silent-action rollers and will include rubber stop buttons inside the pockets so the doors will work without clatter.

FIG. 16-38

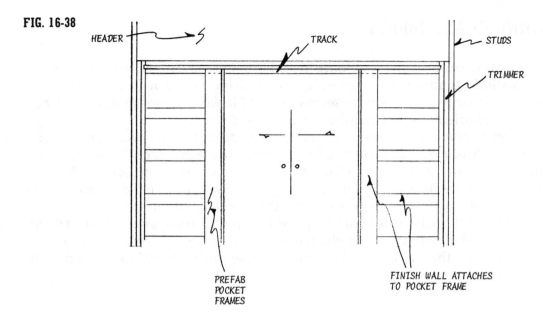

HEADER

TRACK

STUDS

TRIMMER

PREFAB
POCKET
FRAMES

FINISH WALL ATTACHES
TO POCKET FRAME

CAFE DOORS

These units are practical, for example, between a kitchen and a dining area, since they permit passage even though a person's hands are full. They do not provide much privacy, but enough to hide kitchen clutter from the diners.

The doors can be simple or fancy and are hung on special pivot hardware that is attached to conventional jambs (**Figure 16-39**). Check the hardware you buy to be sure the bottom pivot includes a riser so that the doors can be held in an open position.

FIG. 16-39

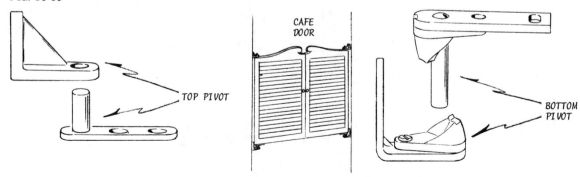

TOP PIVOT

CAFE
DOOR

BOTTOM
PIVOT

SLIDING GLASS DOORS

Sliding glass doors bring the outdoors in and provide easy access to a patio or garden. Installations are done to minimize the division between in and out so the room appears larger and greenery seems a part of it. Such doors should be placed following the orientation principles we discussed in Chapter 2.

Factors to consider are thermal insulative quality, built-in weathertightness, a security locking system, and of course, the appearance you seek. Those made for residential use are framed with steel or aluminum or wood and are available in various stock sizes. It is important these days to check out the glass used and to spend more money if necessary for insulating types that will minimize heat loss.

All units will include at least one fixed and one sliding panel (**Figure 16-40**). They are available with three panels, the center one sliding in one direction, and with four panels, the two center ones moving in opposite directions, as shown in **Figure 16-41**.

FIG. 16-40

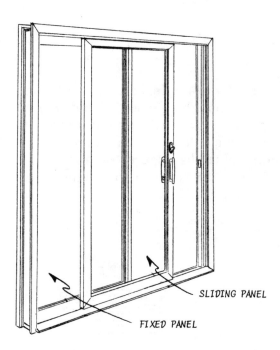

SLIDING PANEL

FIXED PANEL

FIG. 16-41

3-PANEL   4-PANEL

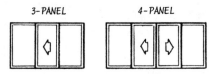

The size of the units and the dimensions of the rough opening will vary depending on the manufacturer, but the design of the opening will not differ from what is required for any door (**Figure 16-42**). You should, of course, have the specifications of the units on hand while you are doing the wall framing.

FIG. 16-42

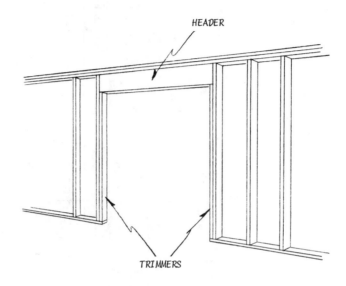

HEADER

TRIMMERS

Some types require finish framing — jambs, etc. — like those you would install for any door, but others are so complete, including sills, that you just slip them into place and secure and weatherseal them following instructions that come in the package.

In all cases, parts must be plumb and square if doors are to slide properly. Cross sections through header, jambs, and sill of a typical installation are shown in **Figure 16-43**.

It's a good idea to cover the installed door with plywood or some other material to protect the glass and frame from damage while other work is going on. The least you should do is place strips of masking tape over the glass so no one will try to walk through it.

GARAGE DOORS

Most modern garage doors will either swing up or roll up. Sliding bypass doors and hinged doors are much less practical. Swing-up doors are most common, probably because they are cheaper than others and, if you wish, can be made on the site with the same materials used for house siding. Often, though, the door

FIG. 16-43

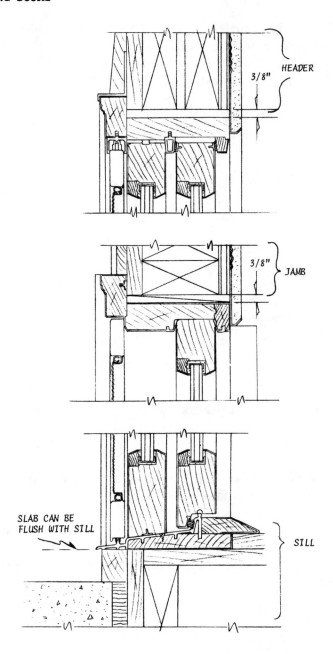

3/8"   HEADER

3/8"   JAMB

SLAB CAN BE
FLUSH WITH SILL

SILL

cover is a thin exterior plywood, since it is easy to work with and results in a lighter door.

A door 8 to 9 feet wide and 6½ to 7½ feet high is okay for a single-car garage, but use a minimum 16-foot width if the garage is for two cars. Rough openings are done in standard fashion, but always with substantial headers because of the wide openings (**Figure 16-44**). Cripples may be used between the header and the top plate if the wall height makes them necessary.

The opening is framed with side and head jambs just like those installed for regular exterior doors except that heavier material, usually 2× stock, is used and integral stops are not required. Inside jambs are often added as shown in **Figure 16-45**, when they are required to support track or other door hardware.

FIG. 16-44

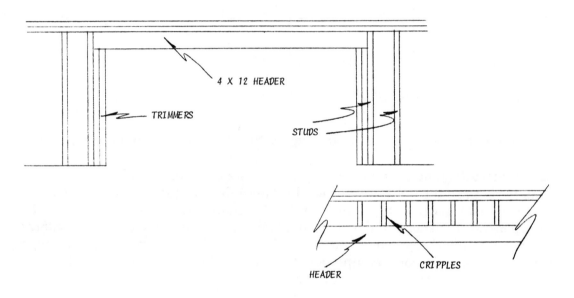

FIG. 16-45

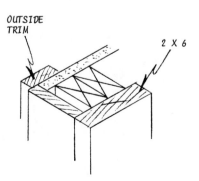

A typical frame that you can make for a swing-up door that will be sheathed with plywood is designed as shown in **Figure 16-46**.

The same type of frame can be done but with nailing strips or additional blocking if other siding materials will be used as a cover. Wide doors should be reinforced to prevent bowing; this can be done by using ready-made, adjustable

FIG. 16-46

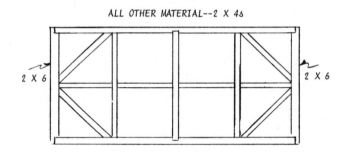

ALL OTHER MATERIAL--2 X 4s

2 X 6   2 X 6

hardware which can be purchased along with other hanging materials you will need.

Counterbalancing devices, usually extension or torsion springs, are included in the package so the door will lift easily. These devices may be considered old-fashioned today because there are so many electric-powered and radio-controlled door openers on the market, which allows you to sit in the car and have the garage door open merely by pushing a button. Check out these units before making a decision. All of them come complete with whatever materials are required for the job plus detailed installation instructions.

17

INSTALLING WINDOWS

Like exterior doors, windows should be installed, or framed for, just after walls are sheathed or otherwise prepared for siding. This sets the stage for getting weatherproof joints in the opening and where siding and frames meet.

Today's builder is fortunate, for he doesn't have to fabricate windows or have them custom-made in a local shop. Modern windows are mass-produced to rigid standards that cover materials and methods of assembly and even the design of such details as drip caps and sills.

Wood is a very popular window material because it has natural insulating qualities and minimizes condensation. It may be painted to match any color scheme or it can be stained or finished in natural tones to match various sidings. Some of the units on the market, like the one in **Figure 17-1** by Pella, are provided

FIG. 17-1   PELLA'S FACTORY WEATHERSTRIPPING

with complete weatherstripping at the head, jambs, and sill to give a positive seal against the elements. Some are double-glazed systems that include aluminum-framed glass panels that can stay in place throughout the year and so eliminate the need for additional storm windows. Insulating glass is also available.

Wood windows are available with outside casing (trim) already attached, as in **Figure 17-2**, so that the basic installation can be done by nailing through the casing into the sheathing or the frame of the rough opening.

FIG. 17-2

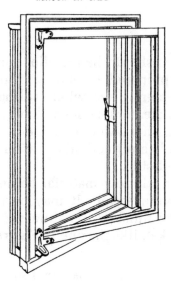

Since metal is strong and rigid, windows made of steel or aluminum have narrower framing pieces than those made of wood. Generally, metal windows cost less, but they do not insulate as well and can have condensation problems. A good argument for aluminum is that it doesn't need painting; for steel, that it is strong. Much research and development work is going on with all windows—accelerated today because of diminishing energy factors—so it's wise to make a good study of what is on the market when you build. It's possible, for example, that by the time you read this, metal windows made of laminations that include insulation may be available. Plastic materials may also have an impact on future designs.

For those who believe in wood construction but want the advantages of a metal exterior that does not require painting, there are products like Pella's Clad windows with exteriors protected by factory-applied and factory-finished aluminum (**Figure 17-3**).

FIG. 17-3

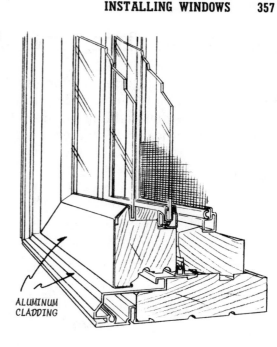

ALUMINUM
CLADDING

FIG. 17-4

ANDERSEN PERMA-SHIELD

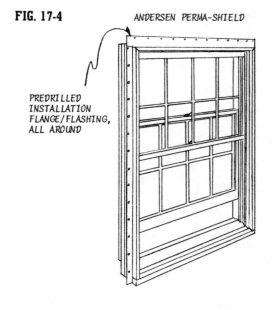

PREDRILLED
INSTALLATION
FLANGE/FLASHING,
ALL AROUND

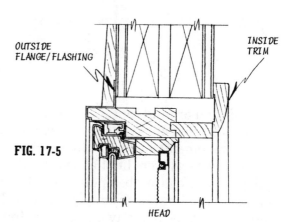

OUTSIDE
FLANGE/FLASHING

INSIDE
TRIM

FIG. 17-5

HEAD

The design of many windows eliminates the need for exterior casings to secure the window to the wall, like Anderson's Perma-Shield (**Figure 17-4**). The perimeter of the units is a predrilled installation flange that also acts as flashing. Attachment nails are driven through the flange (**Figure 17-5**).

WINDOWS CAN SLIDE, SWING, OR STAY PUT

Within the basic categories of windows that slide, swing, or are fixed, as shown in **Figure 17-6**, there is much variety in design and sometimes even in method of operation. The double-hung window is a sliding type, but its two bypassing sashes move vertically. With the exception of the fixed window, it is probably the

FIG. 17-6

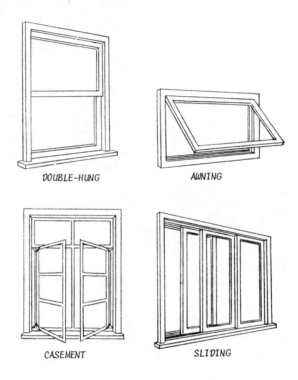

DOUBLE-HUNG

AWNING

CASEMENT

SLIDING

oldest design around, but it is still very popular, as evidenced by the many modern versions that are available. They are adaptable to many architectural schemes and improvements have eliminated most of the original drawbacks. At one time, the moving parts were held at a particular position through a system of cords and pulleys and a counterbalancing weight that was hidden behind the side jambs. Replacing the cord or freeing a stuck weight was a real chore. Today the job of holding either sash at any position is done largely through friction designs or by spring-type hardware. In addition to easier and surer operation, there are improvements in materials and features such as built-in weatherstripping, double glazing, and so on. Some units are made so the sashes can be removed, which solves the old problem of how to clean the exterior surface of the glass while remaining indoors.

A simplified cross section of a double-hung window is shown in **Figure 17-7**. Usually, and especially if the window is not under a protective overhang, a drip cap is advisable (**Figure 17-8**).

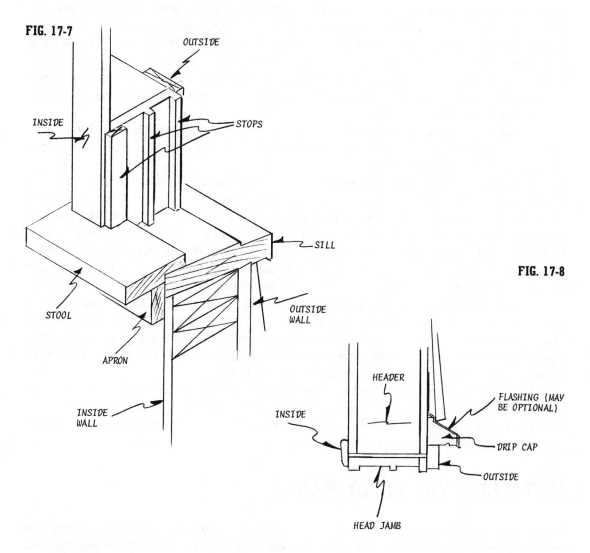

FIG. 17-7

OUTSIDE

INSIDE

STOPS

SILL

FIG. 17-8

STOOL

OUTSIDE WALL

APRON

INSIDE WALL

HEADER

INSIDE

FLASHING (MAY BE OPTIONAL)

DRIP CAP

OUTSIDE

HEAD JAMB

Sliding windows that move horizontally have two or more sash units. The most common designs are two-sash units with one or both of them being movable, or three-sash designs with the center one fixed. A common objection to the sliders has been a lack of security, but recent improvements include better locking hardware and built-in adjustable stops that can limit the distance a movable sash will open.

Extra-wide windows are often made by doing a custom job with a fixed center panel and then installing a two-sash unit at each end.

Casement windows have sashes that are hinges on one side like a door. They swing outward and may be adjusted by a crank or by a push bar which may be locked at any position. Since the movable parts swing to the outside, screens or additional storm sashes must be mounted on the inside. The design may be used throughout the house, but it is especially practical over a counter or similar object you must lean over to reach the window. Older units were difficult to make weathertight, but improvements in design and new weatherstripping materials have minimized the problem.

Do some thinking about the location of casement windows in relation to outside activities. They can cause a traffic problem if they open over a patio or porch or a path that is close to walls.

Awning windows are hinged like casements but along the top edge, so they swing out and up. Some units are made with special hardware that provides a type of pivot action—the top of the sash moves down as you push the bottom outward. A practical feature of the concept is that units can be placed side by side to provide adequate light and ventilation but close to the ceiling, leaving a maximum amount of wall space for placing furniture. Used this way, they can also provide, for example, greater privacy for a bedroom or a bathroom.

Often awning windows are used in combination with fixed windows to provide necessary ventilation. Opening and closing mechanisms include the cranks and push bars used on casement windows, but some are designed so that such hardware is not needed. The outside-clearance consideration mentioned in relation to casement windows applies here as well.

Hopper windows look like awning types but are hinged at the bottom and designed to swing inward. They are most practical for installations near the ceiling. If placed low on a wall they will obviously interfere with drapes and can cause traffic problems inside the room.

INSTALLING THE READY-MADES

If you do a good job on the rough openings, being careful with dimensions and making sure that sills and headers are horizontal and the trimmers are plumb, you will have no trouble placing the windows. Be sure, though, that you check any literature that comes in the package just in case there are special directions that apply to the units you purchased.

Many units are shipped with temporary braces or spacers that guard against distortion. Let these stay in place during installation, but do check first to be sure the units have not been damaged or thrown out of alignment despite the precautions. Cut 8- to 10-inch-wide strips of building paper and tack them in place around the rough opening. It's a good idea to do this in all situations, but it's especially important over board sheathing.

Set the window unit in the rough opening from the outside so the exterior casing or the installation flange overlaps the sheathing (**Figure 17-9**). Since rough openings allow about 1 inch of extra space both vertically and horizontally, you have plenty of leeway to make sure the window is both level and plumb. Best bet, to start, is to rest the unit on the rough sill and then drive a nail partway at one of the top corners. In all cases use rust-resistant nails that are long enough to penetrate framing members by at least 1 inch. Work with casing nails if you are attaching the window through exterior trim (casing) that is part of the unit.

FIG. 17-9   **FIG. 17-10**   **FIG. 17-11**

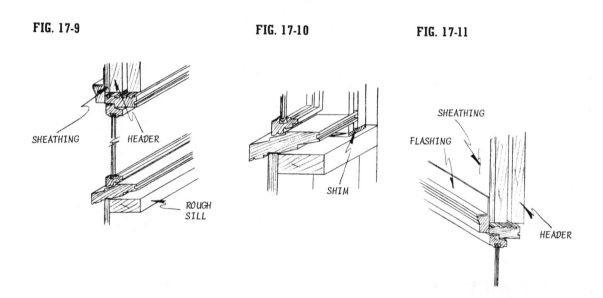

Check with a level to be sure the window is in perfect alignment both vertically and horizontally. If necessary, use shims to make adjustments (**Figure 17-10**). It's a good idea on wide windows to place wedge blocks at intermediate points to guard against any tendency of the sill to sag. When you are sure the window is placed correctly, drive nails at the remaining corners and then check all movable sash for easy operation before completing the nailing. Space nails through casing about 10 to 12 inches apart. The nailing pattern is already set if the unit has a predrilled installation flange. Attach flashing, if any, by nailing through the sheathing (**Figure 17-11**). Don't nail into the head casing or a drip cap.

Caulking is usually done after the siding is up, but it won't hurt to place a bead around the frame of the window right off. Stuff the open spaces around the window on the inside with insulation and then cover the areas with strips of vapor-barrier material (**Figure 17-12**).

FIG. 17-12

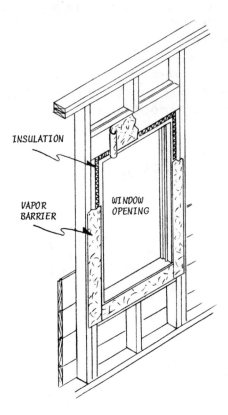

FIXED WINDOWS

Fixed units are windows that do not have movable sash. They are available already framed so you can install them as you would any window. This is often done in combination with units having movable sash that provide the ventilation; the whole becomes a wide picture-window area. Fixed-glass windows, since they are usually on the large side, should be double-glazed (**Figure 17-13**). It's best to work with units of standard size, since having such products made to order can be very expensive. Some types are made with separate pieces of glass that are spaced as much as ½ inch apart. The captive air is specially dehydrated; all edges are hermetically sealed and the perimeter is enclosed and protected with a metal channel.

Double glazing can save on fuel bills and can increase the efficiency of air conditioning. The cost of the double glazing might be offset by the elimination of the need for storm windows.

Many times, glass units are glazed on the site after the rough opening has been framed as shown in **Figure 17-14**. The frames can be preassembled with one set of stops included, as in **Figure 17-15**, so they can be installed in the rough opening as you would any window unit. The frame can be done with heavy material that is rabbeted to receive the glass as shown in **Figure 17-16**. The glass should be set in a nonhardening glazing compound so there will be no contact between the edges of the glass and the frame. Special neoprene spacers are available that clip to the edges of the insulating glass to maintain necessary clearances.

Insulating glass can weigh as much as 7 pounds per square foot, so you want to use substantial material when you make a frame. Work with a dry, high-grade, warp-free wood; using waterproof glue in the joints is not a bad idea.

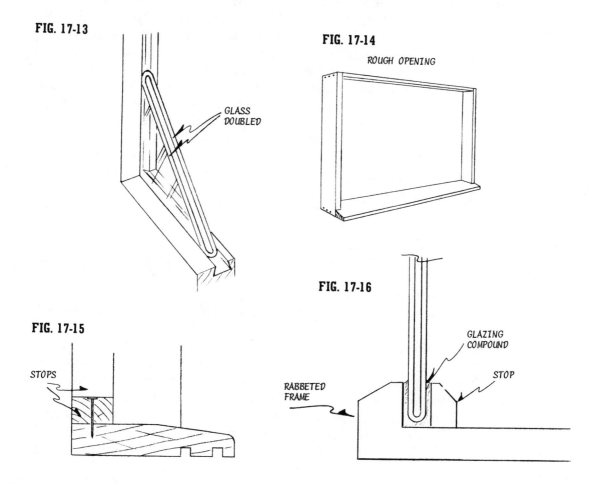

FIG. 17-13

GLASS DOUBLED

FIG. 17-14

ROUGH OPENING

FIG. 17-16

GLAZING COMPOUND

RABBETED FRAME

STOP

FIG. 17-15

STOPS

18

PLUMBING

Most people view plumbing installation with some apprehension. They do not think they can understand what seems to be a complicated assembly of bits and pieces.

It may sound overly optimistic to say the installation is basically simple, but it is. Knowing where and how to place and how to connect ready-made components so water will be directed to points of use or from points of discharge is what plumbing is all about. Waste materials and water that are not consumed must be discharged, so drainage units are introduced. The discharge system and the pipes carrying incoming water should be viewed separately, even though they must work very closely and together to make up the complete plumbing installation.

There can be situations where a third system of pipes is required—for example, when the house has a hot-water heating system. The pipes carrying domestic hot water—what is used in the kitchen or laundry room or tub—and those that carry hot water for house heat are separate installations; water from one should not flow into the other.

The fresh water enters the house under pressure through a single pipe, which may branch off in different directions but eventually arrives at a water-heating device, where it divides and becomes two lines, usually running parallel to each other throughout the house, one line carrying cold water, the other hot (**Figure 18-1**).

At each point of use and at other critical places the systems are interrupted with shutoff valves so you can stop the flow of water completely or at an isolated point.

The assembly of pipes that carry used water and wastes to a sewer system or your own septic tank is the discharge or drainage system. Those pipes that carry off used water are often called *waste* pipes, while the larger ones that carry off discharge from a toilet are called *soil* pipes (**Figure 18-2**). In addition, the installation includes vents and traps. Traps are included so sewer gases and odors will not enter the house. Traps are usually U-shaped connections that always contain water to form a seal between inside air and the air in the drainage lines (**Figure**

FIG. 18-1

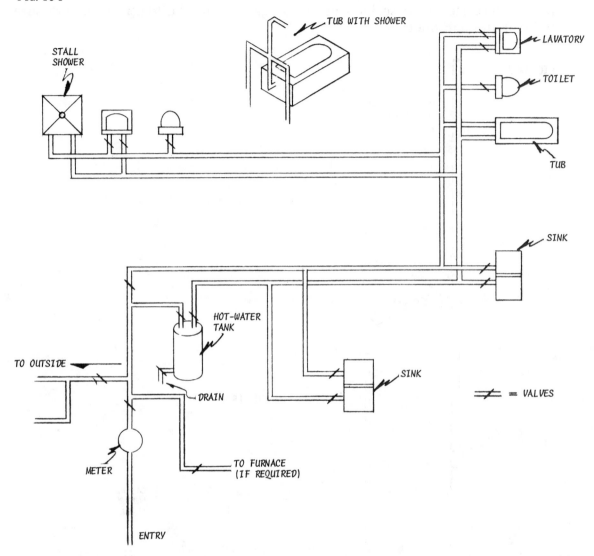

18-3). Toilets have built-in traps, and they work the same way. It's possible for some fixtures to share a trap, but usually each has its own. Whenever water runs down a drain or a toilet is flushed the trap water is replaced, so it is always fresh and can't cause a problem.

Vents are part of a discharge system so gases and odors that can't break through the trap water will be able to escape. Venting also serves to maintain atmospheric pressure, without which the water in a trap could be siphoned off. With the sewer side of the trap lacking a vent, siphoning could easily happen if many drains or several interconnected ones were filled with water at the same time. If the system were sealed — no vents — gases would build up enough pressure to break through the traps.

FIG. 18-2

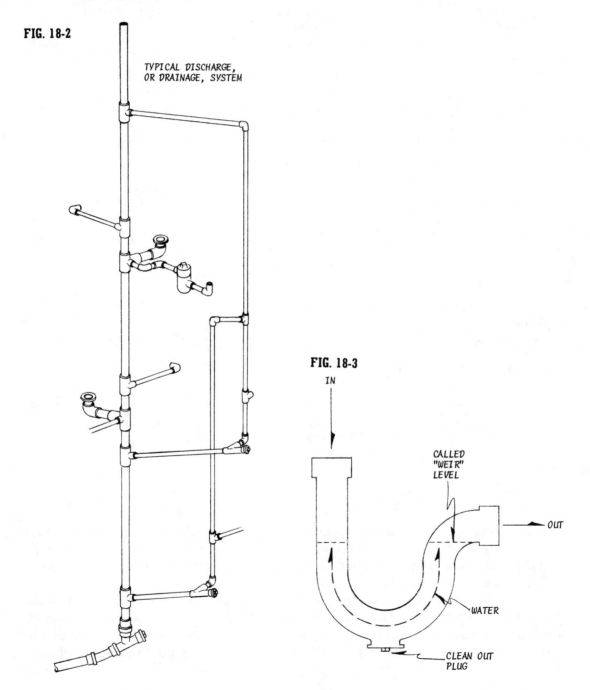

TYPICAL DISCHARGE,
OR DRAINAGE, SYSTEM

FIG. 18-3

IN

CALLED
"WEIR"
LEVEL

OUT

WATER

CLEAN OUT
PLUG

Vents for toilets are called main vents, while those that serve fixtures are secondary. Each vent must pass through the roof and be open to the air, unless the system is organized like the bathroom setup in **Figure 18-4** so that secondary vents will exhaust into a main vent which then passes through the roof.

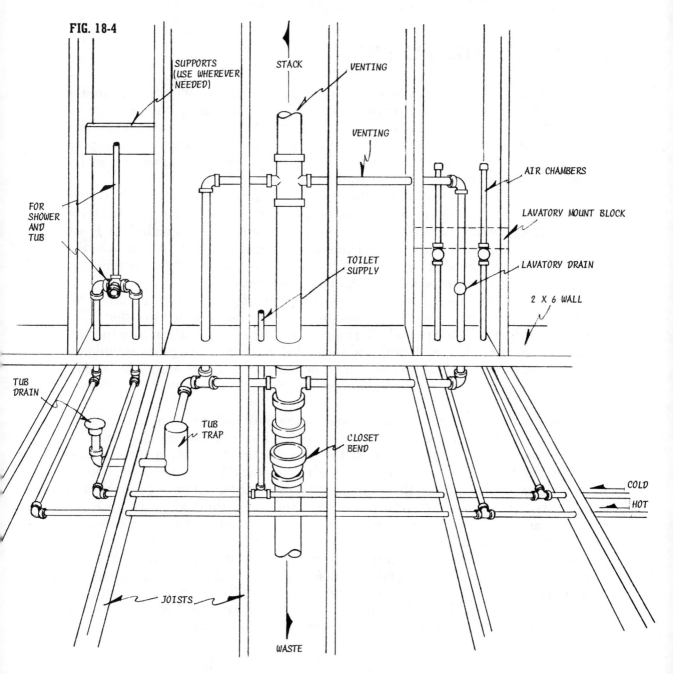

FIG. 18-4

The systems include cleanouts, which are openings with removable plugs so that you can get to the inside of pipes should blockage occur. At least one is provided for each horizontal run.

Since the complete drainage installation includes *drain* lines, *waste* lines, and *vents*, it is usually referred to as the DWV system.

BASICS ON SEPTIC SYSTEMS

Many millions of homes are constructed in areas where a municipal sewer line is not available. There have been a few unacceptable "solutions" to the problem—a cesspool being a common one. But today, health authorities generally agree that the most practical and the safest alternative to a municipal hookup is a septic-tank system. When correctly sized for the house, properly installed and maintained, such a system will provide trouble free service for many years.

As shown in **Figures 18-5** and **18-6** a septic system consists of a subsurface, steel or concrete tank, which receives wastes from the house, and a subsurface drainage field, which is usually made of perforated pipe imbedded in loose gravel. The drainage field disposes of the effluent from the tank. Waste from the house is carried to the tank by means of a sealed sewer line, and connections are made just as they would be to a municipal sewer main.

The tank is actually a settling tank designed to retain solids long enough for maximum decomposition to occur through anaerobic bacterial action (that is, in the absence of air). The matter which is not completely dissolved sinks to the bottom of the tank as a sludge. There, it collects, filling the tank to a point at which it interferes with the tank's action and must be pumped out, preferably by professionals with special trucks designed for the job. It's not really a project for the do-it-yourselfer.

How often? Much depends on the size of the tank and the amount of sewage that enters it. You can conduct a test by removing the tank's access cover and inserting a long stick until it bottoms in the tank (**Figure 18-7**). As a general rule, when the thickness of the scum at the top of the tank and the sludge at the bottom of the tank approaches 2 feet, the tank should be pumped out. I have tried this and other tests, but I prefer to avoid them. Our tank gets pumped every 18 to 24 months—maybe more frequently than necessary. This lets me avoid a nuisance chore, and it's a good precaution against a backed up system.

Tank capacity is noted in gallons. The minimum size tank should not be less than 500 gallons but even for a small house (two bedrooms or less) going to 750 gallons makes sense. Tank sizes go up as the number of bedrooms increases. This, of course, assumes there will be more people using the facilities. When you get to five bedrooms, the minimum tank size should be 1,250 gallons.

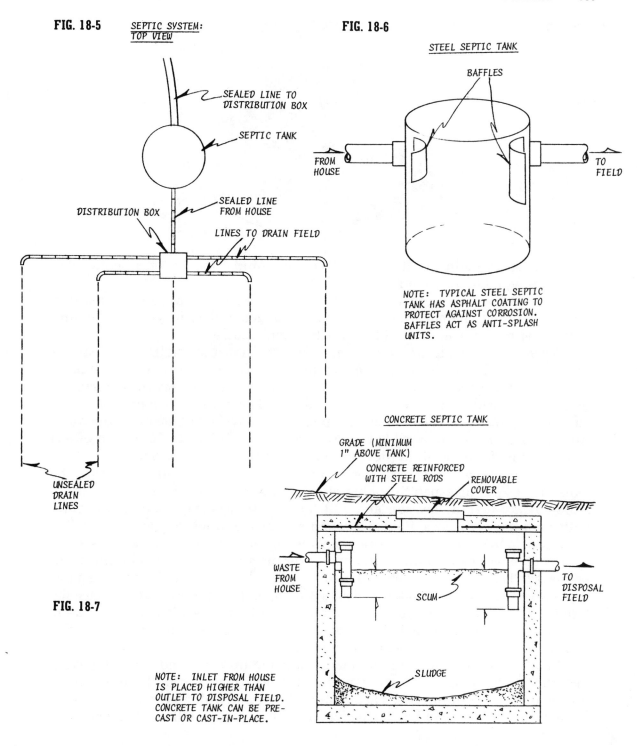

FIG. 18-5   SEPTIC SYSTEM:
TOP VIEW

SEALED LINE TO
DISTRIBUTION BOX

SEPTIC TANK

DISTRIBUTION BOX

SEALED LINE
FROM HOUSE

LINES TO DRAIN FIELD

UNSEALED
DRAIN
LINES

FIG. 18-6

STEEL SEPTIC TANK

BAFFLES

FROM
HOUSE

TO
FIELD

NOTE: TYPICAL STEEL SEPTIC
TANK HAS ASPHALT COATING TO
PROTECT AGAINST CORROSION.
BAFFLES ACT AS ANTI-SPLASH
UNITS.

CONCRETE SEPTIC TANK

GRADE (MINIMUM
1" ABOVE TANK)

CONCRETE REINFORCED
WITH STEEL RODS

REMOVABLE
COVER

WASTE
FROM
HOUSE

TO
DISPOSAL
FIELD

SCUM

FIG. 18-7

SLUDGE

NOTE: INLET FROM HOUSE
IS PLACED HIGHER THAN
OUTLET TO DISPOSAL FIELD.
CONCRETE TANK CAN BE PRE-
CAST OR CAST-IN-PLACE.

FIG. 18-8

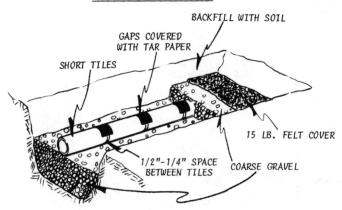

TYPICAL DRAINAGE TRENCH

BACKFILL WITH SOIL

GAPS COVERED
WITH TAR PAPER

SHORT TILES

15 LB. FELT COVER

1/2"-1/4" SPACE
BETWEEN TILES

COARSE GRAVEL

Another guideline is that the minimum liquid capacity of a septic tank should be equal to a 24-hour flow of sewage.

Effluent from the septic tank is only partially treated sewage and so is still in septic condition. Therefore it must not be discharged to the surface of the ground, or directed to a stream or used for irrigation purposes.

A good system calls for the effluent to flow through a sealed line to a distribution box (a small concrete box) so the liquid can be distributed among the number of drainage lines needed. The drain lines can be concrete or clay tiles or various types of pipe made of perforated fiber or plastic. The drain lines *must not be* watertight (**Figure 18-8**). The effluent is supposed to move out of the drain lines into surrounding soil where aerobic bacteria (those that live in the presence of air) complete the decomposition process and render the effluent harmless.

The total linear footage in the drainage/disposal field will depend on the moisture-absorption capability of the soil. A sandy-gravel soil will let water move more quickly than a clay soil and will let a septic system function efficiently with a minimum disposal field. To determine the condition of your soil you can perform a "percolation" test. Use a post hole digger to form about six holes to the depth of the proposed drainage trenches, over the area where the drain lines will be placed. Then fill the holes with water. After about 24 hours, add or remove water from the holes so you have about a 6-inch depth. At this point you must determine how long it takes the water level to drop 1 inch. Based on this percolation test, you can consult charts in local codes to determine the number of square feet of absorption area per bedroom you will need.

Some local codes require that the percolation tests be conducted by health-department technicians. It's also possible that tests may already have been conducted on your building site. Then, you will be told (which is convenient) how big the tank must be and how many feet of drain line you will need. Other code-controlled factors include distance between drain lines, distance of tank from the house, distance between disposal field and streams or wells that might be present, and so on.

Take advantage of any advice or planning aids you can get from city or county planning commissions and local health departments. Don't forget that the engineering or agricultural department of a nearby college or university may have done a soil survey of the area because this can help you select a suitable site for an absorption field.

PLANNING HOUSE PLUMBING

Building codes cover plumbing installations, and there can be enough differences in various localities, even when they almost adjoin, to make some rules and regulations seem like bureaucratic foolery, and some of them might well be. The acceptance or rejection of plastic pipe is a case in point. It has already made an impact on DWV systems and it is definitely the coming thing for general rough-plumbing use. The material is light, as durable as it has to be for the purpose, is easy to cut, and requires only daubs of cement for connections. It's an ideal amateur plumber's product, which, justifiably or not, may be a main reason for its being opposed.

Anyway, building codes should be checked out before you make plans and buy materials. And let's not forget that codes can be very educational, supplying data that will help you do the job correctly. In addition to types of materials, they cover such things as sizes of pipes for various applications, the distance a fixture should be from a stack or a vent, the design of the venting system, and so on.

Generally, pipe sizes run along these lines: Water enters the service through a ¾-inch or 1-inch pipe. The main lines that deliver hot and cold water to points of use are ¾-inch; ½-inch branch lines take off from the main lines to fixtures and sometimes to appliances. Toilets and lavatories are often serviced with ⅜-inch lines. There can be variations, with much depending on what you are going to install. Best bet is to know in advance the tubs, showers, toilets, and so on that you will be using. All such products come with installation instructions that will suggest sizes of connections and tell where the rough plumbing should pierce the wall and, when necessary, the floor. The latter facts are critical, since you must know where to open up for toilet, tub, and shower drains and where to come through the wall with hot- and cold-water lines.

The main drains and vents in the DWV system that services toilets or groups of fixtures are 3-inch or 4-inch. Secondary vents that pass through the roof can be smaller, but check codes. Toilets drain through 4-inch lines, showers through 2-inch lines, tubs, sinks, washers and the like through 1½-inch lines. Lavatory drains can also be 1½-inch, but often 1¼-inch pipe is used. Sewer lines are 4-inch. A good procedure is to plan the installation on paper, using the pipe sizes specified above, but then have your plan checked by the local plumbing inspector before proceeding.

There are two important points to remember. Fresh water enters the building under pressure, which is, or should be, enough to supply requirements at points of use while absorbing pressure losses caused by flow friction as water passes through pipes and fittings and by any vertical travel that might be in the system. About 50 psi (pounds per square inch) is about average pressure for water supplied through a municipal system. This generally works out okay with the pipe sizes specified above. If pressure is lower (as from a well) you can compensate somewhat by using larger pipes, and, of course, you can use a sealed, pressurized storage tank. Remember too that copper tubing causes less pressure loss from friction than galvanized pipe. The difference often permits the use of smaller sizes of copper tubing under normal pressure conditions, and can be helpful in a low-water-pressure situation.

The other important point has to do with the DWV system. Unlike the incoming water supply, which operates under pressure, the drainage installation uses gravity flow to exhaust its contents. That's one reason why larger pipes are used and why the system should be designed in a straightforward manner with runs as direct as possible, horizontal pipes sloped, and angular connections made so the gravity flow can function efficiently.

GALVANIZED PIPE

This is steel pipe, available in various lengths and diameters, that has a rust-resistant coating on inside and outside surfaces. Connections are usually made by cutting threads in the pipe, and since this is done after the coating is applied any exposed threads will show rust. Because it has a long life expectancy even when buried, it has been and still is a popular rough-plumbing product. The principal objection is its roughness, which contributes to water-flow friction and the collection of mineral deposits and sediment that, over the years, reduce the inside diameter of the pipe.

The pipe is available in many lengths already threaded, and there are numerous fittings and connectors, a few of which are shown in **Figure 18-9**. Consequently, a complete water-supply system can be installed with a minimum of on-the-job cutting and threading.

When cutting is necessary it should be done with a pipe cutter (**Figure 18-10**). In addition to making the job easier, the cutter leaves a much cleaner and squarer end than you would get with a saw.

FIG. 18-9

FIG. 18-10

FIG. 18-11

FIG. 18-12

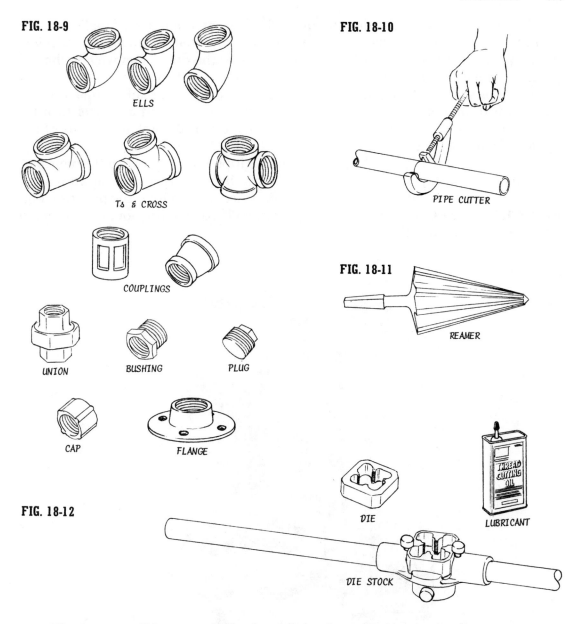

ELLS

TS & CROSS

COUPLINGS

UNION   BUSHING   PLUG

CAP   FLANGE

PIPE CUTTER

REAMER

DIE   LUBRICANT

DIE STOCK

The cutter will leave a slight shoulder or burr on the inside edge of the pipe, and this should be cleaned out, preferably with a cone-shaped reamer. These are made, like the one shown in **Figure 18-11**, to fit an ordinary woodworking brace, or they may be fitted with their own handles.

After it is cut and reamed the pipe is threaded with a stock and dies (**Figure 18-12**). The stock is a holder with a handle, sometimes ratcheted, for turning that can be used with various dies for pipe of different diameters. Stocks usually include a guide which helps you get the threading job started squarely. Cutting is done by turning the stock clockwise, which causes the die to remove a V-shaped piece of metal. To remove the waste so the die won't clog and cause rough threads, the stock is then turned counterclockwise. The threading operation is done by turning the stock to the left (counterclockwise) about a quarter-turn for each half-turn you make to the right (clockwise). Use lubricant at the start of the job and frequently while cutting.

To do the job correctly, the pipe must be gripped firmly to keep it from rotating. You can accomplish this with an ordinary bench vise if you equip it with special pipe jaws, or you can work with a pipe vise designed for the purpose (**Figure 18-13**). All the tools mentioned can be rented if you choose not to add them to your permanent collection.

FIG. 18-13

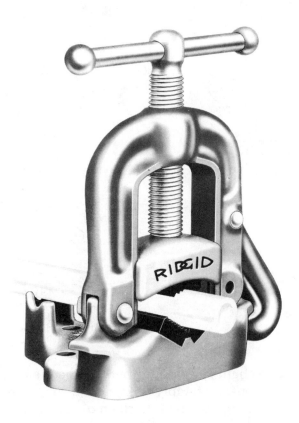

All pipe threads must be coated with a pipe-joint compound before connections are made. The compound assures a watertight joint and makes it easier to disassemble the union should it be necessary. A new product on the market which may be used in place of the compound is a special Teflon tape.

Know the difference between the actual and the visible length of the pipe you must cut (**Figure 18-14**). The length of pipe that enters the fittings at both ends must be added to the distance between the face of the fittings.

The pipe wrench is made for working pipe (**Figure 18-15**). The teeth are sharp ridges that run across the jaws and are designed to dig into the pipe when pres-

FIG. 18-14

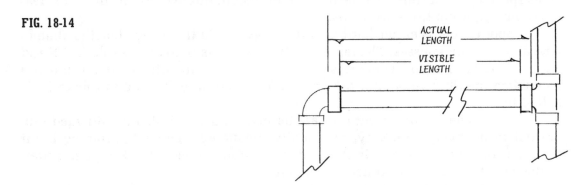

FIG. 18-15

sure is applied. The jaws should be adjusted to make a light contact with the pipe. A pipe wrench is not the best tool to use on polished pipe or on fittings made of soft materials.

COPPER PIPE AND TUBING

Most amateur plumbers find copper tubing easier to work with than other materials. It does cost more and is softer than, for example, galvanized pipe; in some areas special precautions must be taken to protect it. But overall, the desirable features outweigh the little problems.

Copper tubing is either hard or soft and available in various lengths, diameters, and wall thicknesses. The tempered type (hard) is rigid and is sold in 10- and 20-foot lengths. The soft version is flexible and comes in coils of 30, 60, and 100 feet. Often the rigid version of the product is called pipe and the soft material is called tubing.

Wall thicknesses of copper pipe are termed M, L, and K, being thin, medium, or thick respectively. The M type is usually adequate for general plumbing, but if the pipe contacts the soil or is exposed to possible damage, the K type is better. Codes may tell you what to use and where.

Most types of copper fittings and connectors are designed for soldering and are called sweat fittings. The variety in sizes and shapes is extensive; some are shown in **Figure 18-16**. You can work with ready-made parts to suit any design.

FIG. 18-16

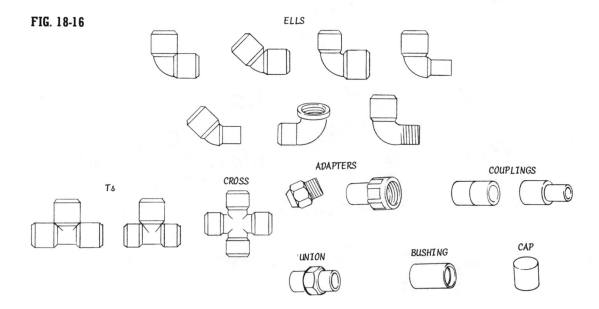

DWV copper pipe is specially made for drainage systems. It weighs and costs less than similar diameters of standard copper pipe, but the fittings, some of which are shown in **Figure 18-17**, are very expensive, so the cost factor imposes limitations even though the simplicity of the soldered joints is appealing.

FIG. 18-17

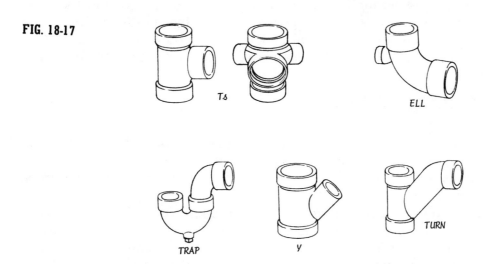

T&

ELL

TRAP

Y

TURN

Soft-temper copper pipe (or tubing) is so flexible that it's possible to do an entire run so that fittings are required only at the ends. Long, gentle curves are easy to form without special equipment, but tight bends, especially in short pieces, should be done with a special spring-type bender that will keep the material from kinking (**Figure 18-18**). These come in various sizes and are used by passing the tubing through them so the tool covers the bend area. You can do the actual bending over your knee, gently, or over a suitable curved surface. The bender is easy to remove if you rotate it as you pull it off.

Both types of copper pipe can be cut with a hacksaw or with a tube cutter, which is a smaller version of the pipe cutter (**Figure 18-19**). Gripping the pipe in a

FIG. 18-18   **FIG. 18-19**

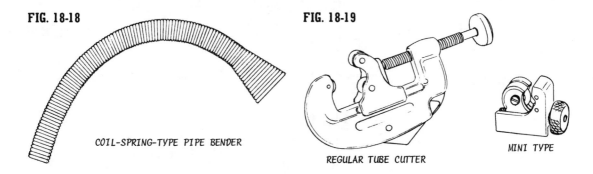

COIL-SPRING-TYPE PIPE BENDER

REGULAR TUBE CUTTER

MINI TYPE

vise is not good practice, since the material is too soft to withstand much pressure. A better way is to rest the part on a flat surface and grip it firmly with your hand, or use a conventional miter box. The handle of the tube cutter, which is turned to bring the cutting wheel against the pipe, also supplies cutting pressure. This should be applied gradually, since excessive pressure can distort the pipe. Getting a smooth, clean cut is more important than doing it quickly. Most tube cutters have a triangular piece of metal attached that you use to remove burrs when the cut is complete. Hacksaw cuts can be cleaned with a smooth round file.

SWEATING COPPER PIPES

All contact surfaces — the exterior of the pipe and the interior of the fitting — must be polished either with steel wool or a fine emery paper. Steel wool works fine, since it readily conforms to contours. Check both the pipe and the fitting to be sure each is round and free of dents.

Soldering is done with a propane torch, soldering flux, and solid-core wire solder. The important thing to remember is that the pipe and fitting are heated enough to melt the solder; the solder is not melted by the flame of the torch.

Use a small stiff-bristle brush to coat all contact points with flux. Be a bit generous, since too much flux does no harm while too little can cause a poor joint. A typical soldering procedure is shown in **Figure 18-20**. Apply the flame to the

FIG. 18-20

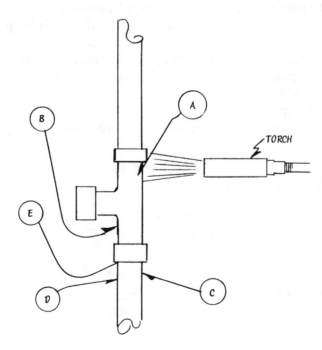

heaviest part, usually the fitting (A), until it is hot enough for the solder to melt when touched at about point B. Then shift the flame to the pipe (C) but keep it about an inch away from the fitting. Heat the pipe until the solder readily melts when placed at about point D. When the parts are hot enough, press the tip of the solder against the joint (E), and capillary action will draw the solder to fill between pipe and fitting.

Hold the solder at that point until you see a full circle form around the joint. It's okay to keep applying heat for a few seconds longer, but don't overdo it or the solder may flow out. Don't touch the joint for a minute or so after the solder loses its brightness. Then, if you must handle the piece, cool it with water first. When installing runs, it's best to do all cutting, polishing, and fitting of joints first; then move from point to point to do the soldering.

Both types of copper pipe can be handled with sweat fittings, but the flexible variety can also be done with flared fittings. This is convenient, involves no soldering, and provides self-sealing joints that can be disassembled at any time with a pair of wrenches. The objection is cost; flared fittings cost more than other types.

The flare is made after cutting and polishing the end of the tube (**Figure 18-21**). This is a critical procedure, since burrs or ridges or ends that are not square will prevent a watertight seal. Don't use a reamer to the point where you bevel the inside edge of the tube. All you must do is remove burrs.

FIG. 18-21   **FIG. 18-22**

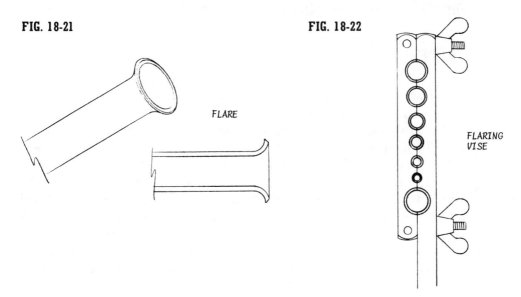

FLARE

FLARING VISE

After it has been slipped through the flare nut, the tube is gripped in a die or flaring vise (**Figure 18-22**). The flaring tool, which has a threaded ram with a polished, cone-shaped tip, is placed on the die as shown in **Figure 18-23**. As you turn

FIG. 18-23   FIG. 18-24

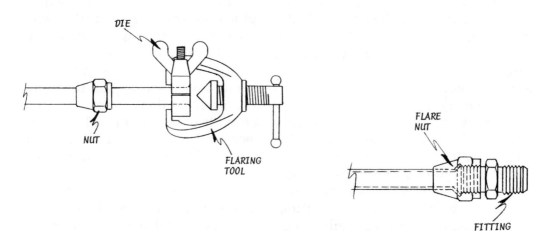

the ram, the cone bears against the tube and forms the flare. Use a touch of wax or oil on the cone and do the turning slowly. Check the flare carefully, and if it is flawed in any way cut it off and do the job over.

Bring the nut up against the flare and put the fitting in place (**Figure 18-24**). Tighten by using two open-end wrenches. In the case shown, the fitting serves as a coupling to join two lengths of tubing.

CAST-IRON PIPE

Cast-iron pipe is heavy and a little awkward to handle, but it is so strong and durable that it is in general use for drain, sewer, and vent lines. Of the two weights available — "service" and "extra heavy" — the lighter is quite adequate for residential use and should be used wherever local codes permit. The common, standard length is 5 feet, but if you search you might find some 10-foot pieces.

Cast-iron soil pipe is hub-shaped at one end and is ridged or smooth at the "spigot" end (**Figure 18-25**). The hub is shaped and sized to receive the spigot end of the length that follows. Pipes are always joined so the hub is at the up side of a flow. Otherwise, solid wastes could lodge in the joint. Double-hubbed pipe is made for use in short lengths; two short lengths, each with the necessary hub, can be made from each pipe. Only one usable length could be cut from single-hubbed pipe.

Like the other pipe materials, there are numerous types and sizes of fittings designed for cast-iron installations (**Figure 18-26**). You must be careful when choosing fittings for the drainage system. While some types of pipe may be used

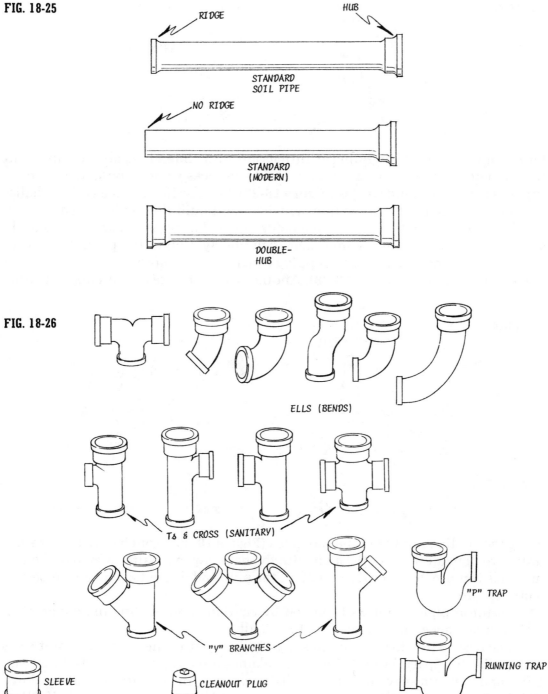

FIG. 18-25

RIDGE

HUB

STANDARD
SOIL PIPE

NO RIDGE

STANDARD
(MODERN)

DOUBLE-
HUB

FIG. 18-26

ELLS (BENDS)

Ts & CROSS (SANITARY)

"Y" BRANCHES

"P" TRAP

RUNNING TRAP

SLEEVE

CLEANOUT PLUG

FIG. 18-27

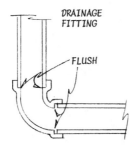

DRAINAGE
FITTING

FLUSH

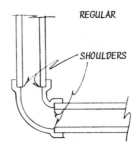

REGULAR

SHOULDERS

for either water supply or drainage, fittings are not so interchangable. Drainage fittings are designed to provide smooth inside surfaces so no shoulders can interrupt the passage of solid wastes (**Figure 18-27**). Regular fittings in a DWV installation must be limited to the vent areas where no liquid flow is involved.

To cut cast-iron pipe, rest it on a block of wood near the cut area. Use a hacksaw to form a slot about 1/16 inch deep completely around the pipe. Then work with a hammer or a light sledge, tapping constantly as you rotate the pipe until it cracks along the slot (**Figure 18-28**). Another way is to use a cold chisel, tapping

FIG. 18-28

HOW TO CUT CAST-IRON PIPE

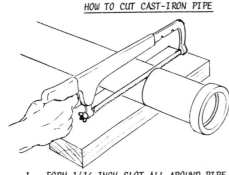

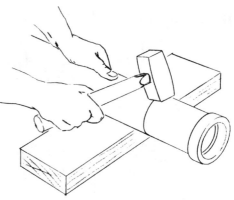

1. FORM 1/16-INCH SLOT ALL AROUND PIPE.   2. CRACK PIPE ALONG SLOT WITH HAMMER OR

along the cut line until the pipe separates (**Figure 18-29**). Rough edges will result, unless you cut all the way through with the hacksaw, no matter which of the two methods you use, but they don't matter; the pipe end doesn't even have to be perfectly square.

Cast-iron pipes are usually joined by caulking and leading the connection. This is a nuisance, but it really isn't difficult.

The first step is to be sure the spigot is centered in the hub and that it seats firmly. Fill the space in the joint with oakum—a ready-to-use material that you buy—and then use an offset chisel, called a yarning iron, to force the oakum down tightly. Repeat the procedure, if necessary, until the space is about half filled. Don't be afraid to really compress the oakum.

FIG. 18-29

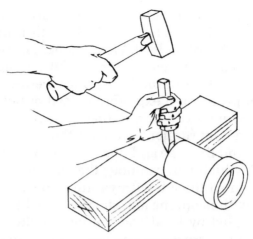

2A.   USE A COLD CHISEL.

The remainder of the joint is filled with molten lead, which, when cool, is tamped down with caulking irons. An inside iron is used first, as shown in **Figure 18-30** and then an outside iron, whose bevel is opposite the one on the inside iron, is used to tamp the outside edge of the lead.

FIG. 18-30

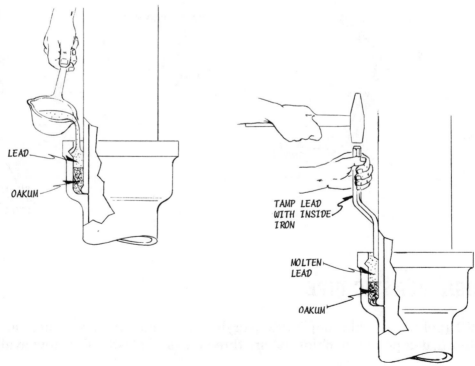

LEAD

OAKUM

TAMP LEAD
WITH INSIDE
IRON

MOLTEN
LEAD

OAKUM

The big job is melting the lead, but a plumber's furnace, which works with propane and includes a cast-iron pot and a ladle with a long handle, can be rented. Don't be careless with molten lead. Wear leather gloves with long sleeves, and keep the ladle well away from *you*. Pour carefully to avoid splatter. Don't pour lead if there is any moisture in the joint, since wetness can turn to steam, which can create enough pressure to explode the lead out of the joint.

Connections in a horizontal run or one that is pitched are done by using a special implement called a joint runner to contain the lead (**Figure 18-31**). When the runner is clamped around the pipe and pushed tightly against the hub, a small opening remains through which you can pour the lead. Remove the runner after the lead has cooled and finish the job with yarning irons.

No-hub joints, done with neoprene and stainless-steel sleeves, are something new, and if they are accepted by local codes, they provide a much simpler installation method. After the sleeves are in place, the clamps are tightened to secure the connection (**Figure 18-32**). It's as simple as that, and the joint can easily be opened for repairs or additions.

FIG. 18-31   **FIG. 18-32**

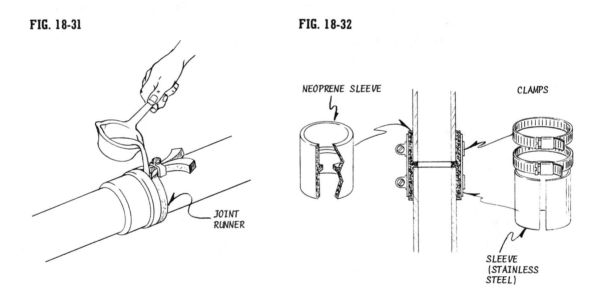

RIGID PLASTIC PIPE

PVC (polyvinyl chloride), ABS (acrylo-nitrile butadiene-styrene), and CPVC (chlorinated polyvinyl chloride) are three types of plastic pipe now available for use in home plumbing systems. All three are usable for cold-water lines, but only the CPVC is rated for use in a hot-water system. As mentioned before, plastic piping is a controversial subject, and while its use is pretty generally permitted in outdoor sprinkler systems, many codes don't permit it within walls, especially when it will carry hot water. Even when it is permitted, connections at the hot-water heater must get special attention. For example, the heater's relief valve must coordinate with the water-pressure and temperature rating of the pipe, and connections at the heater must be through a long brass nipple to keep the plastic away from the heat in the wall of the tank. At any rate, a quick check with authorities will let you know if you can use plastic and, if you can, how it should be installed.

One area where plastic is making an impact is in DWV setups (**Figure 18-33**). You can look through quite a few mail-order catalogs and find complete systems, if the materials are not available locally.

FIG. 18-33

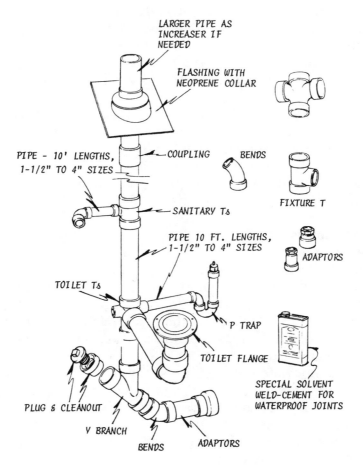

Rigid plastic pipe is easy to cut with any saw that has fine teeth—backsaw, hacksaw, crosscut saw, whatever—but you can also use a special tool that works much like those made for cutting metal pipe (**Figure 18-34**). Be sure that cuts are square and that you remove burrs from inside and outside edges. This can be done with a sharp knife or fine sandpaper.

FIG. 18-34

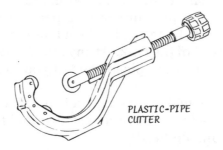

PLASTIC-PIPE
CUTTER

Rigid-plastic-pipe connections are usually done with a special solvent which literally welds the parts together. The solvent works quickly, so you must be sure the parts fit and are aligned correctly. Best bet is to do a dry run first—that is, without solvent—and mark the pieces for accurate repositioning when you get to using the cement. It isn't wise to make a major adjustment after the solvent starts to set.

Read the instructions that come with the solvent very carefully, since the materials and the method of application may vary depending on the type of plastic you are working with. Some are done with a single solvent; others, like CPVC, require using a liquid cleaner first. In all situations, contact surfaces should be dry and free of foreign matter, like oil or grease, that will prevent adhesion. Some workers use fine sandpaper to remove the shine from the insert piece, feeling that it helps the cement to adhere more strongly.

Most instructions say to apply a thin coat of solvent inside the fitting and a heavier coat on the pipe. Use a brush that is wide enough to cover the joint area in a single swipe; you want to apply the cement quickly.

Put the pieces together immediately, using a short rotating action to join the alignment marks you made during the dry run. A good joint will show a bead of cement completely around the pipe at the fitting. Hold the parts tightly together for about 30 seconds, and then wait for three or four minutes before doing more work in that same area. The instructions will tell you how long you should wait before conducting any watertightness tests.

Rigid plastic pipe and metal pipe can be combined in a plumbing system, but special transition fittings or connectors must be used. For example, a plastic-to-steel joint requires an adapter that is threaded on one end for the metal pipe but is shaped at the other end so it can be forced into the plastic pipe, where it is secured with a stainless-steel clamp.

THE PLUMBER AS A CARPENTER

The installation of a plumbing system requires some work with a saw and a drill. Planning should be done to minimize cutting of framing members, but some is inevitable. Try to do hole-forming as much as possible by using drill bits and hole saws. This will result in much neater work and in less weakening of joists and studs. Keyhole saws and saber saws may be used of course, but carefully so the job *won't* look like the one shown in **Figure 18-35**.

FIG. 18-35

Such work is disgraceful, even though it will be hidden. Also, parts are weakened unnecessarily and require additional bits and pieces to make up for the lost strength.

Smooth, clean holes can be cut almost anywhere along a joist as long as they are centered between top and bottom edges, have at least a 2-inch edge distance, and have diameters that are not more than 25 percent of the joist width.

Notches are best cut along the end quarter-sections of joists and shouldn't be deeper than one-quarter of the joist's width. When done elsewhere, the joist should be reinforced with a 2×4 at least 4 feet long (**Figure 18-36**). Another way to compensate for the strength lost by notching is to make braces from joist material and use them as shown in **Figure 18-37**.

When pipes can run under or along joists, in a crawl space for example, notching will not be necessary, but the pipes should be supported at points not

FIG. 18-36

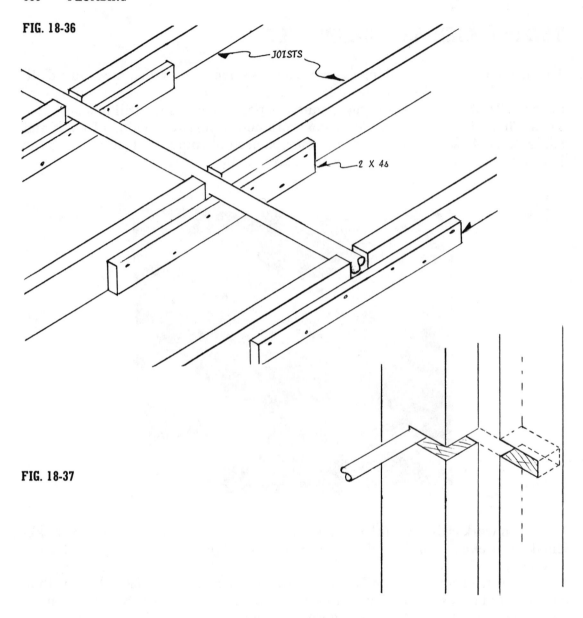

JOISTS

2 X 4s

FIG. 18-37

more than 10 feet apart with pipe clamps (**Figure 18-38**). Use galvanized clamps and nails with galvanized pipe, and copper clamps and nails with copper pipe. Any type of clamp can be used with plastic pipe, but it must fit loosely so that the plastic can expand. Also, plastic pipe should be supported at more frequent intervals—every 4 to 6 feet.

FIG. 18-38

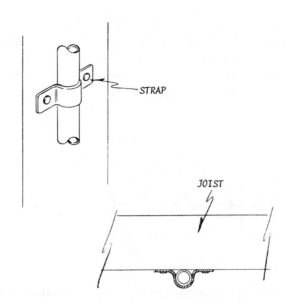

STRAP

JOIST

The depth of notches cut in studs for pipes that run horizontally should be kept to a minimum; no more or less than the pipe needs. It's a good idea to add metal covers over the notches as shown in **Figure 18-39** to protect the pipe during future nailing and is usually required by codes.

FIG. 18-39

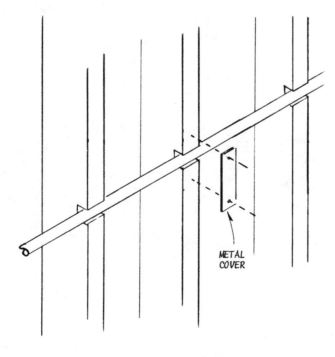

METAL
COVER

Stud-notching in a bearing wall is more critical. Codes usually say that it's okay to notch at the top half of the stud to about half its depth as long as two un-notched studs remain. It's difficult to judge the loss of strength when the rule can't be followed, but when a question remains, the solution is to include additional studs placed as shown in **Figure 18-40**.

FIG. 18-40

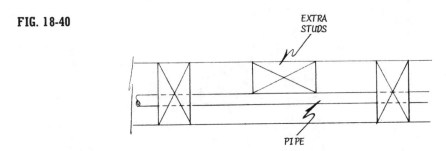

Closet bends, which are the connections that run between the floor flange of the toilet and the soil stack, can create problems because they are bulky and the assembly of connections in the area can be quite heavy (**Figure 18-41**). Whenever possible, position it between joists and use blocking and shims to support it and maintain its alignment (**Figure 18-42**). If the bend can be fitted by notching a joist, make the cut as small as possible and add compensatory braces on each side.

FIG. 18-41   **FIG. 18-42**

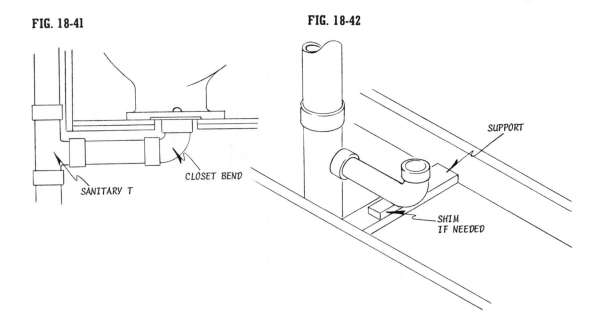

If you must cut away a portion of a joist, then doubled headers and, possibly, trimmers should be added for any opening through a floor frame (**Figure 18-43**).

FIG. 18-43

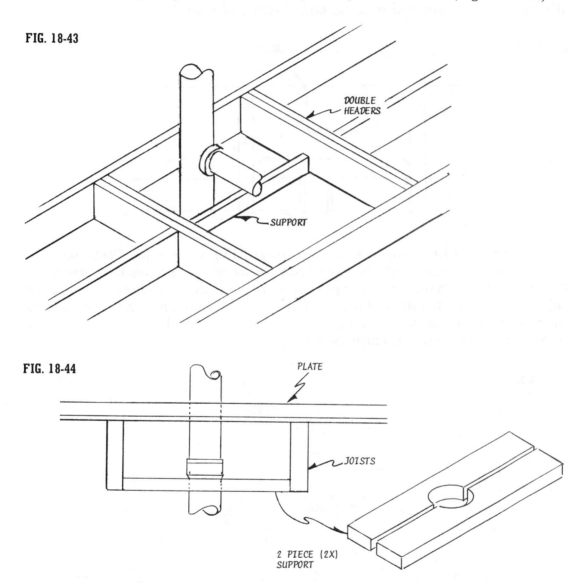

FIG. 18-44

Stacks, especially when they are done with cast-iron pipe, represent a considerable amount of weight which will rest entirely on the house drain unless intermediate support is provided. It's a good idea to transfer some of this weight to each floor frame the stack passes through by installing additional frame pieces along the lines shown in **Figure 18-44**.

Horizontal runs of drain line in exposed situations (crawl spaces and basements) can be supported with special heavy-duty pipe straps (**Figure 18-45**). Usually the straps are secured with lag bolts instead of nails.

FIG. 18-45

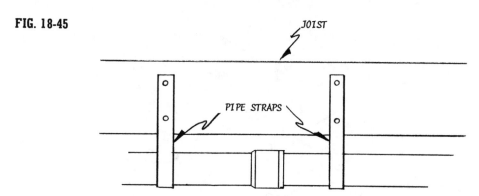

All drainage lines must be pitched toward the sewer pipe. Tests have indicated that a ¼-inch pitch per foot of run allows waste to move most freely. At the other extreme, a pitch of 45 degrees or more poses no problems at all since the solids slide easily. If a straight run can't be done within the maximum pitch of ½ inch per foot, do the major part of the run using the ideal pitch and slope the remainder at 45 degrees (**Figure 18-46**).

FIG. 18-46

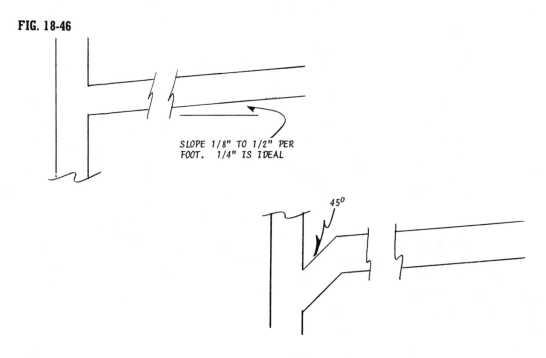

AIR CHAMBERS

Water hammer is the common name for the noise that can result when a rapid flow of water is suddenly halted. It happens frequently when a faucet is turned off by hand or by an appliance such as an automatic washing machine. The solution is to install air chambers in the line, which act as cushions for the water and also generally take strain off the entire system (**Figure 18-47**).

FIG. 18-47

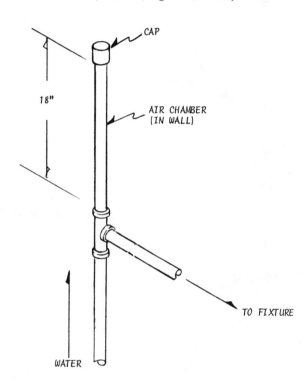

CAP

18"

AIR CHAMBER
(IN WALL)

TO FIXTURE

WATER

It's common today for plumbing installations to include an air chamber in every pipe that leads to a fixture. Make the air chamber at least 18 inches long; 24 inches won't hurt. Use pipe which is as large as that in the supply line. An added safety factor would be to use a larger pipe for the air chamber. This you can do by ending the supply run with a connector that takes different-size pipe at each end.

VENTING

In order for the drainage system to operate correctly it must, as we have said, be vented to admit air and to pass off sewer gases and smells. There are many ways to design vent systems and many pros and cons for each. The principal guidance factor is to do it as simply and economically as possible while still staying within local codes.

The sketches that follow should not be accepted as architectural designs, but they can be the basis for a discussion with the local plumbing inspector. Then you can alter, if necessary, to make him happy and to suit your needs specifically.

Figure 18-48 shows a reasonable DWV installation for a two-story house. Note that while the solid lines show how it can be done, the dotted lines indicate additional venting that local codes may call for.

FIG. 18-48

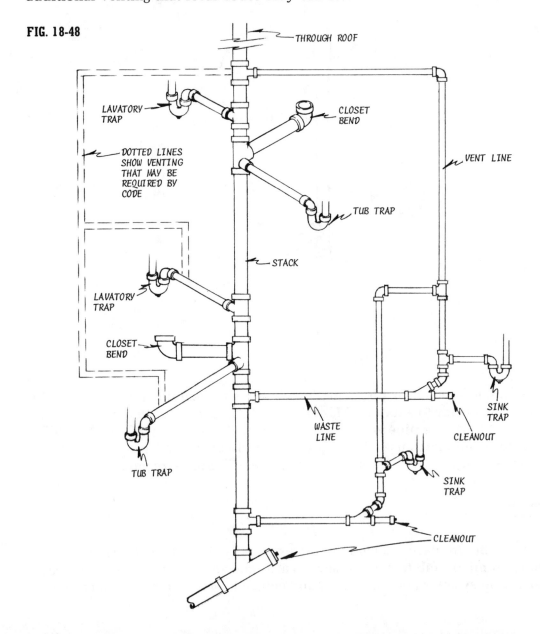

In the bathroom setup in **Figure 18-49**; both the tub and the lavatory will be *back vented* to the main stack. Note that the drain lines are sloped toward the stack so water can flow out more readily. The horizontal vent slopes upward toward the stack.

FIG. 18-49

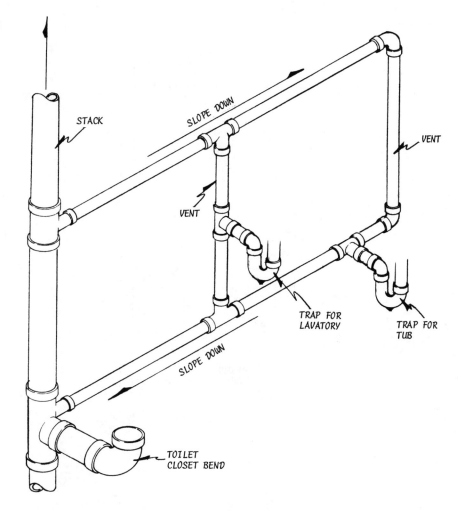

STACK

SLOPE DOWN

VENT

VENT

TRAP FOR
LAVATORY

TRAP FOR
TUB

SLOPE DOWN

TOILET
CLOSET BEND

For a half-bath you can do something along the lines shown in **Figure 18-50**. This is called *wet venting* and is not permitted by all codes, even though it makes for a rather simple installation.

A *dry venting* system can be substituted by installing a through-the-roof vent for each fixture or by venting the lavatory through the main stack (**Figure 18-51**). If the alternate vent is permitted, codes will usually tell you how high above the floor or the fixtures it must be.

FIG. 18-50

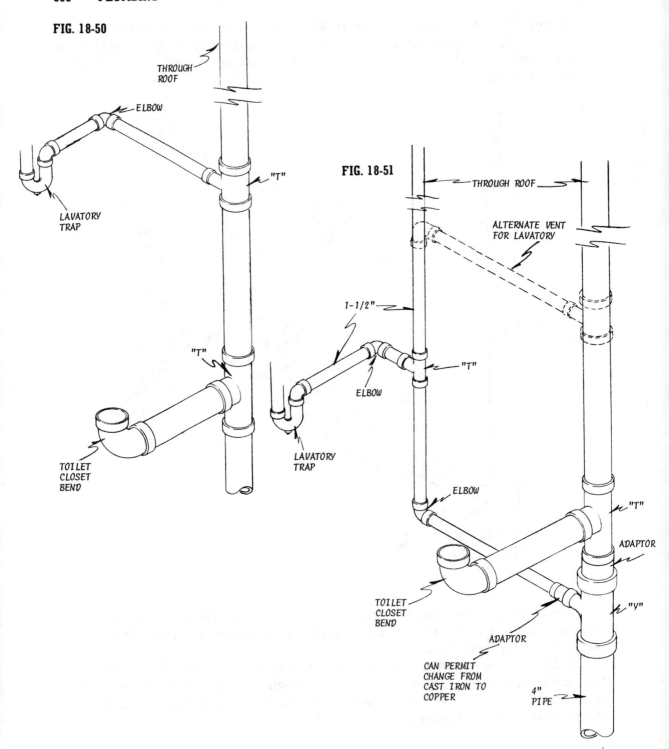

THROUGH
ROOF

ELBOW

"T"

LAVATORY
TRAP

FIG. 18-51

THROUGH ROOF

ALTERNATE VENT
FOR LAVATORY

1-1/2"

"T"

ELBOW

LAVATORY
TRAP

"T"

TOILET
CLOSET
BEND

ELBOW

ELBOW

"T"

ADAPTOR

TOILET
CLOSET
BEND

"Y"

ADAPTOR

CAN PERMIT
CHANGE FROM
CAST IRON TO
COPPER

4"
PIPE

FIG. 18-52

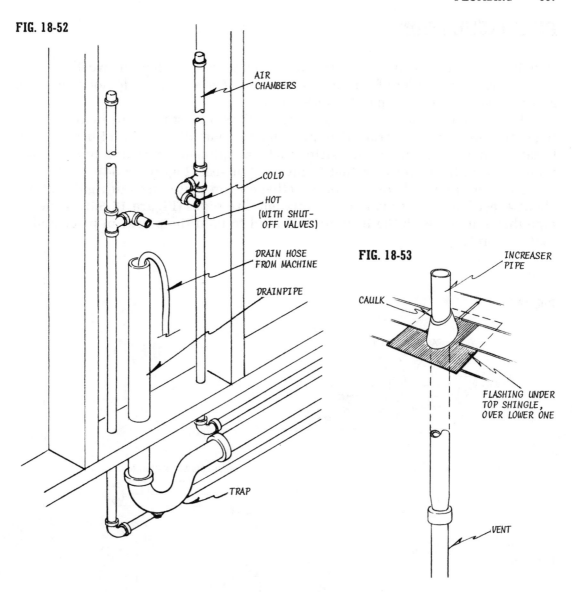

AIR CHAMBERS

COLD

HOT (WITH SHUT-OFF VALVES)

DRAIN HOSE FROM MACHINE

DRAINPIPE

TRAP

FIG. 18-53

INCREASER PIPE

CAULK

FLASHING UNDER TOP SHINGLE, OVER LOWER ONE

VENT

A typical installation for a washing machine is shown in **Figure 18-52**. The drain runs off to a main stack, which might also be the vent for the system, or the drain might be interrupted on the sewer side of the trap with a through-the-roof vent of its own.

There is always the possibility that secondary vents, being smaller than main vents, will become clogged in snowy or wet and cold weather. For this reason, in some areas the size of the vent pipe is increased where it passes through the roof. (**Figure 18-53**).

PIPE INSULATION

Pipe insulation can keep cold-water pipes from sweating when the weather is hot and will reduce heat loss from hot-water pipes. The materials do not cost much and they help minimize maintenance costs.

The job can be done in various ways, but the simplest methods involve working with ready-made materials that are available generally. One is a fiberglass material that comes in 3-inch-wide strips that you spiral-wrap around the pipe and then cover with an overwrap that forms a vapor seal (**Figure 18-54**).

Another is a vinyl-covered polyurethane foam that comes in rigid sections about 4 feet long and is made to fit different size pipes (**Figure 18-55**). A special tape that you buy with the insulation is used to hold the pieces in place and to seal the joints.

FIG. 18-54

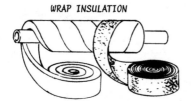

WRAP INSULATION

FIG. 18-55

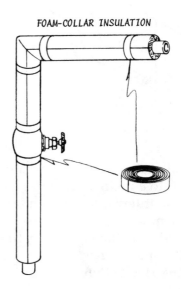

FOAM-COLLAR INSULATION

19

WIRING

Water and electricity shouldn't mix, yet the house systems for each have some things in common. Both make an entrance through a main line, which then branches off to points of use, with interrupters included in the layout so it can be shut down entirely or at isolated points.

The shutoffs in the water line are operated manually. Those in the electrical system can be operated manually, but to provide necessary safety factors they also function automatically, when correctly installed, to prevent dangerous overloads on any branch line or on the system as a whole. Valves do the job in a water line; fuses or circuit breakers do the job in an electrical system.

You might view the incoming current as having to pass through a main gate before it can service smaller gates, which are the controls for branch lines—the individual circuits (**Figure 19-1**).

The larger the water pipe, the greater the flow of water. The larger the wire, the greater the flow of current. Water flow is stated as so many gallons per minute, with pressure as pounds per square inch. In essence, the flow of electric current is stated as amperage, while the pressure is voltage (**Figure 19-2**). Wattage, which is a measurement of the total energy flowing in a circuit at a given time, can't be told by either amperes or volts alone, but by a combination of the two—wattage = amperage × voltage.

Like plumbing, the electrical system in a house is not really difficult to plan or do, but doing it less than perfectly, or with inadequate materials, or without regard for your safety, is folly. If you forget to shut down the water system before working on a section of plumbing, you and the surrounding areas might get wet. If you forget to shut down the electrical system when doing *any* kind of wiring, you can be killed. So, the inflexible rule is—*No current in the lines while you are working!* Remember that if connections have been made at the entry panel, the system to that point will still be *hot* even though you disengage the main house fuse or switch. Black wires are always connected to black wires; white wires to white wires. White wires are continuous—never broken for a connection except in particular circuits, and then the white wire is specially marked to indicate its strange use.

Electrical codes are very particular, and they should be, since faulty installa-

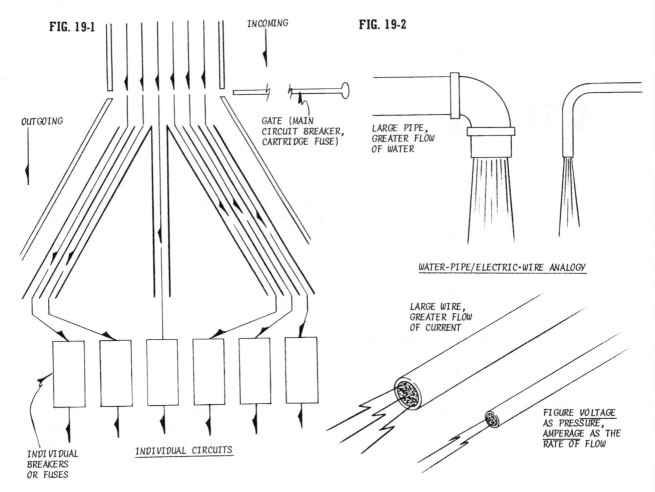

FIG. 19-1

INCOMING

OUTGOING

GATE (MAIN CIRCUIT BREAKER, CARTRIDGE FUSE)

INDIVIDUAL BREAKERS OR FUSES

INDIVIDUAL CIRCUITS

FIG. 19-2

LARGE PIPE, GREATER FLOW OF WATER

WATER-PIPE/ELECTRIC-WIRE ANALOGY

LARGE WIRE, GREATER FLOW OF CURRENT

FIGURE VOLTAGE AS PRESSURE, AMPERAGE AS THE RATE OF FLOW

tions can be so disastrous. There is a National Electric Code in book form which is revised periodically. While it can't be enforced by the people who produce it (the National Fire Protection Association), it is pretty much accepted as a bible by local inspectors. If you don't want to buy a copy, you might be able to borrow one from a library or, maybe, the utility company. It's also important to have a copy of the local code, since that is the code you *must* obey. Check with local authorities to see how you can get a copy. Codes result from experience, knowledge, and testing. They keep up with new materials and new methods, so accept them as an aid, not a hindrance.

Always work with materials that have been tested and are listed by the Underwriters' Laboratories. Such products are stamped or labeled with the UL identification, so you know they have been tested and are listed as meeting minimum quality and safety standards.

GROUNDING

The grounding of an electrical system, at the entrance and continuously through the installation, provides a necessary safety factor. Without grounding, or with improper grounding, there can be much danger from shocks and fire and even damage to motors and appliances. Grounding, essentially, is a deliberate connection of the parts of a wiring installation to a *ground*, which is usually a water pipe that makes considerable contact with the earth or a special metal rod that is driven deeply into the earth.

What we might call the main ground, which will be illustrated later, occurs at the entrance, but also, as most codes say, all receptacles, boxes, switches, lights — all wiring — must be grounded. Grounding wires, while they are run together with current-carrying wires, do *not* carry current *unless* the wiring system or a plugged-in appliance is damaged or becomes defective.

As the sketch of only grounding wires in **Figure 19-3** shows, separate grounding wires are continuous but tie in to all components. If a box is installed for a receptacle — again shown in **Figure 19-4** with grounding wires only — a grounding-

FIG. 19-3

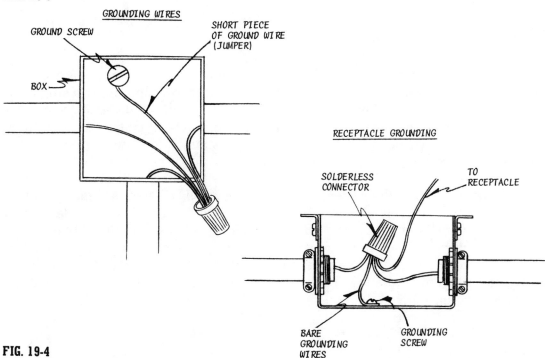

FIG. 19-4

wire connection would be made from the junction to the grounding terminal, usually green-colored, on the receptacle.

The two sketches demonstrate the wisdom of the code. If you were, for example, to remove the receptacle, you would not destroy the continuity of the ground.

Systems that are done through metal conduit or by using cable that has metal armor may not require the extra grounding wires, but the conduit or the armor must be securely grounded to components. That's why when working with either of the two materials it is critical that connectors or locknuts make absolute contact with metal boxes in the system so you get a continuous metallic circuit that is independent of any of the wires that are installed.

Overall, grounding occurs two ways: system grounding done with one of the current-carrying wires (the white or "neutral" wire, sometimes called the ground or continuous wire); and equipment grounding, which has to do with parts of the installation that do not carry current and which is often accomplished, as we have said, with special extra grounding wires.

Check your local codes very carefully for all grounding requirements as well as other electrical-installation specifics. Working with the rules may seem a nuisance, but you'll have the satisfaction of knowing the job will be correct and safe.

WHAT IS A CIRCUIT?

A circuit is a part of the electrical system, with its own wires and its own circuit breaker or fuse, that feeds a particular area of the house or, sometimes, a single appliance. Codes will tell the minimum number of circuits, but the suggestion is based primarily on the square footage of the house, and while the installation may be essentially safe it may not be, as even the code states, adequate for good service and possible expansion.

The smart way is to obey the codes as far as the mechanical installation is concerned but to plan the system to handle *your* needs. There are many advantages in having a considerable number of individual circuits. You can plan lights and outlets so no large area of the house will be in darkness should a breaker trip. With many circuits, the total load can be more equally divided so no one set of wires or breaker will have to carry more than it should. Voltage drop in each circuit will be less, so appliances and lights will operate more efficiently and power waste will be minimized.

What you can carry on a circuit depends on the amperage capacity of the wire, which relates to wire size, and the ampere rating of the breaker. Wire-size capacities are shown in **Figure 19-5**: #14 wire goes with a 15-amp breaker; #12 with a 20-amp breaker, and so on. Generally, #14 wire is usable on circuits for

FIG. 19-5.   AMPERAGE CAPACITY OF WIRE

WIRE SIZE	MAXIMUM CAPACITY
#14	15 amps
#12	20 amps
#10	30 amps
#8	40 amps
#6	55 amps
#4	70 amps
#2	95 amps
#1/0	125 amps
#3/0	165 amps

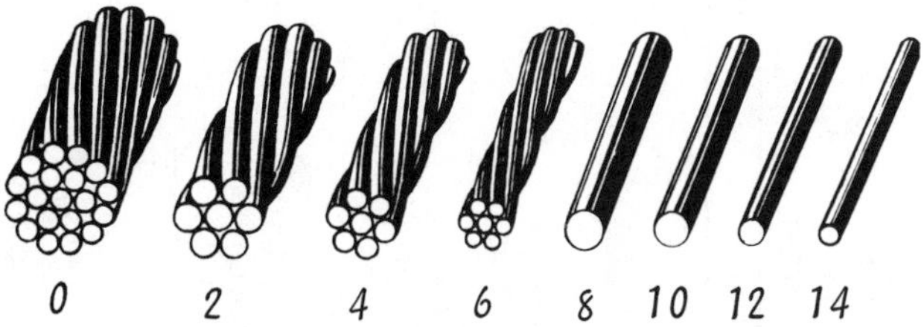

0   2   4   6   8 10 12 14

lighting and general-use receptacles, but the trend is to use #12 wire as a minimum.

To judge wattage, let's assume #14 wire at 115 volts. The number of amps, 15, multiplied by 115 equals 1,725 watts—amperage × voltage = wattage. To find amperes, divide the wattage by the volts. In the above example, 1,725 watts divided by 115 volts equals 15 amps.

THOUGHTS ON PLANNING

Minimum requirements call for a 20-amp general-purpose circuit for every 500 square feet or a 15-amp circuit for every 375 square feet of livable area. The kitchen and dining room should each have two 20-amp circuits; the laundry room, one. In all three cases, the room lights should not be included in the circuits. There should be a convenience outlet for every 12 feet of running wall space, and an outlet every 4 feet along kitchen counters. Remember that the minimum may not be adequate for you. In the case of the laundry room, for example, if you plan to include an electric dryer, an automatic washer, or maybe even a freezer, additional circuits would be needed. Obviously, the system for an all-electric house would have to be designed differently than one for a house that is also supplied with gas.

A good way to start is to title a sheet of paper for each room of the house and to list on it the necessary electrical equipment and those items that will consume the electricity. With such lists and the table in **Figure 19-6**, you can get a pretty accurate idea of demands on circuits and the system as a whole. Since it's not likely, especially in a large house, that all electrical equipment will be in use at the same time, codes include a "demand factor" which may be applied when deciding on service-entrance equipment. Essentially, the demand factor is based on 100 percent of the first so many watts and a percentage of the remaining watts. The factor applies to the service entry only, not to the *number* of circuits. Remem-

FIG. 19-6. GENERAL GUIDE FOR DETERMINING ELECTRICAL CONSUMPTION

AREA	BASIC NEEDS		OTHER	
	Item	Watts	Item	Watts
BEDROOM	ceiling light w/switch at entry	100	electric blanket	150-200
	closet light w/pull chain	60	heating pad	75
	normal wall outlets	(see note)	clock-radio	40-150
	dressing table light w/switch	up to 300	hair dryer	400-1000
			TV	200-400
			water bed/electric heat	500
BATHROOM	ceiling light w/switch at entry	100	ultraviolet sun lamp	275-400
	mirror light w/switch	100	infrared heat lamp	250
	duplex outlet		shaver	8-15
LIVING ROOM	ceiling lights w/entry switch or several switch-controlled outlets for table lamps	100-300	TV	200-400
			Hi-Fi w/tape deck, phono	check equip.
			slide or movie projector	300-500
	normal wall outlets	(see note)	Xmas tree	up to 150
HALL	ceiling or wall light w/3-way switch each end	100		
	duplex outlet			
ENTRY	inside ceiling or wall light w/ switch	100	guest-mirror light w/switch	100
	outside light w/inside switch	30-100		
	duplex outlet			
DINING ROOM	ceiling or wall lights w/dimmer switch	100-300	hot plate (food warmer)	500-1200
	normal wall outlets	(see note)	coffee maker	500-1000
KITCHEN	normal wall outlets	(see note)	toaster	500-1200
	outlets over counters	(see note)	waffle iron	600-1000
	ceiling lights with entry switch or maybe 3-way switches	100-300	frying pan	1000-1200
	built-in counter lights if needed	variable	food blender	500-1000
	wall clock	2-3	knife	100-150
	refrigerator	up to 500	rotisserie	1200-1700
	range top	4000-8000	roaster	1200-1700
	oven	4000-5000	mixer	120-300
	dishwasher	1000-1500		
	sink garbage disposal	500-900		
	coffee maker	500-1000		

UTILITY: LAUNDRY ROOM	ceiling light w/entry switch	100-300	
	normal wall outlets	(see note)	
	washing machine	up to 800	
	dryer	4000-8000	
	water heater	up to 8000	
	freezer	up to 500	
	sewing machine		60-90
	iron		600-1200
PANTRY	ceiling light w/switch	100	
	duplex outlet		
GARAGE	ceiling light w/3-way switches	100	
	outside light w/in-house switch	60-100	
	normal wall outlets	(see note)	
	special circuits if freezer, washing machine placed here		
SHOP GARAGE AS SHOP (best to lead in from special sub-panel)	general overhead lights w/switch	variable	
	workbench lights w/switch	100-300	
	more than normal wall outlets	(see note)	
	240 circuit or circuits		
	¼ hp motor		300-400
	½ hp motor		500-650
	over ½ hp		900-1100 per hp
PATIO	post or wall light w/in-house switch	variable	
	waterproof duplex outlet		
STAIRS	ceiling or wall lights w/3-way switch at top and bottom	variable	
	duplex outlet at top or bottom		
MISCELLANEOUS	portable heater	1000-2000	
	installed-in-wall heater	1000-4000	
	portable fan	50-300	
	room-type air conditioner	800-2000	
	vacuum cleaner	250-1000	
	fluorescent lamps	15-60	
	fuel-fired furnace	800	
	central air conditioner	5000	

NOTE: The number of outlets per area or the spacing of outlets is usually suggested by national or local codes. It's okay to change the spacing should this be more convenient in terms of room layout. It's okay to add more as long as you recognize the possible total wattage in relation to wire size and the amps at the circuit breaker or fuse.

ber, if you follow the code specifically, you will come up with minimum requirements. Working on the plus side, especially at the entrance, leaves you set up, for one thing, for possible future expansion — for example, the unfinished attic or basement that you may wish to make livable later.

DRAWING A PLAN

Don't try to work out your electrical system as you are actually installing the wires. Do it on paper first so you can make corrections with an eraser. The final diagram should be something you won't be ashamed to present to the inspector. It will reveal your sincerity and willingness to abide by codes. This kind of approach is always welcomed and more likely to receive a sympathetic, cooperative reaction.

Professional drawings are neat because they are done with symbols and abbreviations to indicate runs of wire and wiring devices. There are many of these, and you can probably pick up a chart that shows them all, but the ones in **Figure 19-7** have been selected as being most applicable for a residential installation. Another way to go is to use an electrical-symbols template, which you can buy at most stores selling art supplies.

You can, if you wish, make a floor plan of each room on those sheets of paper you used to list electrical requirements. Then you can use the symbols to indicate

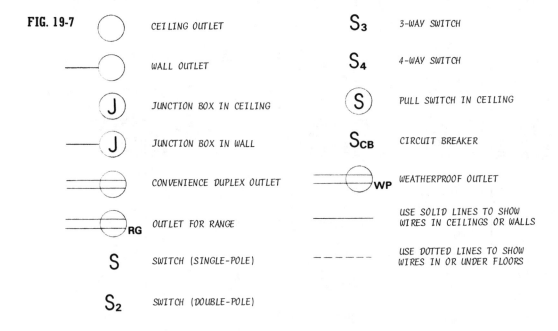

FIG. 19-7

CEILING OUTLET

WALL OUTLET

JUNCTION BOX IN CEILING

JUNCTION BOX IN WALL

CONVENIENCE DUPLEX OUTLET

OUTLET FOR RANGE

SWITCH (SINGLE-POLE)

SWITCH (DOUBLE-POLE)

S_3 3-WAY SWITCH

S_4 4-WAY SWITCH

S PULL SWITCH IN CEILING

S_{CB} CIRCUIT BREAKER

WP WEATHERPROOF OUTLET

USE SOLID LINES TO SHOW WIRES IN CEILINGS OR WALLS

USE DOTTED LINES TO SHOW WIRES IN OR UNDER FLOORS

FIG. 19-8   SAMPLE DO-IT-YOURSELF PLAN

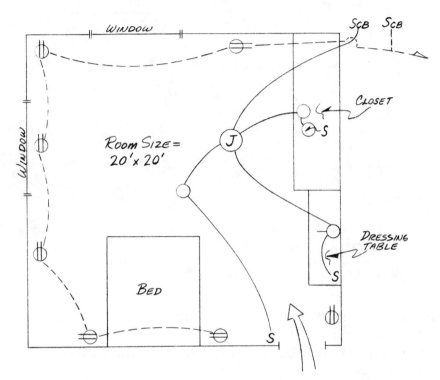

where lights, outlets, and the like should go. You might even include wire runs to see what must enter the room to feed the equipment (**Figure 19-8**).

Eventually, you must produce an overall layout done on a floor plan that is drawn, preferably, to scale (**Figure 19-9**). This will be easier to do if you work on paper that has been squared so each square can represent an actual square foot of the house. If you wish to play safe, do the floor plan on the squared paper but the electrical layout on an overlay of tracing paper. Then you can make as many changes as you wish and, finally, do on the floor plan what you accept and what you will present to the inspector.

Figures 19-8 and **19-9** are *not* designs to follow—they are merely examples of working drawings you can display to an inspector for approval or discussion.

TYPES OF WIRING

There are many types of materials used in wiring systems, but those acceptable by codes for conventional, residential installations are usually nonmetallic sheathed cable, flexible armored cable, or wires that are pulled through thin-wall (or rigid)

FIG. 19-9

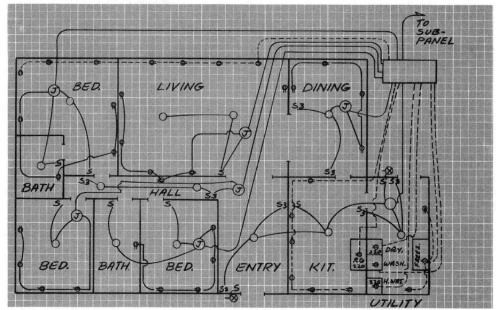

conduit (**Figure 19-10**). Which you use depends on which you are permitted to use. Often, the complete system may involve different materials. For example, entrance wires may have to come through steel conduit while cable may be allowed anywhere inside.

FIG. 19-10   *TYPES OF WIRING*

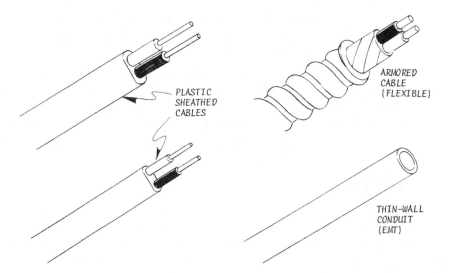

PLASTIC SHEATHED CABLES

ARMORED CABLE (FLEXIBLE)

THIN-WALL CONDUIT (EMT)

Cable is any group of two or more wires that are encased in additional insulation. The number and the size of the wire are stamped on the insulation; for example, 12–2 means there are two #12 wires. If a ground wire is included, it is also noted. All wires are color-coded so you can identify them no matter where you make a cut. A two-wire cable will have a black and a white wire; with three wires, the colors will be black, white, and red. A ground wire, if it isn't bare, will usually be green.

The trend today is for all cable to be sheathed with plastic regardless of whether the conductor (wire) is copper or aluminum. Codes regarding aluminum wiring are a little more rigid today than they used to be. Often a larger size of wire is required if aluminum is used instead of copper. Also, all devices on aluminum-wired circuits must be rated CO/ALR, which simply means that it's okay for use with aluminum, copper, or copper-clad aluminum.

Plastic-sheathed cable is basically of two types, which might be identified as either "indoors" or "dual-purpose." The indoor type is used as the name says — in dry, protected areas. The dual-purpose may be used anywhere in place of the indoor type and can even be used underground and in exposed situations without conduit, unless there is a possibility of mechanical damage. Local codes may call for special *ground fault* protection if the wire is used on an outdoor circuit.

Plastic-sheathed cable should be secured with special straps, as shown in **Figure 19-11**, about every 3 feet along a run and within 12 inches of every box. (Staples are not used because overdriving them can damage the cable.) Make all bends gently and be generous enough with radii so you won't damage the cable. Runs that are made along the side of a framing member like a stud, joist, or rafter, or along the edge of, say, a joist over a crawl space, need no further protection (**Figure 19-12**).

FIG. 19-11   **FIG. 19-12**

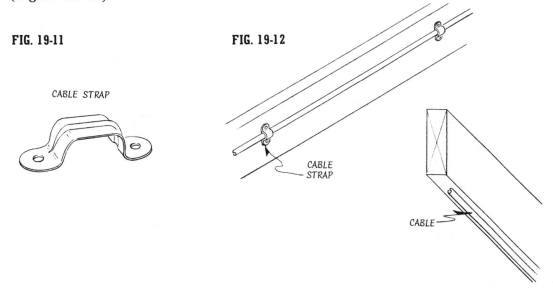

CABLE STRAP

CABLE STRAP

CABLE

FIG. 19-13

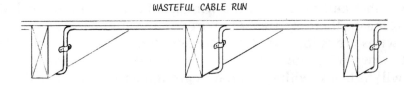

WASTEFUL CABLE RUN

Those exposed runs that cross framing members can follow the structure as shown in **Figure 19-13**, but the method, although often followed, is wasteful of material and contributes to voltage drop because of the longer lengths of cable. Better ideas are shown in **Figure 19-14**. If you drill holes, locate them on the timber's centerline and make them just large enough so the cable won't have to be forced through. Cables that run across the top of ceiling joists in an attic should be protected with strips of wood that are at least thick enough to match the diameter of the cable (**Figure 19-15**).

Entrances into boxes are made as shown in **Figure 19-16**, with special connectors used to secure the cable. Insulation should be continuous to just inside the box.

FIG. 19-14

EFFICIENT CABLE RUN (THROUGH HOLES)

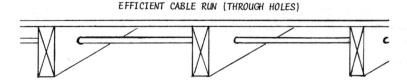

EFFICIENT CABLE RUN (RUNNING BOARDS)

1 X 3 OR
1 X 4

CABLE

STRAP

CABLE

FIG. 19-15   CABLE ACROSS CEILING JOISTS

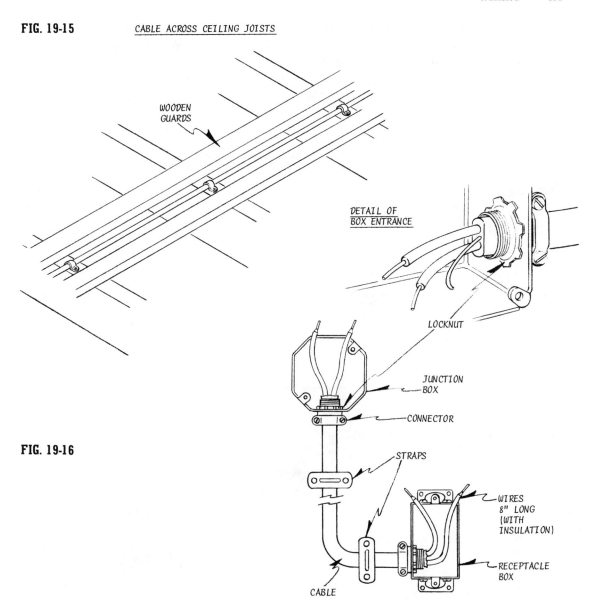

FIG. 19-16

ARMORED CABLE

Armored cable is restricted to indoor wiring and only in dry locations. It is never used in damp areas, or outdoors or underground, and it is always installed with steel boxes. It is possible for a local code to require it, if only for the part of a system that might feed, for example, a stove or a furnace or a large motor.

The spirally wound steel cable contains insulated wires that are wrapped in kraft paper as protection against abrasion, and a bond or ground wire or strip that runs between the paper and the armor.

It takes a little more effort to work with armored cable, but the job will not be difficult if you follow the steps in **Figure 19-17**.

FIG. 19-17

CONNECTING ARMORED CABLE TO BOX

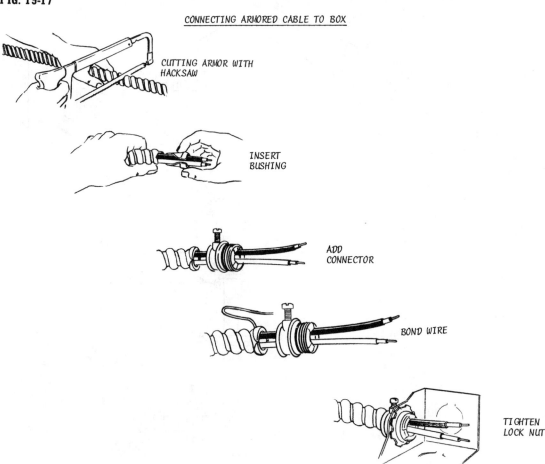

CUTTING ARMOR WITH HACKSAW

INSERT BUSHING

ADD CONNECTOR

BOND WIRE

TIGHTEN LOCK NUT

Hold the cable firmly and use a hacksaw at an angle to cut through one strip of the armor. All you want to do is cut through the thickness of the strip. Grasp the cable on each side of the cut and twist your hands in opposite directions so the armor will "break" and permit you to cut the wires. Repeat the procedure about 10 inches back from the cut end so you can remove the unwanted section of armor and protective paper.

A special bushing (demanded by codes) made of tough and highly insulative fiber is inserted so it seats between the wires and the armor. Push the connector onto the cable as far as it will go and tighten the screw to get solid anchorage. If the cable contains a bare copper grounding strip, bend it back over the armor before inserting the bushing and be sure the strap is squeezed solidly by the connector. If the cable is made with a bonding wire, the wire should make positive contact between the armor and the screw on the connector.

Armored cable may be supported and protected as suggested for nonmetallic sheathed cable. In this case you can use staples instead of straps (**Figure 19-18**).

FIG. 19-18

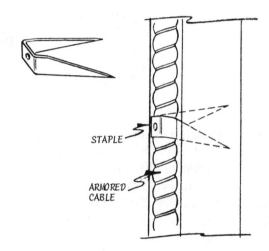

STAPLE

ARMORED CABLE

CONDUIT

Some codes require that you do the electrical system with thin-wall conduit. This differs from the materials we've talked about so far. It is a kind of protective piping system through which you pull individual wires. It may be used indoors or out regardless of whether locations are damp or dry, but check to see if you must use a conduit heavier than thin-wall for an exterior application.

The conduit comes in 10-foot lengths and in various diameters. The diameter is determined by the number and size of wires you plan to pass through (**Figure 19-19**). Like armored cable, it should be used with steel boxes only.

All conduit and boxes are installed before wires are pulled through. You can do the latter job with a long length of galvanized wire, but it's much better to work with a "fish tape" that is ideal for the purpose (**Figure 19-20**). The tape is strong but quite thin and flexible, so you can easily feed it through a run of conduit. Nevertheless, the job is best done with two people—one to feed the wires in at one end, the other at the other end to pull on the tape.

FIG. 19-19.   RECOMMENDED CONDUIT CAPACITIES

SIZE OF CONDUIT	NUMBER OF PIECES OF WIRE CONDUITS MAY CARRY
½"	4 #14 or 3 #12
¾"	5 #12 or 4 #10 or 3 #8
1¼"	4 #6 or 2 #2 or 2 #3 or 2 #4
1½"	3 #1
2"	4 #1/0 or 3 #3/0

FISH TAPE

 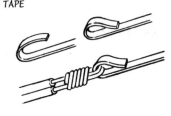

FIG. 19-20

Conduit may be cut with a pipe cutter, but the cutter leaves a very sharp edge, so most professionals use a hacksaw. Of course the sawing leaves burrs, but these are easy to remove with a reamer or with emery.

Bending must be done carefully to maintain the inside diameter of the pipe and so wires can make the turn easily. This is not a job you do over your knee but with a special tool that you can buy or rent (**Figure 19-21**). Codes say that the minimum radius of a bend can be between six and eight times the nominal inside diameter of the conduit and that there can't be more than four quarter-bends (or the equivalent) in any one run. Actually, it makes sense to limit bends as much as possible so it will be easier to pull the wires through. All wires must be continuous. Connections or splices are permitted only at boxes where they will be permanently accessible.

Thin-wall conduit is never threaded; all joints and connections are made with special fittings that grip the material through pressure (**Figure 19-22**). There

FIG. 19-21

CONDUIT BENDER

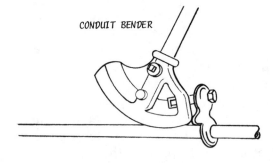

is also a telescopic type of fitting that is used with a special indenting tool (**Figure 19-23**). The tool literally indents the fitting and the conduit after they have been joined to provide a strong joint and one that has low electrical resistance.

Use one-hole or two-hole straps as shown in **Figure 19-24** to support ½-inch or ¾-inch conduit at 10-foot intervals and within 3 feet of every box. Studs may

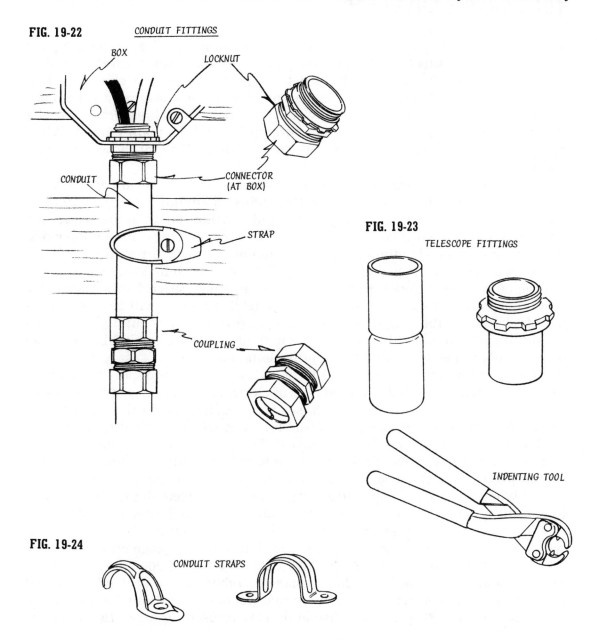

FIG. 19-22   CONDUIT FITTINGS

BOX

LOCKNUT

CONDUIT

CONNECTOR (AT BOX)

STRAP

COUPLING

FIG. 19-23

TELESCOPE FITTINGS

INDENTING TOOL

FIG. 19-24

CONDUIT STRAPS

be notched to receive conduit following the same rules that were outlined for water pipes. Be sure to use metal cover plates so the conduit will be protected from damage should someone later attempt to drive a nail there (**Figure 19-25**).

THE SERVICE ENTRANCE

You can view the service entrance as the heart of your electrical system. A good installation will assure full voltage to the panel so that sufficient current will flow through all circuits in the house. It consists of the utility company's service wires, the house entrance wires, a meter that measures what you consume, a system disconnect so that you can shut down power at any time, circuit breakers or fuses for the system as a whole and for individual circuits, and the main ground. While designs may differ, the finished job appears something like that shown in **Figure 19-26**.

Regardless of your own tabulations concerning possible electrical consumption, it's not likely that you'll be permitted to install anything less than 100-ampere service, unless you're putting up a small vacation home or are requesting temporary service so that you can have electricity while doing construction work. The 100 amps is a code *minimum* and may not be adequate, especially if you are planning for electrically powered ranges, water heaters, clothes driers, air conditioning, and so on. Under such conditions, 150-amp service is a reasonable minimum; 200-amp service is practically a must if the house will be heated electrically. If you are thinking of a supplementary private power supply (perhaps a windmill), you should still install a conventional entrance that will adequately meet the house needs without the supplementary source.

Check the power company (and local codes) before you do any work at all. Good, free advice is available on the type of electric service, sizes of wire, design of the installation, and so on. Check to see how much of the work will be the responsibility of the utility company. Usually, wires will be carried to the side of the house or to a yard pole if one is needed. There will be three wires, one black and one red (or maybe two black) plus a white or bare neutral ground wire. Don't let the word "neutral" make you careless; all three wires should be considered "hot." When you have completed your part of the service-entrance installation, let the utility company make the final connection between their wires and yours.

Sizes of wire will be governed by code, but usually they run like this: For a 3-wire service, rated for 150 or 200 amps, you'll need heavy-gauge wire with RHW insulation. This will be either #0 for 150 amps or #000 for 200 amps. The RHW initials indicate that the insulation is made of rubber and a protective braided covering that make the wire suitable for either wet or dry locations. Smaller diameter wires (#2 or #3) with RHW insulation may be used for a 3-wire service en-

FIG. 19-25

STUD

METAL
COVER
PLATE

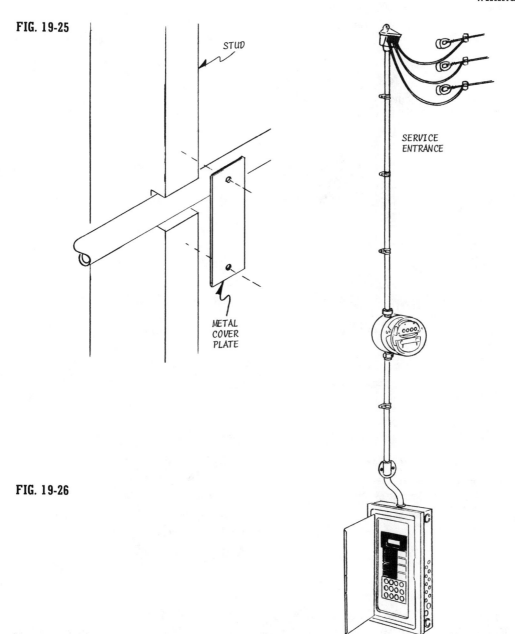

SERVICE
ENTRANCE

FIG. 19-26

trance rated for 100 amps. Codes determine whether the electric service installation can be done with service cable or with wires fed through conduit.

A typical cable installation is shown in **Figure 19-27**. The entrance head
should be at least 10 feet above the ground with the service-cable wires extending

FIG. 19-27   TYPICAL CABLE INSTALLATION

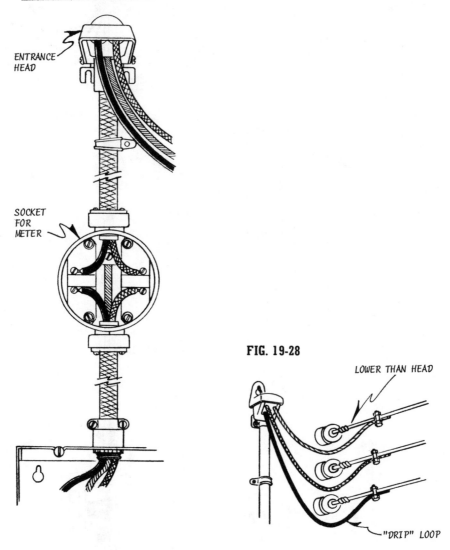

ENTRANCE
HEAD

SOCKET
FOR
METER

FIG. 19-28

LOWER THAN HEAD

"DRIP" LOOP

about 3 feet beyond it so there will be plenty of length for connections to the power lines. The power lines are sometimes secured to the building with insulators, and these should be high enough so they can't possibly be damaged by trucks and the like.

The entrance head must be higher than the top incoming wire, and connections should be made as shown in **Figure 19-28** so water won't drip off exposed lines and enter the service system.

The entrance head for a conduit service entrance may be like the one in **Figure 19-29** or it may be done with a *mast*, which is nothing more than a conduit extension through the roof installed in the manner shown in **Figure 19-30**. It's a neater-looking installation and especially suitable when the house has a low roofline which makes it difficult to get the necessary height for incoming wires by using other means.

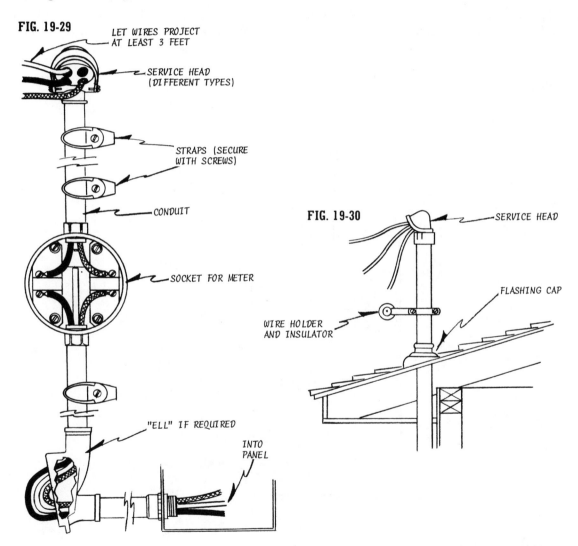

FIG. 19-29
LET WIRES PROJECT AT LEAST 3 FEET
SERVICE HEAD (DIFFERENT TYPES)
STRAPS (SECURE WITH SCREWS)
CONDUIT
SOCKET FOR METER
"ELL" IF REQUIRED
INTO PANEL

FIG. 19-30
SERVICE HEAD
FLASHING CAP
WIRE HOLDER AND INSULATOR

There are other code-governed facts that you should check out. These will tell you about minimum height above sidewalks and driveways, minimum distances from windows and doors, and so on.

The three wires that come into the service panel are connected as shown in **Figure 19-31**. The neutral wire goes directly to the neutral busbar; the black and the red wire (or the two black wires) go to lugs on the system disconnect, which are usually identified. The busbar will also have a connection place for the ground wire and plenty of terminals for the neutrals of the house circuits. Figure 19-31 also shows how a 120-volt circuit is wired. The difference between this circuit and one for 240 volts, in addition to the ampere rating of the circuit breaker, is in the number of wires that connect to the breaker. A 3-wire 240-volt circuit has both a black and a red wire connected to the circuit breaker.

FIG. 19-31

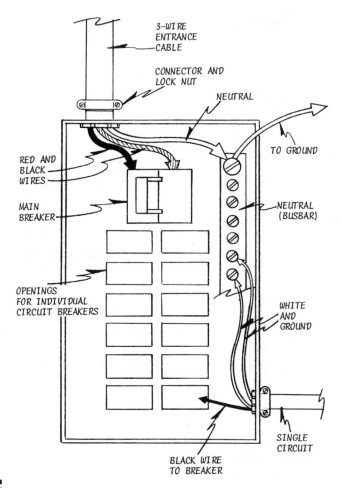

GROUNDING

The usual method of installing the ground wire (check code for adequate size) is to run it the shortest distance possible to a cold-water pipe that makes considerable contact with the earth. Making the connection on the street side of a water meter is wise, since the ground will still be effective if the meter ever has to be removed. If you can't do it that way, then install a jumper as shown in **Figure 19-32**. Use clamps that match the material of the pipe—for example, a steel clamp on a steel pipe.

FIG. 19-32

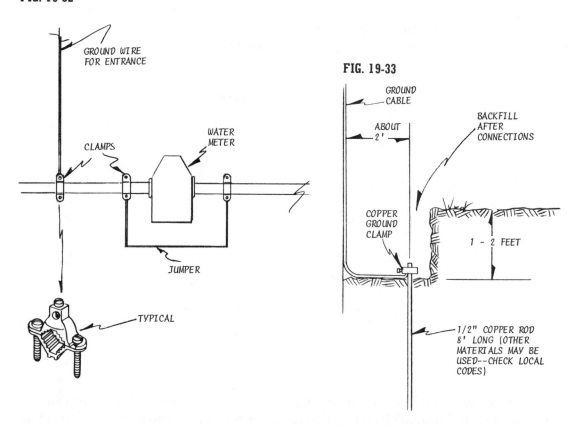

FIG. 19-33

Another permitted method of grounding is shown in **Figure 19-33**. This is called a *made electrode* and is common in rural areas that lack conventional water-piping systems.

Grounding is an extremely important part of the electrical system, so don't approach it casually. Work with the codes to be sure the job will be done correctly.

Don't do any work on the service entrance while there is current coming in. The safest procedure is to make all connections *before* you ask the power company to make the final hookup. Once the connection is made, it will be foolhardy to work on the panel without throwing the main switch. Even then, remember that the incoming lines will still be *hot*.

BASIC CONNECTIONS

The two connections you will be involved with mostly when running circuits are those made to a screw terminal and those where ends of wires are joined together. In each case, the insulation must be stripped from the end of the wire to lay bare only as much of the conductor as you need to make the connection. The minimum is best, since you want no bare wire exposed. Stripping can be done with a knife, but it's easier and better to work with a special tool which can be used to cut the wire as well as strip it (**Figure 19-34**).

FIG. 19-34   **FIG. 19-35**

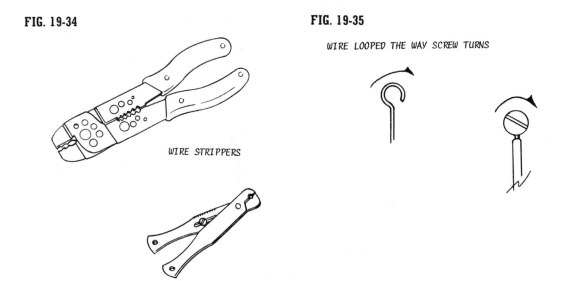

WIRE LOOPED THE WAY SCREW TURNS

WIRE STRIPPERS

Wire ends should be looped, neatly, to fit under a screw, and the loop should be placed so it tends to close as you tighten the screw (**Figure 19-35**). The loops can be made with pliers only, but if you force a small piece of tubing over one jaw and then use the pliers as shown in **Figure 19-36**, you'll get more uniform results.

Some of the more modern receptacles have holes in the back so you don't have to use screws at all. In this case, the stripped end is pressed into the hole, where it locks and makes the connection (**Figure 19-37**). Some of these items have a gauge stamped on the back so you can tell exactly how much of the wire should be bared.

Most wire ends are joined today with solderless connectors, the most common of which are plastic twist-ons, used to join wires in circuits to each other and to fixtures. They come in various sizes, so you can choose in relation to the num-

FIG. 19-36

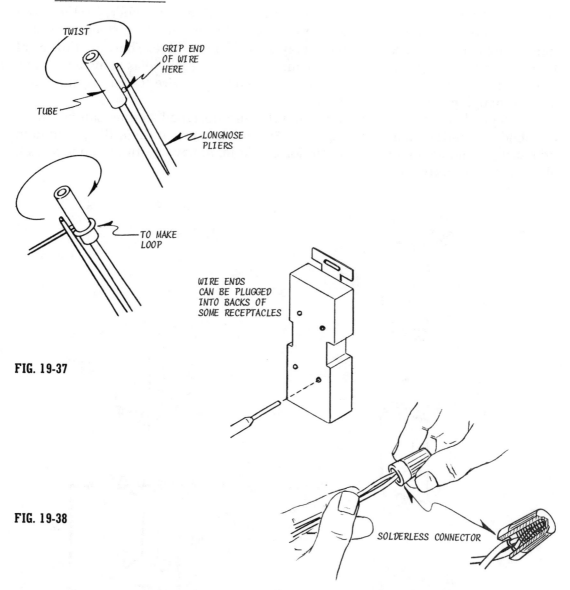

HOW TO LOOP WIRE ENDS

TWIST

GRIP END
OF WIRE
HERE

TUBE

LONGNOSE
PLIERS

TO MAKE
LOOP

WIRE ENDS
CAN BE PLUGGED
INTO BACKS OF
SOME RECEPTACLES

FIG. 19-37

FIG. 19-38

SOLDERLESS CONNECTOR

ber of wires and the size of the wires you are working with. The wires are held parallel and close together and the connector is screwed on tightly (**Figure 19-38**). Check, by holding the connector and tugging on each wire, to be sure the joint is sufficiently tight.

BOXES

Boxes must be used to enclose wire-to-wire connections and wherever there is a switch, outlet, ceiling or wall light, or the like. They are made of steel or plastic, but be sure to check to see if the latter may be used and where. There are dozens of types and sizes, but the most common ones might be classified as being receptacle boxes for switches and outlets, and junction boxes for wire-to-wire connections and for hanging ceiling or wall lights.

Receptacle boxes are usually rectangular and designed so they can be firmly attached to framing members (**Figure 19-39**). With many of them it's possible to do a gang assembly should you wish, for example, to have more than one switch at a location (**Figure 19-40**).

FIG. 19-39

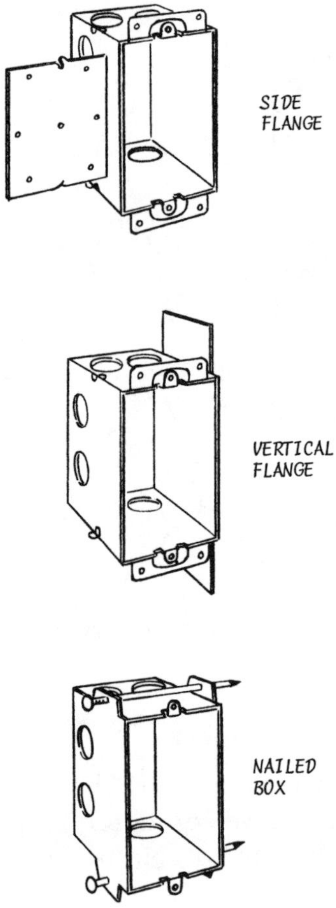

SIDE
FLANGE

VERTICAL
FLANGE

NAILED
BOX

FIG. 19-40

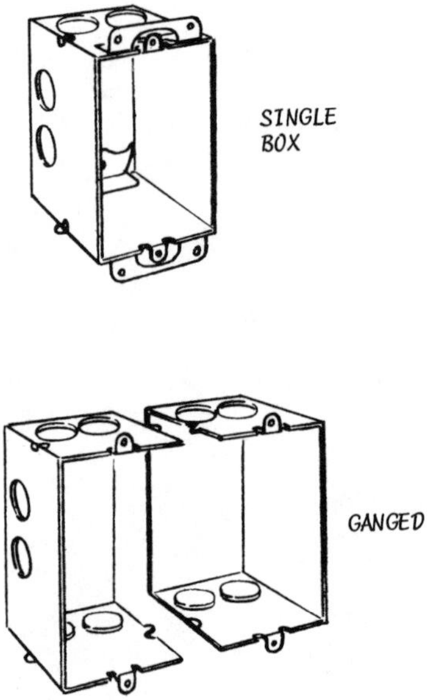

SINGLE
BOX

GANGED

Boxes may also be larger to take more than one switch or outlet or for a particular connection that requires extra room (**Figure 19-41**). This can occur when you need an oversize outlet, and the heavy wire that goes along with it, for something like an electric clothes drier.

Exterior outlets or switches must be installed in special boxes and with covers that are weatherproof (**Figure 19-42**).

When a box must fall between framing members it must be adequately supported by using ready-made metal straps or by installing additional wooden supports (**Figure 19-43**).

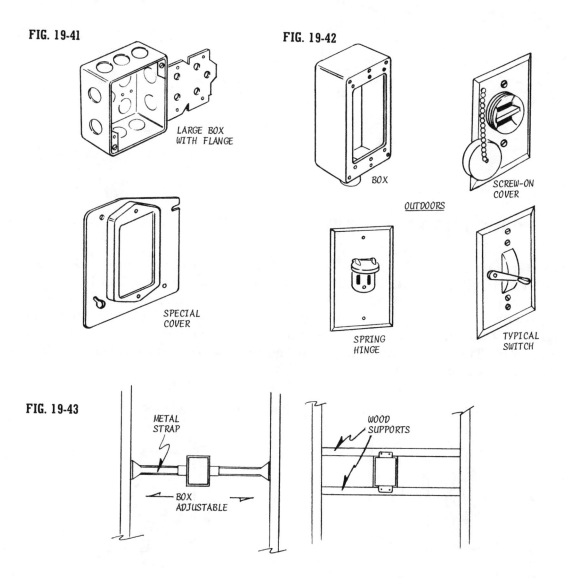

FIG. 19-41

LARGE BOX
WITH FLANGE

SPECIAL
COVER

FIG. 19-42

BOX

SCREW-ON
COVER

OUTDOORS

SPRING
HINGE

TYPICAL
SWITCH

FIG. 19-43

METAL
STRAP

BOX
ADJUSTABLE

WOOD
SUPPORTS

Junction boxes are usually octagonal (**Figure 19-44**). Like other boxes, they must have secure support when they must be located away from a stud or a joist (**Figure 19-45**).

The boxes may have prepunched, circular pieces that you knock out for cable or conduit clamps, or they may contain built-in cable clamps.

FIG. 19-44   **FIG. 19-45**

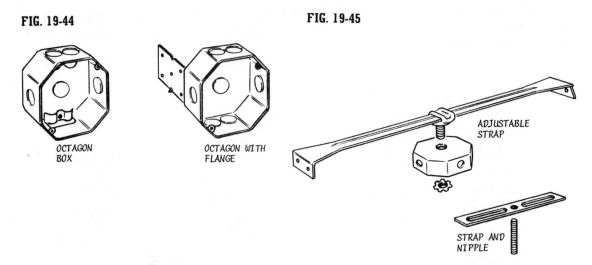

OCTAGON BOX

OCTAGON WITH FLANGE

ADJUSTABLE STRAP

STRAP AND NIPPLE

SWITCHES AND RECEPTACLES

There are many types of switches, but their purpose is always to turn a circuit or part of a circuit on or off. Switches are rated for what they can handle. For example, a rating of 10 amps, 125 volts, means the item can handle 10 amps as long as the voltage is not more than 125. Most designs are of the toggle type shown in **Figure 19-46**, silent or audible, but there are those that operate with a push-button or a rocker. The dimmer type shown in **Figure 19-47** controls a rheostat so you can control a room light from dim to full bright.

FIG. 19-46   TOGGLE SWITCH AND COVER   **FIG. 19-47**   DIMMER SWITCH

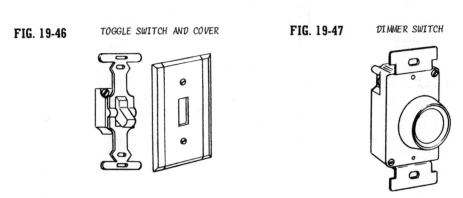

Three-way switches are installed when you wish to control lights from two different locations. It's also possible to use four-way switches so lights can be turned on or off from three different locations.

Outlets, or receptacles, are those items used to plug in radios or TVs or lamps and the like. **Figure 19-48** shows the common duplex type, which has two sets of holes so it can accommodate two plugs, and a special type you can install for a wall clock.

Usually, receptacles (outlets) are wired so they are always hot, but if the overall design of the system suggests it, selected ones may be wired so they can be controlled through a wall switch.

FIG. 19-48

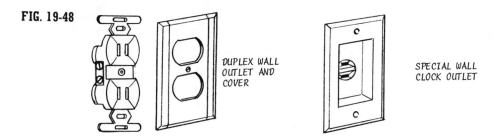

DUPLEX WALL OUTLET AND COVER

SPECIAL WALL CLOCK OUTLET

SOME BASIC WIRING DIAGRAMS

It's best to do the wiring after all boxes have been installed, since you can then plan the shortest routes for cable or conduit. If conduit is used, you are in a position to pull through the exact number of wires, of correct color, for any circuit. Since the number of wires in plastic-sheathed cable is limited, there are situations where the run of the neutral wire *is* interrupted. This does not change the very important basic rule that the neutral wire must be continuous and never connected to a hot (black) wire. When an exception does occur, the white wire must be painted black at both ends of the connection so the strange use will always be evident.

The color of the terminals or lugs on receptacles, sockets, and the like will tell what wire must be connected to them. Hot wires go to copper or brass terminals, and terminals that are whitish in color receive the *grounded neutral* wire. Green terminals are for *grounding* wires. Ordinary switches have copper or brass terminals, since they receive only hot wires. The "common" terminal on a 3-way toggle switch has a special color to identify it.

The diagrams that follow show installations done with plastic-sheathed cable and with screw-on solderless connectors. If separate grounding wires are not shown, it's only so the picture will be clearer. In all situations, the grounding rules that apply to the system you are installing *must* be obeyed. Again, be sure there is no current in the wires you are working on. The safest procedure is to shut down the entire system at the main disconnect. Short of that, be sure the circuit you are working on is not connected at the entrance panel. When you do make the final hookup to the circuit breaker or fuse, be sure the main disconnect is off.

THE JUNCTION BOX

All connections made that route current in different directions must be done in a junction box (**Figure 19-49**). Note that black is connected to black, white to white, and that the grounding wires are continuous and also connected to the box. The boxes must not be overcrowded; codes will tell the number and size of the wires you can connect in a particular size box. Pack the wires neatly inside the box after you have made the connections and then screw down the box's cover.

FIG. 19-49   BASIC JUNCTION

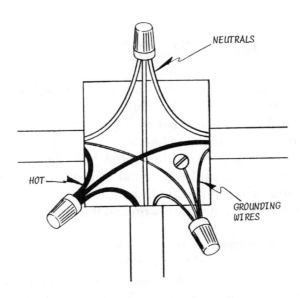

LIGHTS

The wall switch in **Figure 19-50** controls a ceiling light that is located at the end of a run. The neutral wire must be painted black at the points marked A, since, used this way, it becomes a hot wire. This is one of the neutral-wire rule exceptions we mentioned above.

FIG. 19-50   WALL-SWITCH/CEILING LIGHT

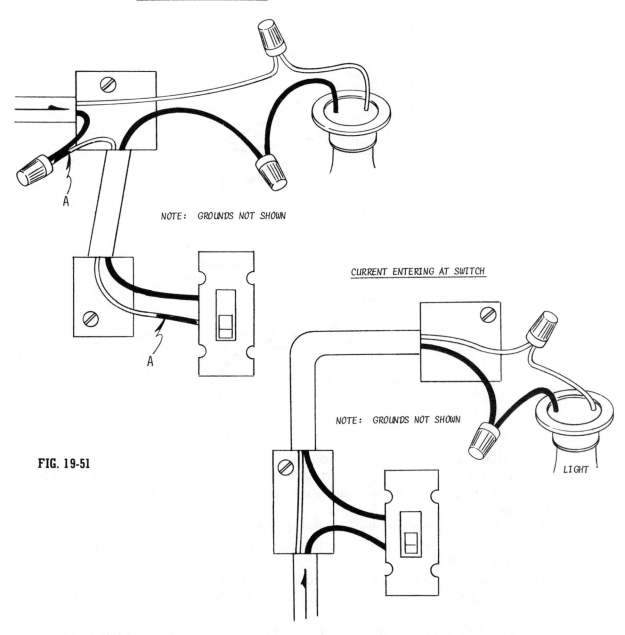

NOTE: GROUNDS NOT SHOWN

A

CURRENT ENTERING AT SWITCH

FIG. 19-51

NOTE: GROUNDS NOT SHOWN

LIGHT

In the situation in **Figure 19-51**, the current is entering at the switch. The black wire is broken·to connect to the switch, but the white wire is continuous and therefore is *not* hot. **Figure 19-52** shows the same situation but with grounding wires included.

FIG. 19-52

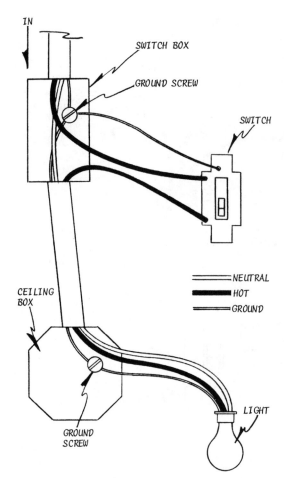

A run that is interrupted to provide a switch that will control more than one light can be wired as shown in **Figure 19-53**. Here, since the neutral wire connects to the switch at one end and to a black wire at the other end, it must be painted black at the points marked A.

A run that is interrupted by a switch-controlled light is wired as shown in **Figure 19-54**. The neutral wire from the switch becomes hot; the current continues beyond the light.

The wiring is done as shown in **Figure 19-55** when you interrupt a run to install a light that has its own switch. The current will still continue beyond the light.

The simplest kind of light to install is one that has its own switch and is located at the end of a run. It is shown in **Figure 19-56** both with and without grounding wires. Such fixtures may have screw terminals, in which case connections are made directly to the screws, or lead wires, in which case they are made by using solderless connectors to join wires.

FIG. 19-53

ONE SWITCH FOR TWO LIGHTS

RUN CONTROLLED
BY SWITCH

LIGHT

LIGHT

A

NOTE: GROUNDS NOT SHOWN

A

INTERRUPTED RUN

FIG. 19-54

A

NOTE: GROUNDS NOT SHOWN

A

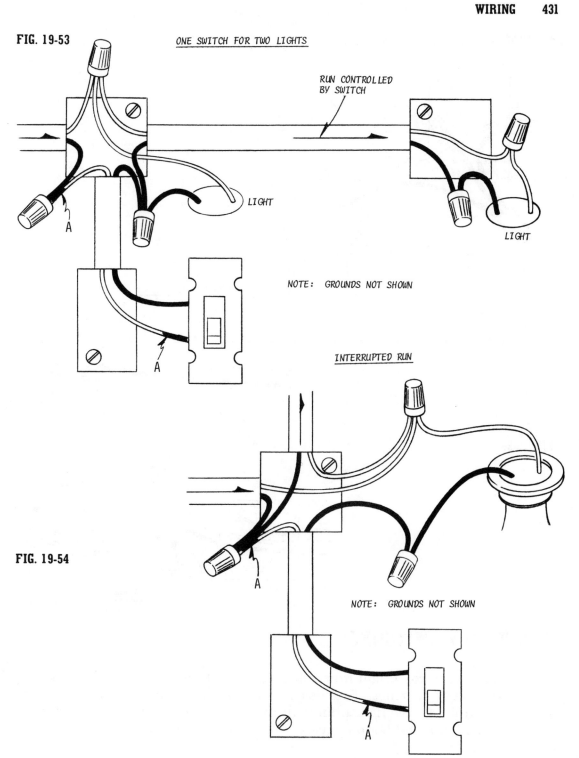

FIG. 19-55   *LIGHT WITH OWN SWITCH*

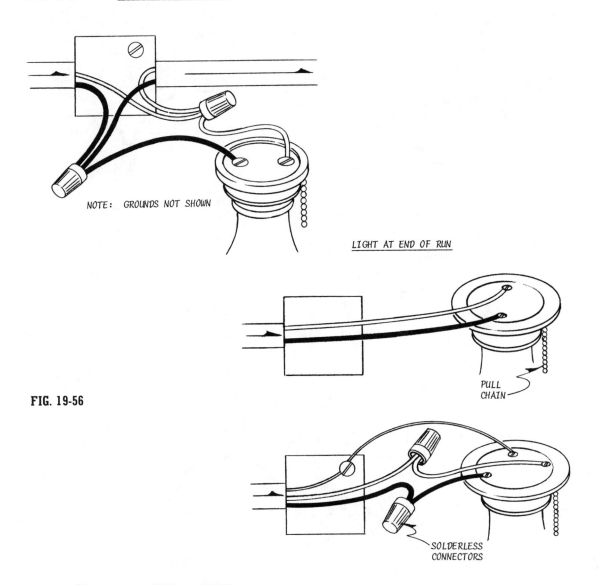

NOTE: *GROUNDS NOT SHOWN*

LIGHT AT END OF RUN

FIG. 19-56

PULL
CHAIN

SOLDERLESS
CONNECTORS

THREE-WAY SWITCHES

If you wish to control a light from more than one location you must work with special three-way switches. The idea is very practical, for example, when you wish to control a garage light from both the house and the garage or a stairway light from both top and bottom levels.

The three-way switches look like ordinary toggle switches but they usually have three terminals, one of which is specially color-coded as the "common" terminal. Be sure you recognize which one it is, since it's an important factor in three-way-switch installations. A basic rule to remember is that the black wire from the current source goes directly to the common terminal on one of the switches. A black wire must also run from the common terminal of the *second* switch to the fixture, where it is connected to the fixture's black wire or brass screw, whichever is present.

The installation is done with three wires, the colors being black, white, and red. The wires at the fixture must always be black to black and white to white. The relative ease or complexity of the wiring depends much on where the current enters the installation, but if you work carefully and slowly and follow the colors, you should have no problems.

A typical installation with current entering through a switch is shown in **Figure 19-57**. The incoming black wire connects to the common terminal on the first switch. The incoming white wire runs directly to the fixture. The black wire that

FIG. 19-57   LIGHT CONTROLLED AT TWO SWITCHES

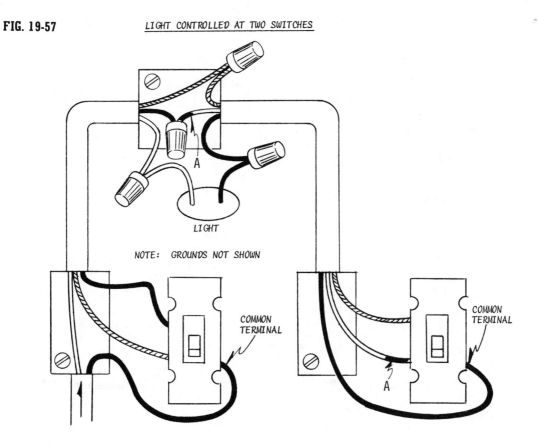

connects to the common terminal on the *second* switch is continuous to the fixture. The white wires should be painted black at points A.

When the current is picked up from a junction box, the wiring is done as shown in **Figure 19-58**. The diagram shows a separate box for the fixture, since it may be required because of the number of connections that must be made in the junction box.

When the current enters the box for a fixture that occurs ahead of the switches, the wiring is done as shown in **Figure 19-59**. Note that a white wire takes over for the black at one of the switches and that it must be painted black at the points marked A.

If the switches are installed to control a light at the end of a run and the current enters through the first switch, the wiring is done as shown in **Figure 19-60**.

FIG. 19-58

LIGHT SWITCHED THROUGH JUNCTION

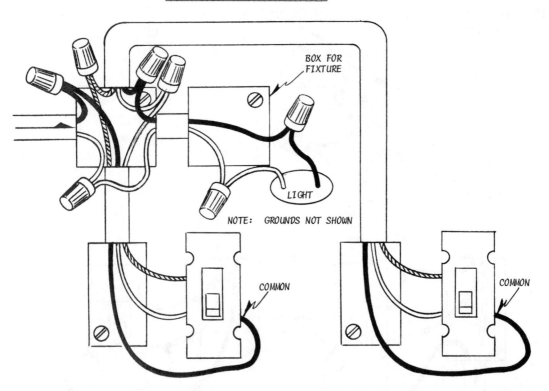

BOX FOR FIXTURE

LIGHT

NOTE: GROUNDS NOT SHOWN

COMMON

COMMON

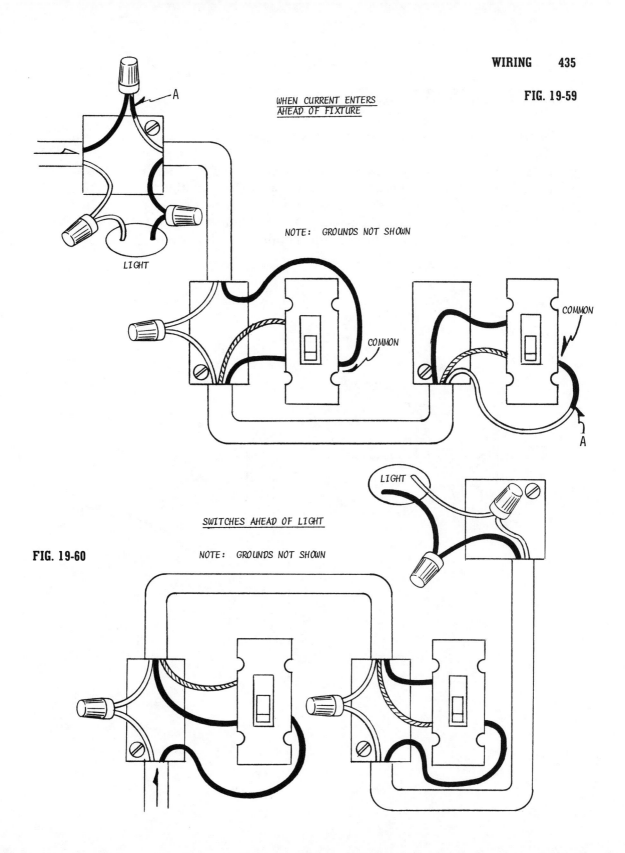

FIG. 19-59

WHEN CURRENT ENTERS
AHEAD OF FIXTURE

NOTE: GROUNDS NOT SHOWN

LIGHT

COMMON

COMMON

A

FIG. 19-60

SWITCHES AHEAD OF LIGHT

NOTE: GROUNDS NOT SHOWN

LIGHT

RECEPTACLES (OUTLETS)

The wiring of routine receptacles is pretty straightforward, as shown in **Figure 19-61** with one receiving current from a junction box. The black wire goes to a copper or brass screw, the white to a silver terminal. You can add other receptacles on the same run by continuing the wires and making similar connections, as shown in **Figure 19-62**, which also shows how grounding wires should be installed. The grounding wires will be either green or bare and should be connected to the green terminal on the receptacle and make firm contact with a grounding screw in the box, itself.

FIG. 19-61   BASIC RECEPTACLE

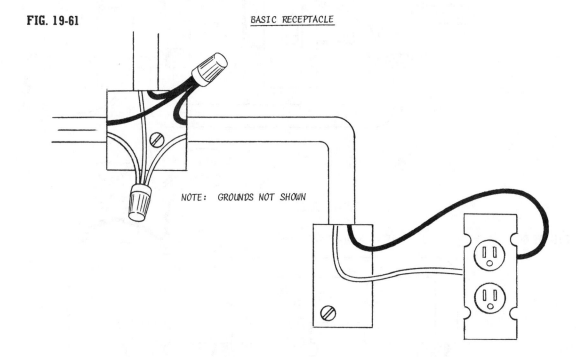

NOTE: GROUNDS NOT SHOWN

FIG. 19-62   ADDED RECEPTACLE

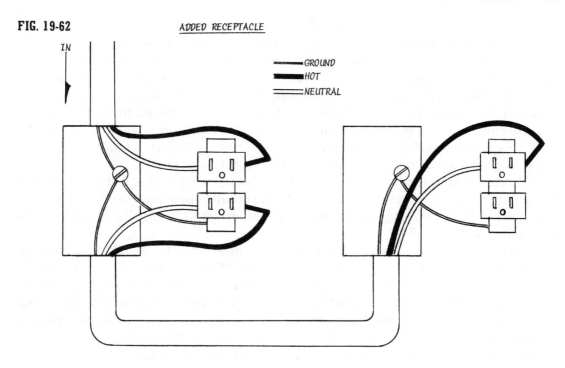

DO SOME TESTING

Because the electrical system is so important and can be dangerous if done incorrectly, it's a good idea to test the circuits before you consider the job complete. This should be done before wall coverings are up and before fixtures are installed. There are many types of testers you can buy, but to prove out circuits you can easily make a reliable one by connecting two dry-cell batteries and a door bell as shown in **Figure 19-63.** Before testing *and without current in the lines,* twist together all wires in boxes as they would have to be to complete the circuits. Do *not*

FIG. 19-63   TWO DRY-CELL
                        BATTERIES

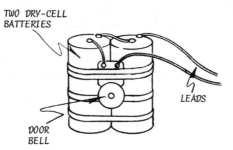

connect the wires that connect to a fixture or receptacle, but bend them so they are well apart. *Do* connect the wires that normally go to a switch.

To test conduit or armored cable, touch the neutral busbar in the service panel with one of the test leads and the black wire terminal of the individual circuits with the other lead. The bell should *not* ring. If it does, a short circuit is probably occurring between a black and a white wire somewhere in the circuit. It's also possible that the insulation has been abraded from a black wire and the bare conductor is making contact with a box or the conduit. At any rate, something is wrong and a complete check of the circuit is in order.

You can test the ground of conduit or armored cable in this fashion. Connect the battery wires to the neutral wire and to the conduit or armor. A correct ground is indicated if the bell rings when you touch its leads to the white wire at outlets and to the box.

Another tester you should have is an inexpensive little gadget that has two prongs at one end and a light bulb at the other (**Figure 19-64**). If you place the prongs in the holes of a receptacle as you would a plug, the bulb will light to indicate the receptacle is hot. No light, no current. By the same token you can touch the prongs to the wires that normally attach to a receptacle and read the results the same way. Here, of course, you are testing with current in the lines and you want to be certain that the prongs, and not your fingers, touch the wires.

FIG. 19-64

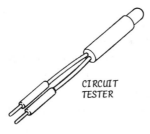

CIRCUIT
TESTER

The tester also supplies a good safety factor, since you can use it to test wires in a circuit after you have shut down a system just to be certain that it is shut down.

There are more sophisticated pieces of test equipment you can buy or, maybe, rent. Always be sure to read and understand accompanying instructions.

INSTALLING INSULATION

House walls and roofs are sometimes like heat exchangers. Heat will flow through any solid body — by *conduction* — because of molecular interaction. Denser materials are usually more heat-conductive. Air is less dense at higher temperatures, so any cold air that is present is heavier and *convection* currents occur that cause a heat flow from one area to another. Air itself can be a good insulator if it is confined to a small space where the convection flow is minimal or nonexistent. Heat can also travel through the air, and while the waves of the *radiation* process do not heat the space they move through, much of the energy is absorbed or reflected when they meet a colder surface. All three methods contribute to some degree to the heat flow through floors, walls, and ceilings of a house.

If inside and outside temperatures could be equal and constant there would be little to fret about. But they are not, and the greater the difference the faster the heat flow. The purpose of insulation is to minimize the transfer of heat and so contribute to adequate inside temperatures with minimum use of energy to operate heating and cooling equipment. Total insulation, especially today, is not only smart, it's almost a duty.

TWO BASIC TERMS

The U-value tells the amount of heat loss through a building section in terms of Btu's (British thermal units) per square foot for each hour when there is a 1-degree difference between inside and outside temperatures. Insulating values go up as U-values go down. A lath-and-plaster ceiling might have a U-value of .61. Add 3 inches of insulation and the U-value will decrease to .078. A conventionally framed wall can have its U-value reduced from about .20 to about .08 merely by adding 2- or 3-inch-thick blankets of insulation between the studs.

R-value is a measurement that tells how a material *resists* the flow of heat. All building materials have an R-value: ½-inch drywall might have an R-value of .45;

½-inch sheathing, 1.32. This also applies to insulation materials. The higher the value, the more efficient the material is. It's a more important point than the thickness of the material. If several products differ in thickness but have the same R-value, they will resist heat flow equally well.

The R-value also has to do with the direction of the heat flow. As shown in **Figure 20-1**, the label on the package should tell the R-value in relation to whether the heat-flow is up, down or horizontal.

FIG. 20-1

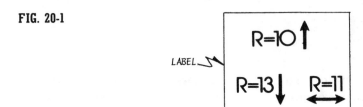

AMOUNT OF INSULATION

How much you need will depend to some extent on the design of the house the construction materials, and, of course, where the house will be located. Many old R-value recommendations should be considered obsolete primarily because they are based on comfort standards, often in minimum terms, and pay little attention to the diminishing supply and escalating costs of conventional fuels.

What we must think about, in addition to our desire for comfort, is amounts of insulation in relation to energy supply and price. The overinsulating you do today will probably be the standards of tomorrow.

A recent survey by Owens-Corning has resulted in recommendations that make a lot of sense, considering the fuel shortages we now face. The study covered 71 cities and the recommendations indicate amounts of insulation that are practical to install with today's products and also realistic in relation to escalating heating and cooling costs.

Even if you design for near-total independence from commercial suppliers of energy, it makes sense to work so that insulation will minimize energy needs regardless of the source.

In developing the new, economically justified levels of insulation as shown in this map in **Figure 20-2**, Owens-Corning used the following definition: "Optimum insulation is the last increment of R-value that will, through reduced operating costs, pay for the initial cost of that unit of R-value plus yield a reasonable return on the investment."

Optimum R-values depend on the cost of energy and on climatic factors, plus

FIG. 20-2   RECOMMENDED R-VALUES

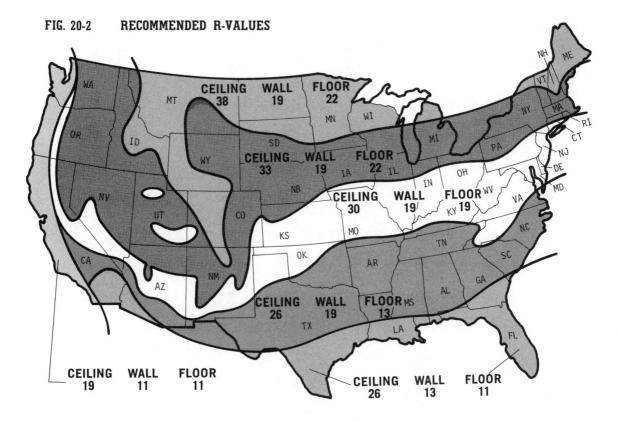

the cost of the insulation. A suitable insulation value for one area will differ from that of another.

To arrive at their conclusions, Owens-Corning considered the following computer-calculated factors:

● Insulating costs
● Weather data
● Today's heating and cooling energy rates
● Projected rate increases (conservatively averaging about 7 percent a year)
● Investment return over a 20-year period with future savings discounted 10 percent a year.

KINDS OF INSULATION

There are many types of insulation, but they are generally classified as being flexible, reflective, loose, or rigid (**Figure 20-3**). Flexible blankets come in rolls of various lengths and in widths that make them suitable for placement between studs

FIG. 20-3

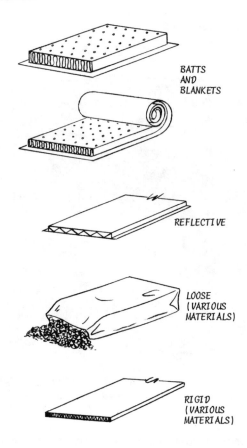

BATTS
AND
BLANKETS

REFLECTIVE

LOOSE
(VARIOUS
MATERIALS)

RIGID
(VARIOUS
MATERIALS)

and joists. The insulation, which may be 1 to 3 inches thick, is usually wrapped in paper, one side of which may be a vapor barrier or a reflective sheet such as aluminum foil. Flexible batts are available in greater thickness (up to 6 inches) and come enclosed in paper, like the blankets, or with just one surface veneered with foil. The batts come precut to specific lengths, usually 24 or 48 inches.

The most effective type of reflective insulation is a bright metal foil that comes flat, in rolls, but is expanded like an accordion when installed to provide air spaces between its surfaces. Reflective insulations also come as flat metal foil or other materials that have a foil veneer. Its insulating value is based on its reflective surface, not the thickness of the material. In order for it to function efficiently, there should be ¾ inch or more air space facing the reflective surface.

Loose insulation can be various types of fibers, vermiculite, granulated cork, or other materials. It's purchased by the bag; the label should tell you the R-value of the contents when used to cover so many feet to a depth of so many inches. It is easy to place, say in an attic, and is ideal for filling the cavities in concrete block.

It is also a good choice when irregular joist spacing or many obstructions would interfere with the placement of batts or blankets. Many times, in order to build up R-value it is poured over blankets in an attic.

Rigid insulations are made by reducing the fibers of wood, cane, and other materials to a pulp and then using the pulp to form lightweight and low-density sheets that are structurally strong. We have already discussed some of these as sheathing and roof decking, but other uses include subflooring, plaster base, perimeter insulation for slab floors, and so on. Sheets of foamed plastic—polystyrene—are being used more and more with masonry construction because of their high water resistance.

CONTROLLING MOISTURE

Part of a good insulation job is a vapor barrier placed on the warm side of the wall to prevent condensation. Usually, the barrier is aluminum foil or some other material which is already applied to the *warm* side of batts and blankets. If the insulation is not so designed, you should cover the walls with a separate material such as vapor-barrier paper, aluminum foil, or polyethylene film. Polyethylene is a popular choice because it comes in widths that make it easy to form a tight cover with a minimum of laps (**Figure 20-4**).

FIG. 20-4

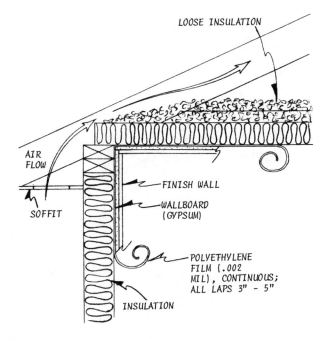

Note the sub-base of wallboard. This is not essential, but it does contribute some R-value and adds to the solidness of the wall, especially if the final cover will be a thin paneling.

Vapor barriers should be installed carefully and after all in-the-wall elements are in place. It's best to apply it just before finishing the walls so there will be little chance of puncturing it while other work is in progress. If you are adding separate vapor barriers, cover everything, including doors and windows. Trim out the openings after the walls are covered with finish materials.

HOW TO PROVIDE VENTILATION

Proper ventilation is needed to help keep a house cool in the summer and to prevent condensation during winter when interiors are heated. Vapor barriers help, but they are not infallible, so something must be done to provide an airflow, especially through an unheated attic or any area above a ceiling if the roof is flat or has little pitch.

Gable roofs are easy to ventilate if you install ready-made units like those shown in **Figure 20-5** at each end. The size of the ventilators is based on a guide

FIG. 20-5   GABLE VENTS

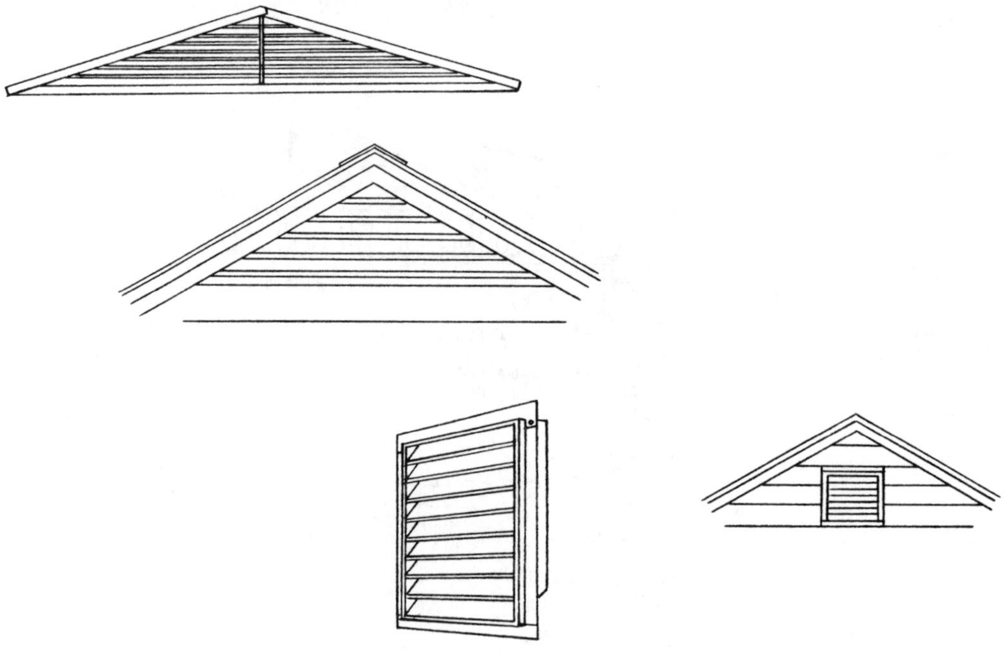

which says the *total* ventilating area should be equal to about ¹⁄₃₀₀ of the total square footage of the ceilings. Since ventilation, when provided this way, depends on air movement, the method is most effective when soffit vents are included (**Figure 20-6**).

FIG. 20-6

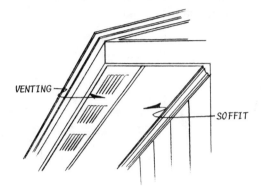

VENTING

SOFFIT

Other types of prefab ventilators are available, so you can make a choice in relation to roof design. Some are designed for placement along a ridge (**Figure 20-7**). Others may be used anywhere on a roof (**Figure 20-8**). There are also types with built-in breeze-activated fans.

FIG. 20-7

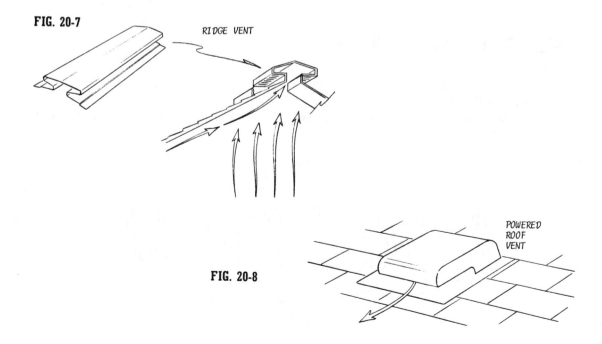

RIDGE VENT

POWERED ROOF VENT

FIG. 20-8

Special units come in the form of louver-covered tubes that are installed merely by forcing them into holes (**Figure 20-9**). They have the special advantage of being installable even after the structure is complete. Thus you can provide compensation ventilation if necessary. The vent diameters range up to 4 inches.

All vents should be screened to exclude insects, but don't use screening so small (less than #16) that it will easily clog.

FIG. 20-9

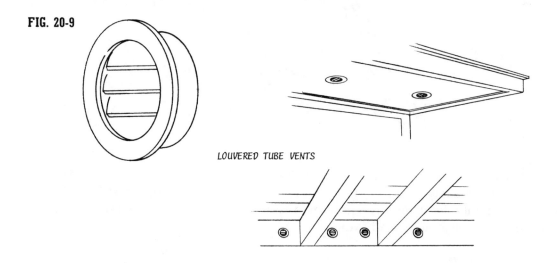

LOUVERED TUBE VENTS

USING BATTS OR BLANKETS

The material may be installed by stapling its flanges to the sides of studs as shown in **Figure 20-10** to provide an air space or to leave the front edges of the studs uncovered. Professional installers of drywall often request this method since it

FIG. 20-10

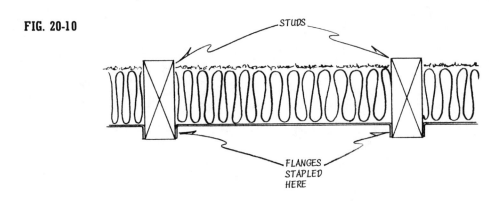

STUDS

FLANGES
STAPLED
HERE

eliminates the possibility of wrinkles and folds on stud fronts that can occur when a job is done carelessly. Usually, though, and especially when the insulation has its own vapor barrier, the flanges are stapled to stud or joist edges as shown in **Figure 20-11**, since this method provides a more effective continuous seal. Use extra staples where necessary to be sure flanges lie flat and smooth. Some flanges you must make yourself, but this is just a matter of cutting the material longer than you need to fill a cavity and then removing a small amount of the insulation from the end or ends requiring a flange.

FIG. 20-11

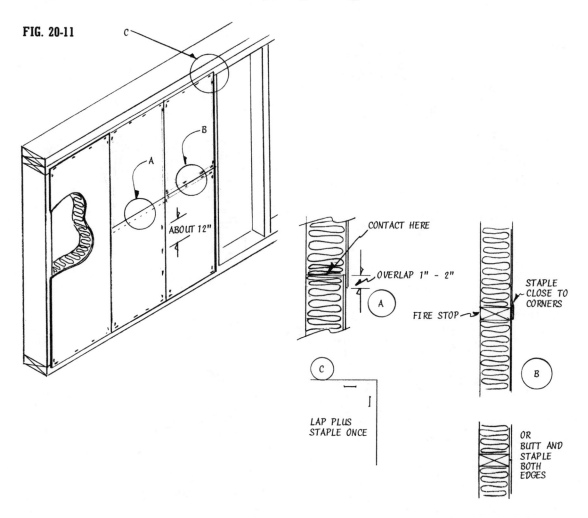

Thread insulation behind plumbing and electrical boxes. If gaps occur, use extra pieces of vapor barrier to cover them. If the insulation is behind a cold-water pipe, add a separate sheet of vapor barrier on the warm side of the room.

Ceilings can be done from below or from the attic. If the attic space is very tight and you are working with blankets or batts it will probably be better to work from below, doing the job along with the walls. Be sure the insulation is carried over the top wall plates and that the corner joints are lapped as shown in **Figure 20-12**.

If the ceiling is open, the insulation should be carried along the rafters as shown in **Figure 20-13**. Note that it is necessary to leave an air space between the insulation and the roof sheathing.

FIG. 20-12

FIG. 20-13

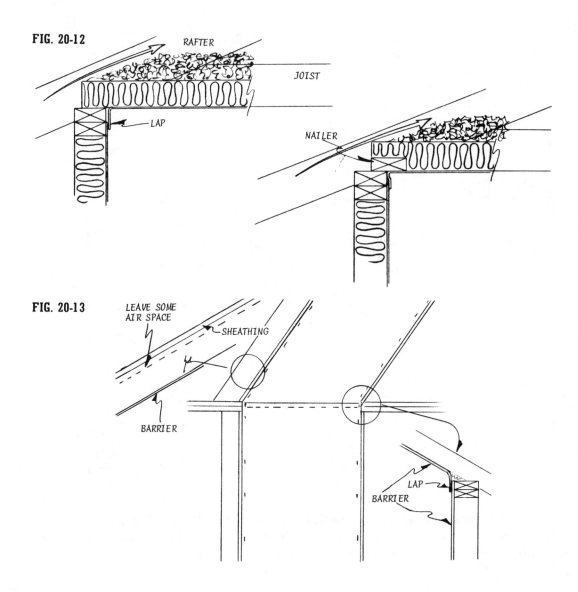

Don't neglect the gaps that exist around doors and windows. Fill the areas with insulation and be sure to provide an adequate vapor barrier (**Figure 20-14**).

FIG. 20-14

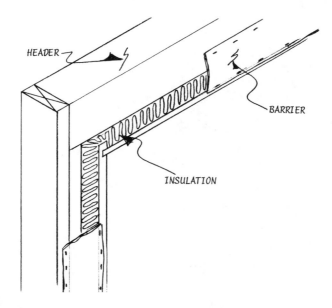

Blanket insulation is easy to cut if you use a sharp knife and a straight piece of wood as a straightedge. Place the wood on the line of cut and then kneel on it to compress the material as much as possible. Then run the knife along the guide, more than once if necessary. Cutting may also be done with snips. Fibers of insulation can be irritating, so you may want to wear gloves and even a respirator.

THE ATTIC IS CRITICAL

Warm air rises, so a considerable amount of heat passes through the attic. Tests reveal that just *minimum* insulation can save up to 20 percent on heating bills and comparable amounts on the cost of air conditioning.

The job can be done with blankets or batts, or with loose fill, or by combining both as in **Figure 20-15**. The table in **Figure 20-16** tells the R-values you can get with various materials of different thicknesses.

If you use loose fill and the ceiling below does not have a vapor barrier, you can provide one by installing polyethylene film between the joists as shown in **Figure 20-17**. Be sure to use a retainer at all vent areas, but do not install it so it becomes a seal that will prevent air flow (**Figure 20-18**).

If there are light fixtures recessed between the joists, it's a good idea to leave some space between the enclosure and the insulation (**Figure 20-19**).

FIG. 20-15

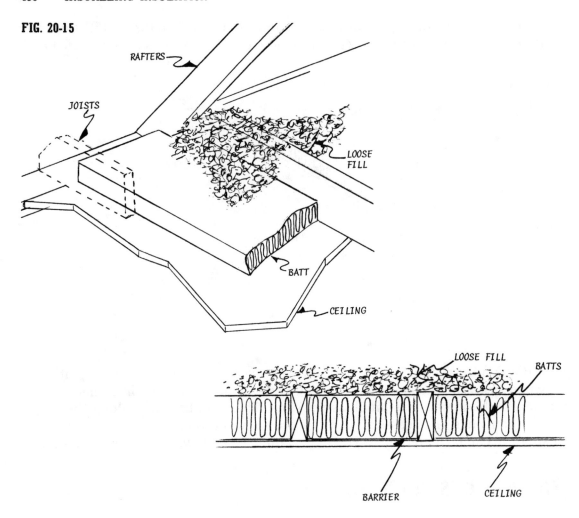

FIG. 20-16.   R-VALUES OF INSULATING MATERIALS

THICKNESS OF INSULATION REQUIRED FOR R-VALUE OF	WITH BATTS OR BLANKETS OF		WITH LOOSE FILL OF		
	Glass Fiber	Rock Wool	Glass Fiber	Rock Wool	Cellulosic Fiber
R-11	3½"-4"	3"	5"	4"	3"
R-19	6"-6½"	5¼"	8"-9"	6"-7"	5"
R-22	6½"	6"	10"	7"-8"	6"
R-30	9½"-10½"	9"	13"-14"	10"-11"	8"
R-38	12"-13"	10½"	17"-18"	13"-14"	10"-11"

FIG. 20-17

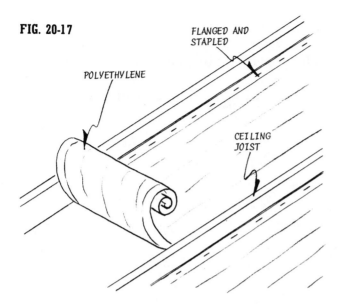

POLYETHYLENE

FLANGED AND
STAPLED

CEILING
JOIST

FIG. 20-18

FIG. 20-19

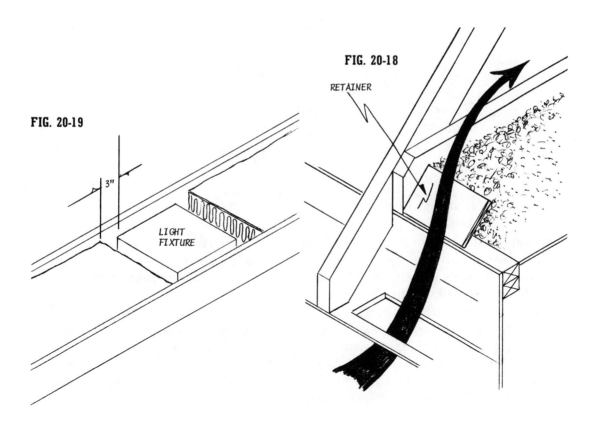

RETAINER

3"

LIGHT
FIXTURE

Loose fill is easy to use in an attic if you have the type that you simply pour from a bag. The material is dumped between joists and then leveled off with a long, straight piece of wood. If the thickness of the insulation is greater than the depth of the joists, erect some "headers" at right angles to the joists to act as retainers and to act as a height guide (**Figure 20-20**). The spacing of the headers is not critical.

FIG. 20-20

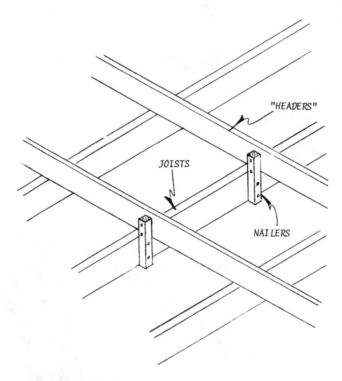

"HEADERS"

JOISTS

NAILERS

THE CRAWL SPACE

The temperature in a crawl space, since it must be vented, is about the same as outside temperature, so the floor area should be insulated like any other part of the house. Blankets or batts may be placed between joists, but remember that the vapor barrier must face the floor (**Figure 20-21**).

This makes it difficult to use stapling flanges in normal fashion, so the job is usually done with wood strips or, preferably, wire mesh to support the insulation. When you encounter bridging, cut the installation in one of the two ways shown in **Figure 20-22**. If you make straight rather than ragged cuts, use small pieces of insulation to fill in. It's also a good idea to cut pieces of insulation so you can fit them against joist headers (**Figure 20-23**).

FIG. 20-21

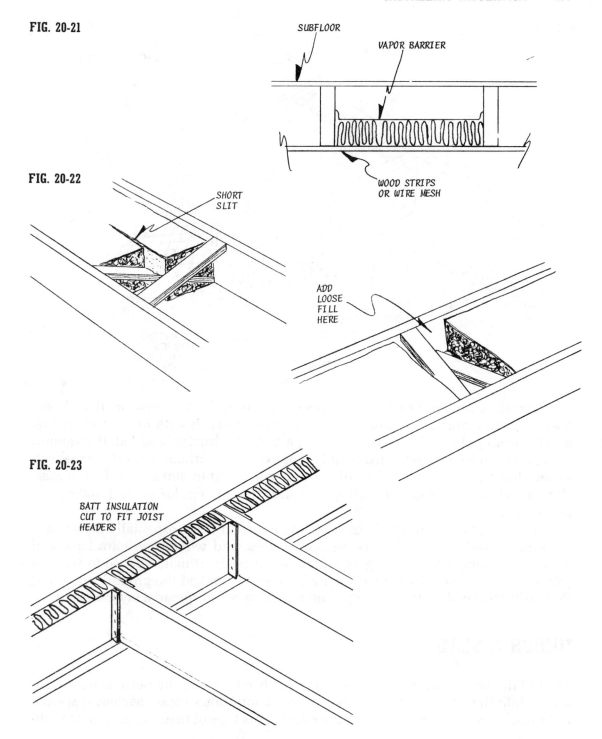

SUBFLOOR

VAPOR BARRIER

WOOD STRIPS
OR WIRE MESH

FIG. 20-22

SHORT
SLIT

ADD
LOOSE
FILL
HERE

FIG. 20-23

BATT INSULATION
CUT TO FIT JOIST
HEADERS

FIG. 20-24

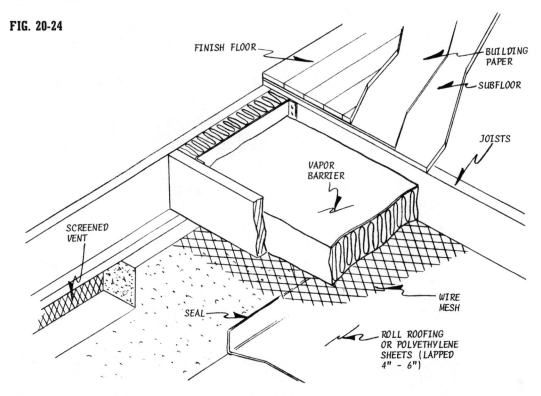

FINISH FLOOR

BUILDING PAPER

SUBFLOOR

JOISTS

VAPOR BARRIER

SCREENED VENT

WIRE MESH

SEAL

ROLL ROOFING OR POLYETHYLENE SHEETS (LAPPED 4" - 6")

Overall, a properly done crawl-space job should look something like **Figure 20-24**. Ground moisture is controlled by covering the soil with heavy roll roofing or with 4-mil polyethylene sheets. If such a moisture barrier *is* added, the amount of venting can be reduced considerably. For example, without a moisture seal you should have about 1 square foot of vent for every 150 to 200 square feet of area. With a seal, the requirements call for 1 square foot of vent for every 1,500 square feet of area.

Installing the seal properly calls for some careful work, especially if you have to work around posts and piers. In such places, and where there are laps, and where the material folds up against foundation walls, it makes sense to seal the joints with a mastic. Place the barrier after you have raked the ground smooth. If that is difficult to do, remove lumps and high ridges and spread a layer of sand.

UNDER A SLAB

Most of the heat loss through a concrete slab occurs along the perimeter, so usually, while the entire slab is poured over a continuous vapor barrier, a special rigid insulation is used only along the edges. This type of insulation, able to withstand contact with wet concrete, comes in thicknesses that range from 1 to 2 inches, so you can make a choice depending on the type of heating you will use and prevailing temperatures. The insulation can be installed in various ways (**Figure 20-25**). Extensions under a slab or vertically on a foundation wall should be at least 2 feet.

FIG. 20-25

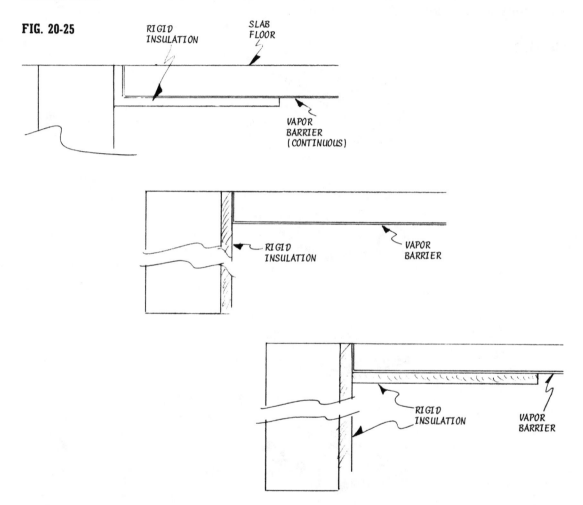

OTHER SEALS

A good deal of heat can be lost through small openings that might be present around doors and windows. If heat can get out, then cold air, and dust and noise, can get in. The guards you can use against such infiltration come under the classification of weatherstripping. The number and types that are available have kept pace with the interest in making homes tight (**Figure 20-26**). Modern, factory-built windows usually come equipped with efficient weatherstripping, but this does not apply to exterior doors unless a special order is placed. So usually exterior doors require special attention to seal them against sneaky drafts. Most products are do-it-yourself orientated, so they come in packages that contain detailed instructions on location and installation.

FIG. 20-26

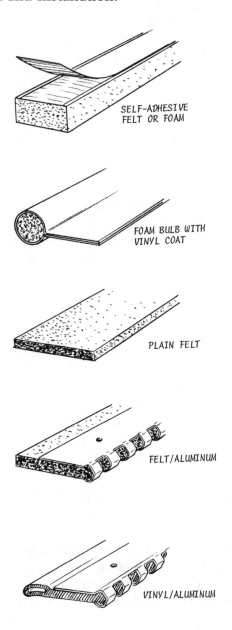

SELF-ADHESIVE FELT OR FOAM

FOAM BULB WITH VINYL COAT

PLAIN FELT

FELT/ALUMINUM

VINYL/ALUMINUM

A basic material is spring or cushioned metal strips which are nailed to jambs against the door stops (**Figure 20-27**). When you close the door, the metal is compressed to form a seal. More sophisticated types are two-piece designs that interlock (**Figure 20-28**). Some are easier to install because they are surface-mounted. Others require a little more work but are neater since they are not visible when the door is closed.

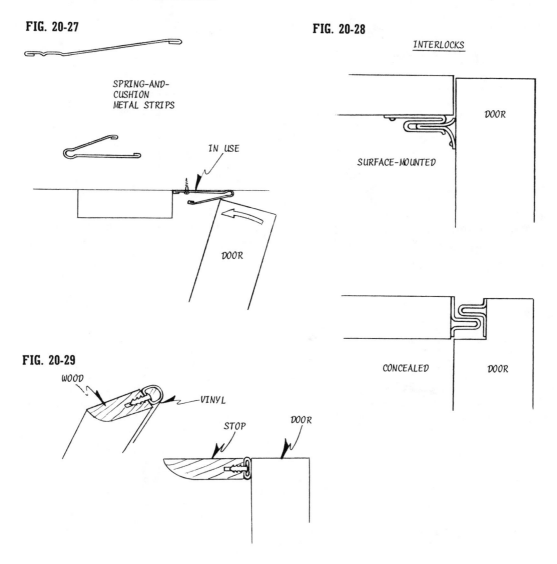

FIG. 20-27

SPRING-AND-CUSHION METAL STRIPS

IN USE

DOOR

FIG. 20-28

INTERLOCKS

DOOR

SURFACE-MOUNTED

CONCEALED   DOOR

FIG. 20-29

WOOD

VINYL

STOP   DOOR

Some products, which include a type of built-in sealing strip, are made to be used in place of regular wood stops (**Figure 20-29**).

Pay special attention to the joint at the bottom of the door. A widely used threshhold is made of metal and includes a vinyl insert (sometimes replaceable) that presses against the bottom of the door (**Figure 20-30**). Since it isn't a totally reliable waterseal, it's a good idea to add a drip cap on the outside of the door.

FIG. 20-30

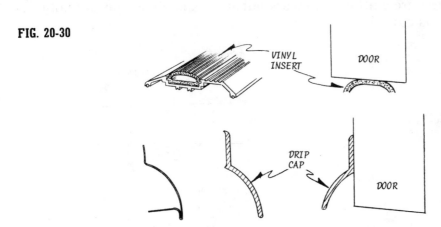

A newer type places the vinyl on the bottom of the door and has an integral drip-cap (**Figure 20-31**). Since the vinyl moves with the door, it doesn't take a beating by being walked on.

FIG. 20-31

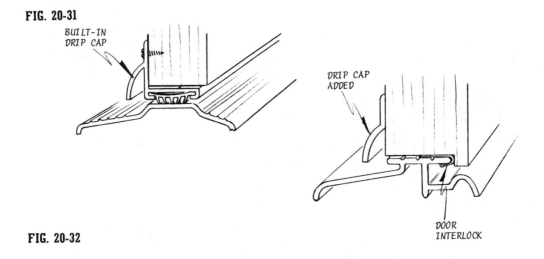

FIG. 20-32

An interlocking type is shown in **Figure 20-32**. Note that the interlock element is nailed to the bottom of the door and that the door is rabbeted.

A most basic kind of seal for a door bottom is a felt or vinyl strip that is nailed or screwed in place as shown in **Figure 20-33**.

FIG. 20-33

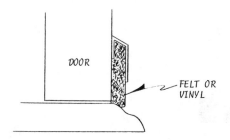

This can be done on any door, but is especially useful when you don't want the job to be too complicated—for example, on a garage door. Some are made so the seal strip moves up or down automatically.

21

BUILDING STAIRWAYS

The construction of a stairway can be simple or complex, with much depending on where the unit is located. Service or utility stairs—those coming from a base-ment or into an attic—can be basic and easy to assemble, especially if they are hidden; a necessary but purely functional element. Often, in the case of an attic, the means of access can be greatly simplified by installing a ready-made folding unit which appears only when needed (**Figure 21-1**). It does its job efficiently and does not consume floor space.

FIG. 21-1

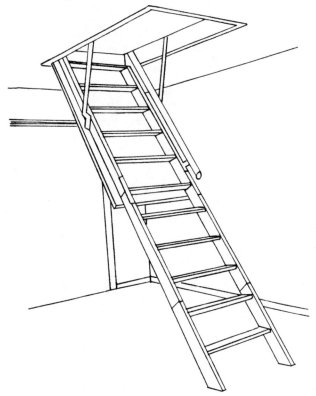

On the other hand, visible main stairs are often architectural features, as in **Figure 21-2**, and deserve the attention and the quality of construction you would ordinarily apply to a fine piece of furniture.

FIG. 21-2

Most professional builders do not require the carpenter to construct a main stairway from scratch right on the job. Specifications are given to a mill whose people specialize in such work; parts are fabricated and assembly is done at the site. Using this method imposes very few limitations on design choices, since available "ready-mades" range from conventional wood structures to more modern units, like that in **Figure 21-3,** which combine metal supports with treads of wood or metal or even concrete.

FIG. 21-3

All stairways can be open or closed (**Figure 21-4**). Or the design might combine the two ideas. The open design in Figure 21-4 shows a wall-like safety barrier, but a railing, of course, can be used just as well. The wall idea does have an advantage, since it provides a surface for decorations such as pictures or plants.

FIG. 21-4

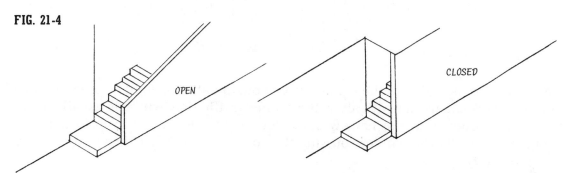

OPEN

CLOSED

Stairways that are straight runs are the easiest to build, but often, for esthetic or practical reasons, the run is broken by a landing, which is a platform that permits a directional change. When a landing is included, through necessity or choice, and space permits, it can be designed attractively so it makes a contribution to pleasant surroundings (**Figure 21-5**). You can actually gain much by being a little more generous with space than a basically practical solution would demand.

FIG. 21-5

Winder-type stairs are used when space is very tight, but they can be unsafe because of the narrowness of the treads at the convergence point (**Figure 21-6**).

FIG. 21-6

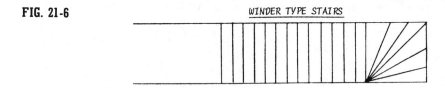

WINDER TYPE STAIRS

When they must be used, the major design consideration should be to maintain a uniform and safe tread width along the normal traffic line, which is usually about 15 to 18 inches from the narrow end. One way to do this is to have the convergence point of the winders *outside* the construction rather than at the corner (**Figure 21-7**).

FIG. 21-7

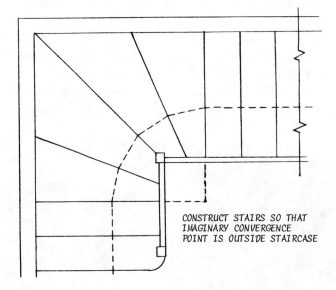

CONSTRUCT STAIRS SO THAT IMAGINARY CONVERGENCE POINT IS OUTSIDE STAIRCASE

THE FUNDAMENTALS

The easiest way to get from one level to another is to use a low-pitched ramp, but since this isn't practical inside a house a series of elevators—individual steps— are assembled to form a staircase. The steps have a tread and a riser and are supported by sawtooth stringers which decide the slope of the project (**Figure 21-8**).

FIG. 21-8

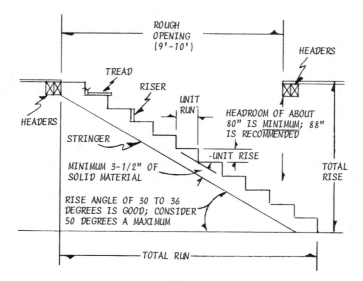

Headroom, rise angle, and all the other factors must be considered when doing initial planning so that the stairway will be safe and convenient to use.

The relationship between the tread and the riser is critical and some guides have been established which relate to what you might call the average climbing stride.

If you study **Figure 21-9** you will discover the importance of applying all three of the rules. For example, using a riser of 4 inches and a tread of 14 inches, rule #2 results in 18 inches which, apparently, is acceptable. But check with rules 1 and 3 and the result is far from the ideal. 2(4) plus 14 = 22; 4(14) = 56.

FIG. 21-9. STAIR TREAD AND RISER GUIDELINES

	THE BASIC RULES	IDEALLY, Y = 7½″, X = 10″
1	2(Y) plus X should = about 25	2(7½) plus 10 = 25
2	Y plus X should = 17 to 18	7½ plus 10 = 17½
3	Y(X) should = about 75	7½ × 10 = 75

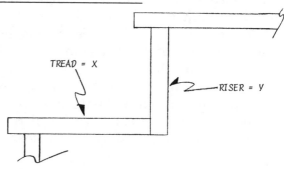

To use the ideal figures would mean having to design around the staircase so total rise and total run would be suitable. This isn't always possible, but you should try to work within reasonable tolerances. For a main stairway, risers can change from 6½ inches to 7¾ inches; treads from 10½ inches to 11 inches. You can see that the range does permit adjustments, so you can fit a stairway in available space without sacrificing safety.

As riser height increases, fewer riser treads are required. The fewer treads there are, the less space is required for the stairway. That's why attic and basement stairs are often built without regard for safety, which is poor thinking really. Excessively high risers and narrow treads should be avoided.

HOW WIDE?

Recommended minimum widths (clear of the handrail) are 32 inches for main stairs and 30 inches for utility stairs. If you agree that stairs should be wide enough so that people can pass and furniture can be transported with a minimum of trouble, then consider 36 inches as a minimum and 42 inches, if space permits, as ideal. The extra inches are especially wise if the stairway is closed or makes a turn or involves winders.

A WAY TO PLAN

Divide the total rise, say 9 feet, by 7. The answer, in this case, is 15.429, but only the whole number is used to tell the number of risers that will be required if each is to have an approximate height of 7 inches. Divide the total rise (108 inches) by the number of risers (15) and you get the actual individual riser height of 7³/₁₆-inches. Work with rule #2 of **Figure 21-9** and you will see that a tread about 10 inches wide is compatible with the riser height. Since the number of treads is always one less than the number of risers, multiply 10 by 14 to find the total run. Adjustments can be made by changing the riser and tread dimensions as long as you don't violate the basic rules.

You can see why it is wise to consider a stairway installation when you are doing the house framing. Changing the height of a ceiling to accommodate a flight of steps might be a bit extreme, but it is no trouble to plan the length of the rough opening so that it will accept a stairway you can design with a minimum of fuss.

The table in **Figure 21-10**, prepared by the Colonial Stair and Woodwork Company, shows suitable riser-tread relationships for installations of various heights and may be used as a guide for your own project.

FIG. 21-10.   TREAD AND RISER RELATIONSHIPS

TOTAL RISE	NUMBER OF RISERS	RISER HEIGHT	NUMBER OF TREADS	TREAD WIDTH
7'-8"	12	$7^{21}/_{32}$"	11	9" to 9¾"
8'-0"	13	7⅜"	12	9½" to 10"
8'-4"	14	7⅛"	13	10" to 10½"
8'-9"	14	7½"	13	$10^5/_{16}$"
9'-0"	15	$7^3/_{16}$"	14	10" to 10½"

THE CONSTRUCTION

The stringer, often called the *carriage* or *rough horse*, is the main structural member and should be made of strong rigid material. Sometimes 1× stock is used, but 2×10s or 2×12s are recommended. If you view the profile of a straight-run flight, you will see that it is a right triangle, the hypotenuse representing, roughly, the length of the stringer. You can pick this up on the job merely by stretching a line from the top to the bottom of the stairs, but add a foot or so to this to find the length of the stringer material to start with.

Another way is to make a full-scale layout of one or two tread-riser cuts and then multiply the bridge dimension, shown in **Figure 21-11**, by the number of risers. Add a foot or so to the result and you will have a good stringer length to start with.

FIG. 21-11

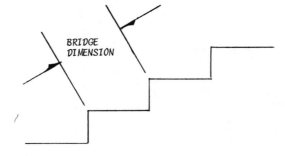

BRIDGE DIMENSION

All stairway material should be kiln-dried and warp-free. Hidden stringers or those used in a basement don't have to be pretty, but visible ones on a main, open stairway should be carefully selected for appearance as well as strength.

Usually, stringers are notched to form a sawtooth pattern that supplies a

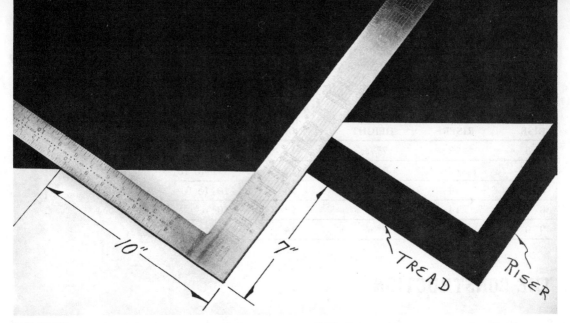

FIG. 21-12

nailing surface for each riser and tread. The pattern can be laid out by using a square as shown in **Figure 21-12**, or by making a special template and using it as shown in **Figure 21-13**.

FIG. 21-13

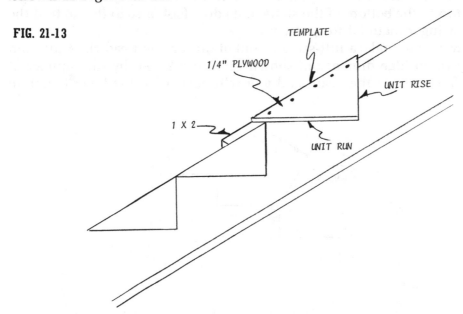

The latter method is advisable, since it minimizes the possibility of human error.

The cutting can be done entirely with a handsaw, or you can use an electric cutoff saw too, moving as far as possible into the cut and then finishing with a handsaw. Similar stringers can be clamped together so two (or three) can be notched at the same time. If sawing through such thicknesses bothers you, then form one and use it as pattern to mark the others. The cut stringers should look like **Figure 21-14**.

FIG. 21-14

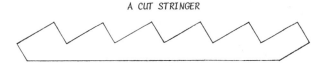

A CUT STRINGER

A widely used type of stairway—usually prefab—is called housed construction and consists of stringers, risers, and treads that are assembled with glue and wedges as shown in **Figure 21-15**.

FIG. 21-15

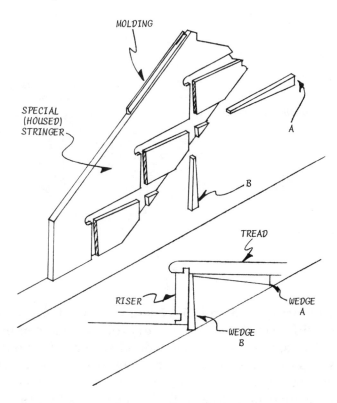

MOLDING

SPECIAL
(HOUSED)
STRINGER

A

B

TREAD

RISER

WEDGE
A

WEDGE
B

This is top-quality work and when correctly installed results in a handsome stairway that is strong, dustproof, and squeakless. This kind of thing can be done on the job, but not without a router and a special adjustable template, which you can probably rent, that serves as a guide for forming the tapered grooves.

Built-up stringers can be formed by cutting triangular blocks with the base and altitude representing tread and riser widths. When the stairway is against a wall, the blocks and a plain piece of wood used as a backing stringer can be done as a subassembly which is then spiked securely to the studs (**Figure 21-16**). Outboard stringers have the blocks nailed to a 2×4 or a 2×6 or whatever is suitable (**Figure 21-17**).

FIG. 21-16

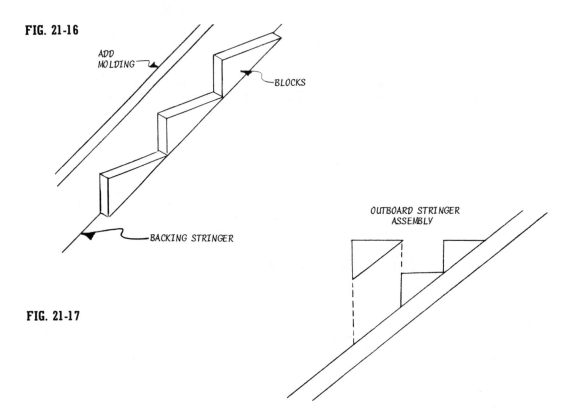

ADD
MOLDING
BLOCKS
BACKING STRINGER

OUTBOARD STRINGER
ASSEMBLY

FIG. 21-17

If you study **Figures 21-16** and **21-17** you will see that if you do a sawtooth-type stringer you will be cutting out triangular blocks that can be used to make a built-up stringer. Thus, you can achieve added strength without extra cost.

Open type stairs—no riser boards—are often used in basements or outdoors. Here, stringers can be formed by using cleats to provide tread support. A more sophisticated way is to form dadoes in the stringers. As **Figure 21-18** shows, riser boards are easily added if you wish to close in the steps.

Open stairs can also be done as shown in **Figure 21-19,** which presents the opportunity to be a bit more creative. But don't let your imagination run to the point where you neglect the safety factors.

FIG. 21-18

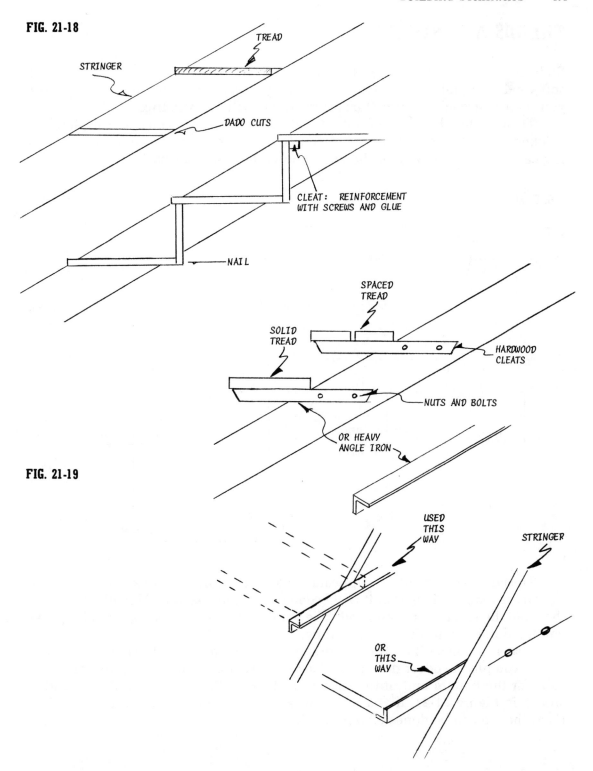

STRINGER

TREAD

DADO CUTS

CLEAT: REINFORCEMENT
WITH SCREWS AND GLUE

NAIL

SPACED
TREAD

SOLID
TREAD

HARDWOOD
CLEATS

NUTS AND BOLTS

OR HEAVY
ANGLE IRON

FIG. 21-19

USED
THIS
WAY

STRINGER

OR
THIS
WAY

TREADS AND RISERS

Stair treads are strongest if they are made of hardwood or a vertical-grain softwood. Flat-grain softwoods are not really suitable for a main stairway unless you plan to use an additional cover material such as carpeting.

Tread material with the front edge already shaped as a nosing is available in thicknesses that may be $1\frac{1}{16}$ inches or $1\frac{1}{8}$ inches. If you make your own, the nosing can be shaped in any of the ways shown in **Figure 21-20**.

FIG. 21-20   SHAPES OF TREAD NOSINGS

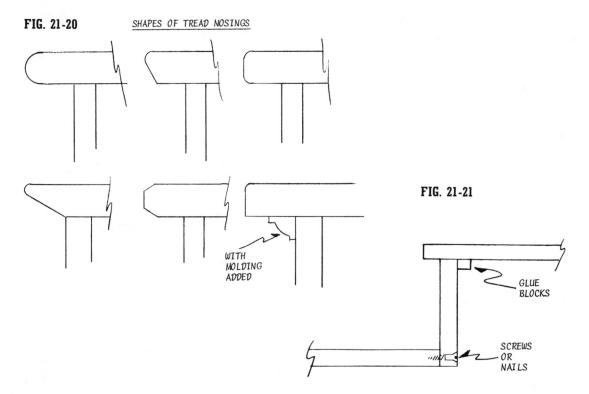

WITH MOLDING ADDED

FIG. 21-21

GLUE BLOCKS

SCREWS OR NAILS

The nosing contributes to appearance, but it should never extend beyond the riser more than $1\frac{1}{2}$ inches or it will become a tripping hazard. Actually, the wider the tread the less nosing you need. The width of the nosing is *not* included when you are doing the initial dimensioning.

Risers can be $\frac{3}{4}$ inch thick and should be made of tread-matching material unless you plan to cover the stairs. Good construction, when butt joints are used, calls for the bottom end of the riser to mate with the back edge of the tread as shown in **Figure 21-21**. If you use screw-reinforced glue blocks at the top of the riser, the job can be done wihout visible nails.

Another way to do the job, which takes some time but results in premium quality, is to use dado and rabbet joints as shown in **Figure 21-22**.

FIG. 21-22

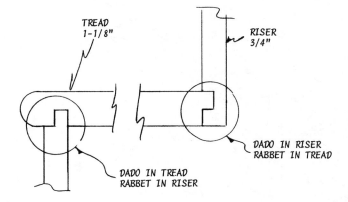

TREAD
1-1/8"

RISER
3/4"

DADO IN RISER
RABBET IN TREAD

DADO IN TREAD
RABBET IN RISER

INSTALLATION

Stairway installation can be started early in the construction by considering the stringers as part of the rough framing. Framing for a straight-run stairway is shown in **Figure 21-23**, and framing for a stairway with a landing in **Figure 21-24**.

FIG. 21-23

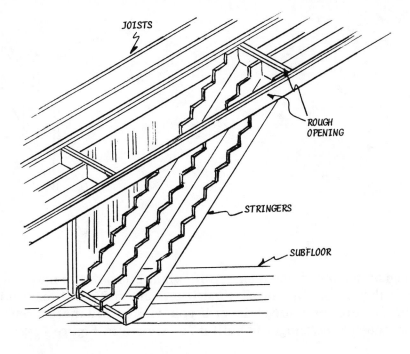

JOISTS

ROUGH
OPENING

STRINGERS

SUBFLOOR

FIG. 21-24

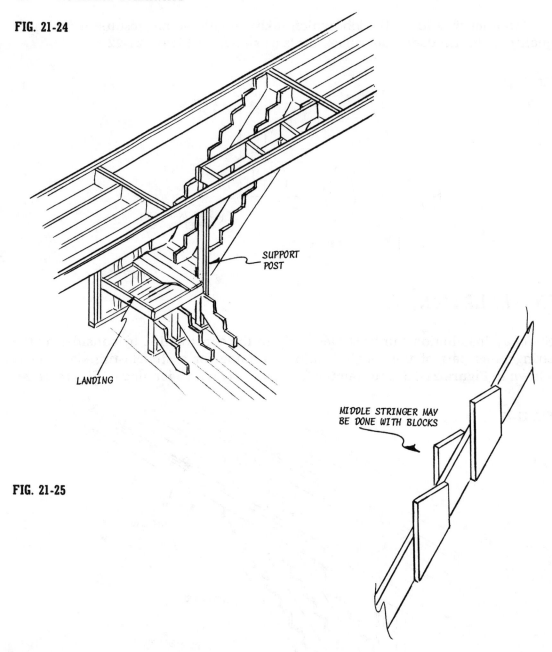

SUPPORT
POST

LANDING

MIDDLE STRINGER MAY
BE DONE WITH BLOCKS

FIG. 21-25

This is a good method, since it's not difficult at this point to make adjustments in the size of the rough opening. Two stringers are sufficient for minimum-width stairs, but use three on wide stairs or if you wish to add more strength and rigidity. Often, the center stringer is built up as shown in **Figure 21-25**.

One way to provide some additional headroom at the bottom of the stairs when space is tight is to install auxiliary headers (**Figure 21-26**). This, in effect, merely provides for a slight extension of the floor above while still allowing adequate headroom for the stairs.

FIG. 21-26

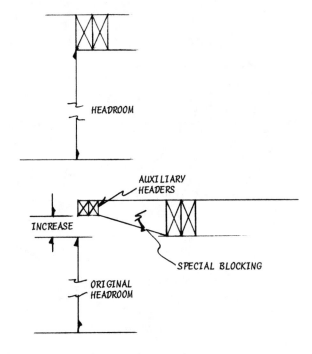

Stringer atttachments at the top end can vary considerably, with much depending on whether the top tread will be flush with the finish floor. If it will be, you can add considerable strength while making installation easier by notching the stringers and including a support ledger (**Figure 21-27**).

FIG. 21-27

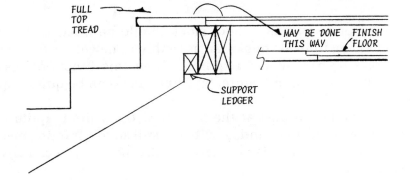

If the top tread is below the finish floor, then you can use either of the methods shown in **Figure 21-28**. Such attachments are especially important on an open stairway. On a closed stairway, the outboard stringers are securely spiked to the wall frames, which results in a very strong installation.

FIG. 21-28

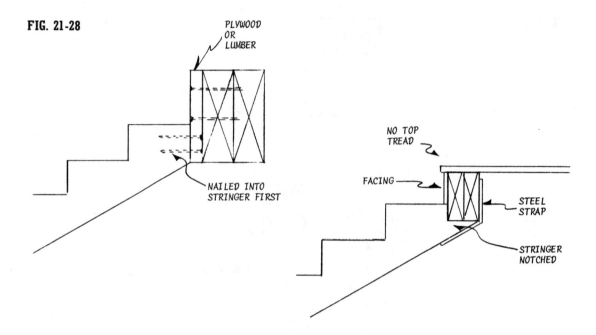

The attachment at the bottom is much simpler, since it doesn't require more than braces, which, if possible, are nailed through the subfloor into joists. **Figure 21-29** shows the stringer braces and also some of the finish work.

THE BALUSTRADE

Stairways that are open on one or both sides should be lined with safety rails. Since these can't hang from skyhooks, an assembly of newel posts, handrail, and intermediate posts or balusters is required. Often this functional assembly is emphasized as an architectural feature. So balustrades can be quite impressive if you choose.

Structurally, the newel post at the bottom of the stairs is quite important. Often it passes through the floor and is bolted or spiked to a framing member (**Figure 21-30**). If a framing member is not conveniently located you can always add a header between joists.

FIG. 21-29

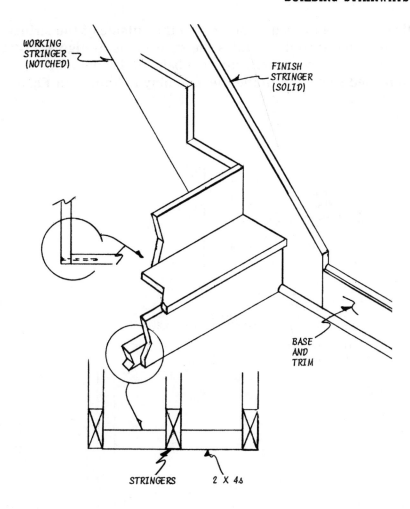

WORKING
STRINGER
(NOTCHED)

FINISH
STRINGER
(SOLID)

BASE
AND
TRIM

STRINGERS   2 X 4s

FIG. 21-30

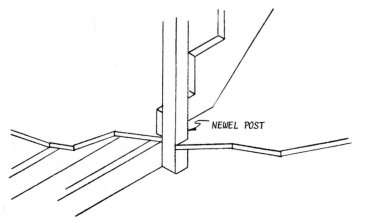

NEWEL POST

Other ways are to secure the post to the outside of the stringer, as in **Figure 21-31**, or to notch it so it straddles the stringer, as in **Figure 21-32**.

You can, of course, make your own balustrade components by working with clear kiln-dried stock and doing the assembly as shown in **Figure 21-33**.

FIG. 21-31

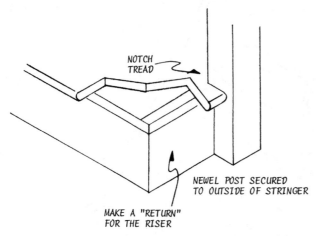

NOTCH TREAD

NEWEL POST SECURED TO OUTSIDE OF STRINGER

MAKE A "RETURN" FOR THE RISER

FIG. 21-32

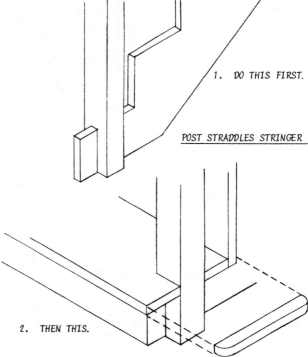

1. DO THIS FIRST.

POST STRADDLES STRINGER

2. THEN THIS.

FIG. 21-33

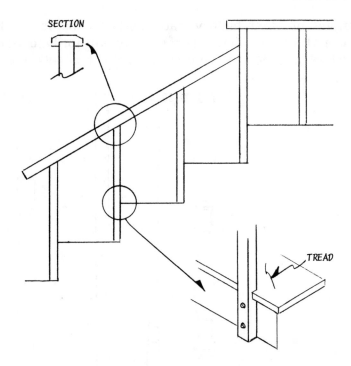

SECTION

TREAD

Handrails are most convenient when they are placed 30 inches above steps and 34 inches above landings (**Figure 21-34**). On wide, open stairs and on closed stairs, stock handrails are often used. These are easily attached to walls by using ready-made brackets designed for the purpose (**Figure 21-35**).

FIG. 21-34

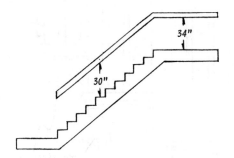

34"

30"

FIG. 21-35

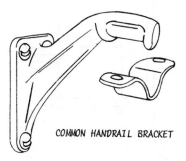

COMMON HANDRAIL BRACKET

A few examples of other factory-made parts are balusters, **Figure 21-36**; newel posts, **Figure 21-37**; complete sets of components for balustrades of various designs, **Figure 21-38**; and decorative pieces such as brackets, **Figure 21-39**.

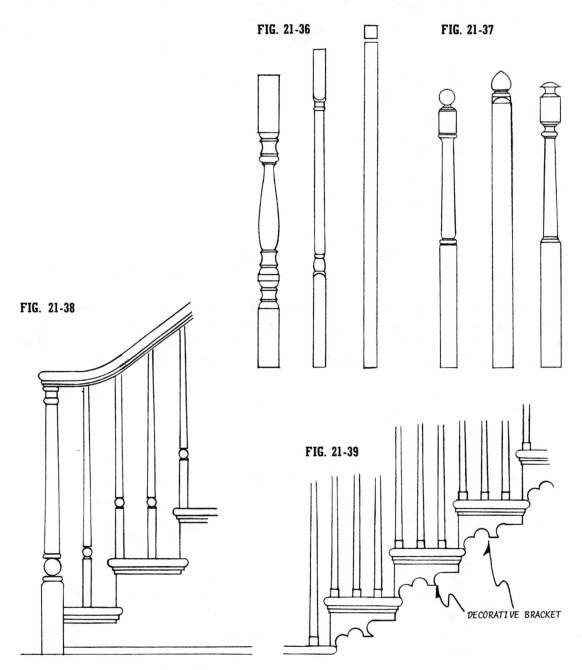

FIG. 21-36

FIG. 21-37

FIG. 21-38

FIG. 21-39

DECORATIVE BRACKET

22

CHIMNEYS FOR FIREPLACES AND WOODSTOVES

Home is where the hearth is. At one time a fireplace was a necessity because it was the only source of heat for warmth and cooking. It became less essential with the advent of more modern heating and cooking systems, so it became an option; included in a house more for appearance and atmosphere than for utility. But these days we're getting back to seeing the fireplace as a heat source, if only a minor supplemental one.

The sad truth about a conventional fireplace is that radiated heat—what enters the room—is slight when compared to what heat is lost up the chimney. Air passes through the fire, is heated, and then escapes up the flue. Simultaneously, outside air with a lower temperature is drawn into the room. People sitting within the radiation zone of the fire will feel its heat. Those who are farthest away may feel a cold draft. Sometimes, a few people stand directly in front of the fire and absorb all the heat so none is radiated into the room.

Despite the drawbacks, a well-designed and properly constructed fireplace is desirable because of its cheerfulness and charm, and it does provide *some* auxiliary heat. In some situations, say in a small vacation cabin or an outbuilding, it might well be adequate as the *only* source of heat, especially in milder climates.

An important consideration is that the heat loss of an ordinary fireplace can be reduced by designing the unit to include a convection system. This can be incorporated in a full-masonry installation by working along the lines shown in **Figure 22-1**. Cool air is drawn from the outside, passes through tubes in a special heating chamber, and is ejected into the room.

Usually, however, the air inlet is located in the room close to the floor. This can be done by installing ductwork in the masonry or by building the project around prefabricated metal units, which are merely set in place and concealed with a construction of your choice (**Figure 22-2**).

FIG. 22-2

FIG. 22-1

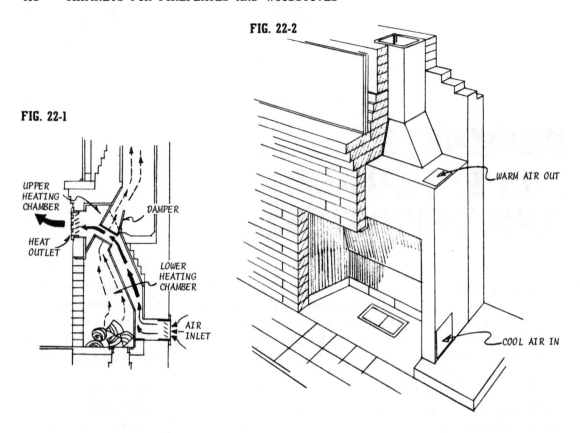

Tests have proved that appreciable amounts of convected heat can be produced by a good recirculating fireplace when it is properly installed and operated. Not only will it heat the room it is located in, but a proper duct installation can direct the heated air to adjoining or second-floor rooms.

BASIC FIREPLACE NOMENCLATURE

Figure 22-3 shows a conventional fireplace in cross section. The *firebox* is the chamber in which the fire is built. It is usually lined with special firebrick and is designed so that its sides splay outward while the upper part of its back slopes toward the room. The idea is to reflect the maximum amount of heat into the room. The *smoke chamber* serves as a funnel that directs smoke to the flue. There is a damper-controlled passage, called the throat, between the firebox and the smoke chamber. The *smoke shelf* is a barrier to downdrafts so they will eddy and be lifted upward by rising air currents. The *flue* carries off by-products of the fire. Its size in relation to the size of the firebox is very important if it is to carry off smoke the way it should.

FIG. 22-3

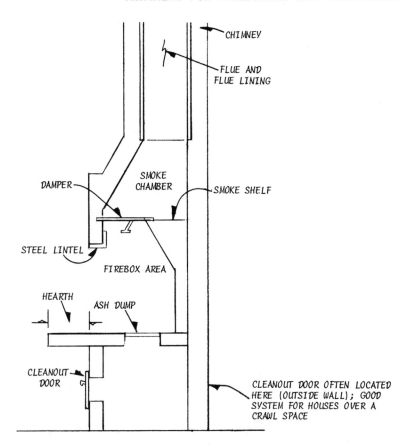

Not all fireplaces include an *ash dump*, probably because construction is easier without one. It's a question of convenience; with an ash dump, ashes are dumped into a fireproof subchamber to be toted off when convenient. Without one, ashes must periodically be shoveled into a bucket and carried out through the house.

Of course an ash dump can't be installed in the conventional manner if the house is built on a slab. But it can be included by designing a raised hearth with an ash pit beneath (**Figure 22-4**). Or a ready-made steel lift-out box can be sunk into the hearth (**Figure 22-5**).

BASICS OF MASONRY CONSTRUCTION

The masonry fireplace and chimney should be an independent structure that does not receive support from or provide support for other building components. If you were to remove the house, the fireplace and its chimney would still stand and be

FIG. 22-4

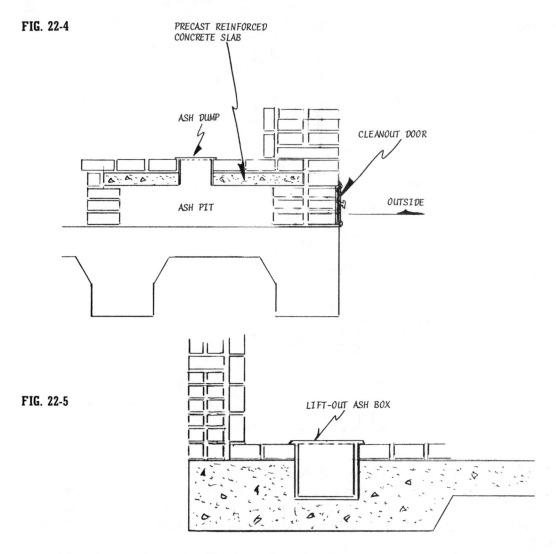

PRECAST REINFORCED CONCRETE SLAB

ASH DUMP

CLEANOUT DOOR

ASH PIT

OUTSIDE

FIG. 22-5

LIFT-OUT ASH BOX

operable. This applies whether the unit is inside the walls or is part of an outside wall as shown in **Figure 22-6**.

This means a concrete base, which may be poured along with the regular foundation, that is capable of supporting the project so it won't settle or crack. Generally, the base, or footing, should be deeper than the frost line and 6 inches greater in width and length than the project itself. The climate of the area will affect how thick the footing must be, so check with local codes before you start. Even when there is no subgrade frost line, the footing must still support considerable weight (about 130 pounds per cubic foot for brick), so you don't want to skimp on initial construction.

FIG. 22-6

THE SIZE FOR THE ROOM

The size of a fireplace should be appropriate for the room it is in. A firebox width of 30 to 36 inches is okay for a room roughly 17 or 18 feet square. As a general guide, you can figure that the fireplace opening should be about 5 square inches per square foot of floor space. This formula is most applicable for small rooms; the ratio may be reduced in large rooms.

THE SIZE OF THE FIREBOX

General guides say that the height of the opening should be about ⅔ to ¾ of its width. The splay of the side walls and the height of the vertical part of the back wall are shown in **Figure 22-7**.

There is a critical interrelationship in all dimensions of a fireplace, including the cross-sectional area of the flue. If the size of the flue is too small for the size of the opening facing the room, the fireplace will not be drafted correctly and will

FIG. 22-7

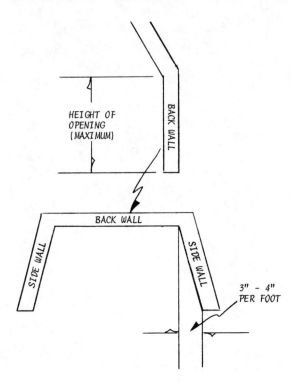

probably be smoky. A larger flue than necessary will merely result in too much downdraft and an excessively generous entrance for rainwater, unless you mount a chimney roof on top.

A fireplace proportioned to function efficiently can be designed by following the rules set forth in the diagrams and table in **Figure 22-8**, which were originated by the Majestic Company, makers of fireplaces.

CONSTRUCTION MATERIALS

Brick is a good choice for fireplace and chimney construction because it is readily available and "comfortable" to work with. The units are light and you can pause at any time during the installation without losing anything unless you mix too much mortar. The rule is to mix only as much as you can easily apply in an hour or so.

Common brick, bonded with regular masonry mortar (1 part Portland cement, 1 part hydrated lime, 5–6 parts sand), can be used for outer walls, but the firebox must be lined with special firebrick that is bonded with fireclay (**Figure 22-9**).

FIG. 22-8. **FIREPLACE SIZING TABLE IN INCHES** (Keyed to drawings below)

| | Finished Opening | | | | | | Rough Masonry | | | | |
A	B	C	D	E	F	G	H	I	K	Inside of Brick Flue	Standard Flue Lining
Width	Height	Depth	Back	Throat	Width	Depth	Smoke Shelf Height	Smoke Chamber	Vertical Back		
24	28	16	16	9	30	19	32	11	14	$8\frac{1}{2} \times 8\frac{1}{2}$	$8\frac{1}{2} \times 8\frac{1}{2}$
26	28	16	18	9	32	19	32	11	14	$8\frac{1}{2} \times 8\frac{1}{2}$	$8\frac{1}{2} \times 8\frac{1}{2}$
28	28	16	20	9	34	19	32	11	14	8×12	$8\frac{1}{2} \times 13$
30	30	16	22	9	36	19	34	11	15	8×12	$8\frac{1}{2} \times 13$
32	30	16	24	9	38	19	34	11	15	8×12	$8\frac{1}{2} \times 13$
34	30	16	26	9	40	19	34	11	15	12×12	$8\frac{1}{2} \times 13$
36	31	18	27	9	42	21	36	11	16	12×12	13×13
38	31	18	29	9	44	21	36	11	16	12×12	13×13
40	31	18	31	9	46	21	36	11	16	12×12	13×13
42	31	18	33	9	48	21	36	11	16	12×12	13×13
44	32	18	35	9	50	21	37	11	17	12×12	13×13
46	32	18	37	9	52	21	37	11	17	12×16	13×13
48	32	20	38	9	54	23	37	$15\frac{1}{2}$	17	12×16	13×18
50	34	20	40	9	56	23	39	$15\frac{1}{2}$	18	12×16	13×18
52	34	20	42	9	58	23	39	$15\frac{1}{2}$	18	12×16	13×18
54	34	20	44	9	60	23	39	$15\frac{1}{2}$	18	16×16	13×18
56	36	20	46	9	62	23	41	$15\frac{1}{2}$	19	16×16	18×18
58	36	22	47	9	64	25	41	$15\frac{1}{2}$	19	16×16	18×18
60	36	22	49	9	66	25	41	$15\frac{1}{2}$	19	16×16	18×18

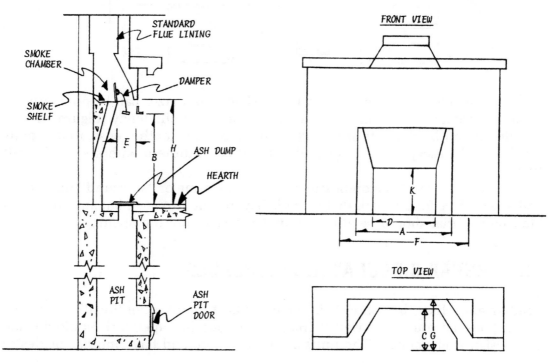

STANDARD FLUE LINING

SMOKE CHAMBER

SMOKE SHELF

DAMPER

E

B

H

ASH DUMP

HEARTH

ASH PIT

ASH PIT DOOR

FRONT VIEW

K

D

A

F

TOP VIEW

C G

FIG. 22-9

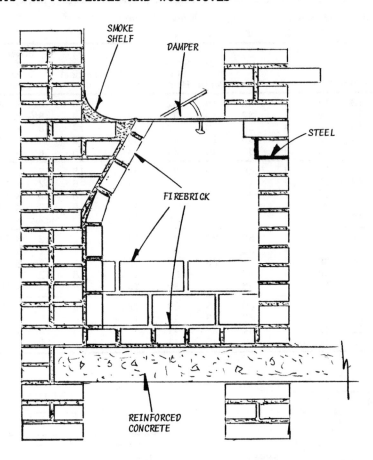

Ready-made steel dampers of various shapes and sizes can be purchased to make construction easier and to guarantee more draft efficiency (**Figure 22-10**).

Be sure all brick joints are packed tight and are smooth. Loose, rough joints can let smoke escape where it shouldn't and can interfere with a smooth draft flow.

You can use any noncombustible material to frame or "surround" the firebox and as a surface for a hearth that extends beyond the fire area, but any combustible trim material should be at least 6 inches away from the opening.

THE PREFAB CIRCULATOR FIREPLACES

Ready-mades like that shown in **Figure 22-11**, with scientifically designed proportions, let you do a masonry fireplace without the paperwork required when you design from scratch. The units are hollow-core and have interior blades as

FIG. 22-10

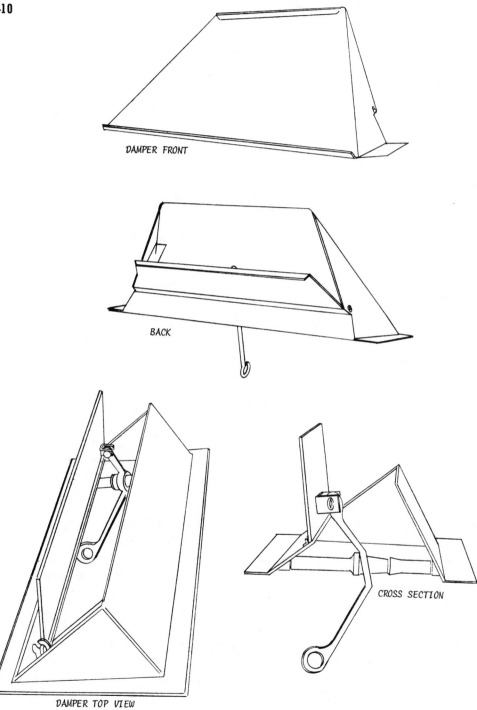

DAMPER FRONT

BACK

DAMPER TOP VIEW

CROSS SECTION

FIG. 22-11

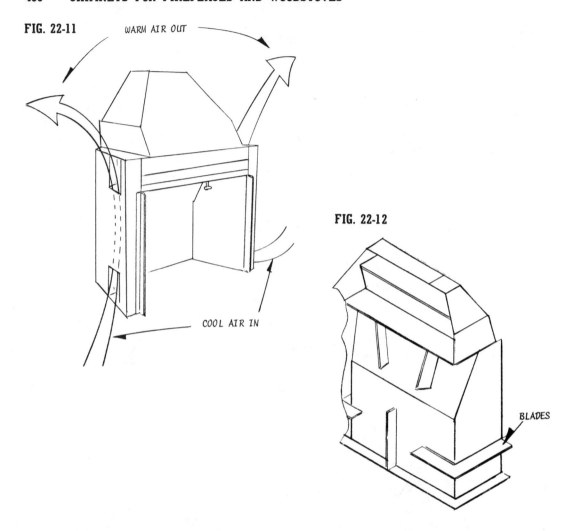

WARM AIR OUT

COOL AIR IN

FIG. 22-12

BLADES

shown in **Figure 22-12** placed to direct air over the hottest parts of the fireplace and to increase radiating-surface and heat-absorbing areas.

The products supply the radiant heat you get from any ordinary fireplace plus warm-air convection currents that can be directed into the fireplace room or into adjacent rooms. This is accomplished with ductwork and inlet and outlet grills. How the grills are placed depends on where the fireplace is located. If it is outside a wall, then the grills must be placed on the face side. If the fireplace projects into the room, then the grills may be on the front or at the sides, or there can be a combination of front and side positions. The inlets should be close to the floor line, but the location of the outlets is optional, as shown by the sample installations in **Figure 22-13**.

FIG. 22-13

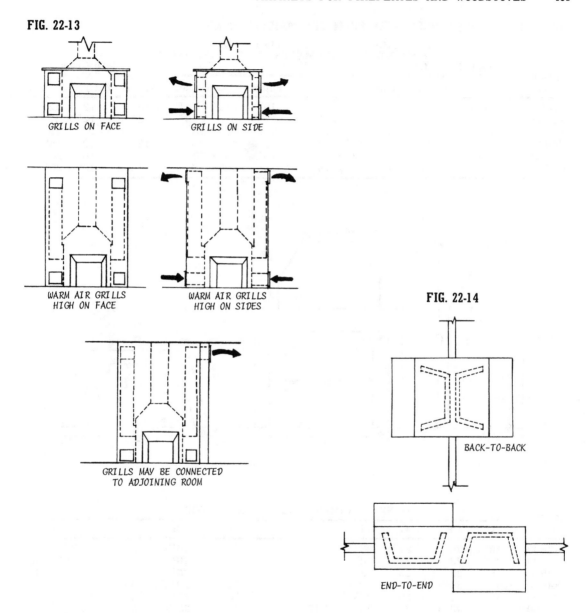

GRILLS ON FACE

GRILLS ON SIDE

WARM AIR GRILLS
HIGH ON FACE

WARM AIR GRILLS
HIGH ON SIDES

GRILLS MAY BE CONNECTED
TO ADJOINING ROOM

FIG. 22-14

BACK-TO-BACK

END-TO-END

Working with a prefab circulator doesn't restrict fireplace location. Actually, by using several units you can do back-to-back and end-to-end designs, so adjoining rooms can each have a fireplace (**Figure 22-14**).

Typical sizes are shown in **Figure 22-15**. The model numbers were taken from the Majestic Company's catalog, but they should be checked out to be sure no changes have occurred since this writing.

FIG. 22-15.   MATED FIREPLACES FROM THE MAJESTIC COMPANY

MODEL NO.	AVERAGE ROOM CAP. (CU. FT.)	FIREPLACE FINISHED OPENING		FLUE SIZES CHIMNEYS OVER 20'		CHIMNEYS UNDER 20'		MODULAR FLUE SIZES, RECTANGULAR
		Wide	High	Rectangle	Round	Rectangle	Round	
R2800	3520	28″	22″	8½″ × 13″	10″	8½″ × 13″	10″	12″ × 12″
R3200	3850	32″	24″	8½″ × 13″	12″	8½″ × 13″	12″	12″ × 12″
R3600	4565	36″	25″	13″ × 13″	12″	13″ × 13″	12″	12″ × 16″
R4000	4950	40″	27″	13″ × 13″	12″	13″ × 13″	12″	12″ × 16″
R4600	5720	46″	29″	13″ × 13″	12″	13″ × 18″	15″	16″ × 20″
R5400	7040	54″	31″	13″ × 18″	15″	13″ × 18″	15″	16″ × 20″

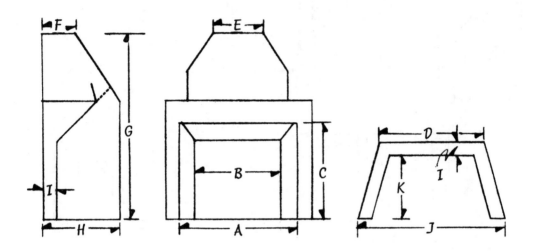

FIG. 22-15A.

GENERAL DIMENSIONS OF CIRCULATOR

MODEL NO.	A	B	C	D	E	F	G	H	I	J	K
R2800	28″	20″	22″	25″	12″	8″	44″	18¼″	3″	34¾″	15″
R3200	32″	23″	24″	28½″	12″	8″	47½″	19¼″	3″	38½″	16″
R3600	36″	27″	25″	32″	12″	12″	51½″	20½″	3″	43″	17″
R4000	40″	30½″	27″	36½″	12″	12″	55″	20¾″	3½″	47¾″	17″
R4600	46″	36″	29″	42½″	17″	12″	60″	22½″	4″	55″	18″
R5400	54″	43¼″	31″	50½″	17″	12″	65″	23½″	4″	62½″	19″

CHIMNEYS AND FLUES

Most chimneys today, because of codes, are lined with fireclay flues that have a minimum wall thickness of ⅝ inch. Using these flues is sensible even when it is not demanded, since they make a straight, smooth channel that lets smoke pass rapidly, and since the flue sections provide an excellent form around which you can place the masonry.

Flue-lined chimneys should be at least 4 inches thick. Chimneys without fireclay flues should be at least 8 inches thick and should be lined with firebrick bonded with fireclay mortar. Be sure to check out the latter method against local building codes.

Most codes will permit installing more than one flue in a single chimney when exhaust systems are needed for more than one fireplace, or for a fireplace, stove, furnace, or whatever; the constructions may differ. It may be necessary to separate the flues by the width of a brick, as shown in **Figure 22-16**, or you may be able to use a fireclay flue of double construction, as shown in **Figure 22-17**.

FIG. 22-16

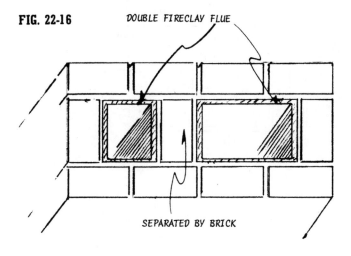

DOUBLE FIRECLAY FLUE

SEPARATED BY BRICK

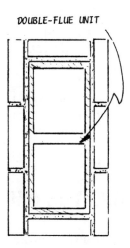

DOUBLE-FLUE UNIT

FIG. 22-17

If you need two flues and are permitted to install two adjoining liners so that only the wall thickness of the liners separates the flues, be sure to stagger the joints in the flue sections by a least 7 inches.

The best way to do the job is to set up sections of liner as you brick up the chimney. Start the first liner at a point about 8 inches away from the throat of the fireplace. Use fireclay mortar to make tight joints that are smooth on the inside.

ENLARGING THE CHIMNEY

Chimneys are often made larger from a point starting 6 to 12 inches below the roof frame. The technique is called *corbeling* and is done to bulk the exposed portion of the chimney for the sake of appearance and greater resistance to weathering. The projection from course to course should not be more than 1 inch, as indicated in **Figure 22-18**.

FIG. 22-18

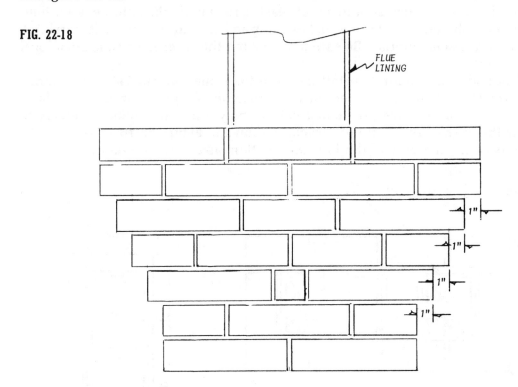

ENDING THE CHIMNEY

The chimney emerges from the roof through the rough opening you provided when you did the roof framing. Remember that combustible materials (wood framing members) should not be closer than 2 inches to the chimney walls. The gap should be filled with a noncombustible insulating material (**Figure 22-19**). The joint between the chimney and the roof is waterproofed by the flashing methods discussed in Chapter 13.

The top of the chimney should be high enough above the roof to be less exposed to the downdrafts that can occur when wind passes over the roof or nearby

FIG. 22-19

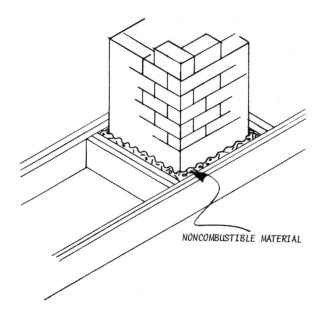

NONCOMBUSTIBLE MATERIAL

structures. The general rule is for the chimney to be 2 feet higher than any ridge closer than 10 feet when the distance is measured horizontally. A flat roof calls for additional chimney height, Code requirements are shown in **Figure 22-20.**

FIG. 22-20

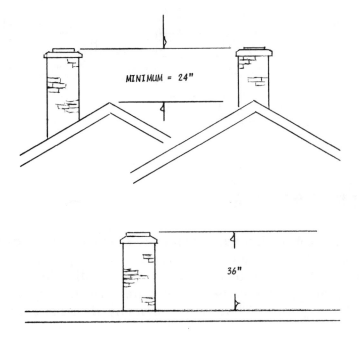

MINIMUM = 24"

36"

Adding a cap such as shown in **Figure 22-21** finishes the chimney nicely, but the cap has practical functions as well. The sloping top deflects outside air upward, and this helps promote a good draft. The cap is easy to cast in place with mortar if you set up some temporary formwork (**Figure 22-22**). Let the cap extend beyond the walls of the chimney so that it will serve as a drip ledge. The ledge, plus the cap's slope, will do much to drain water away from the chimney top.

FIG. 22-21

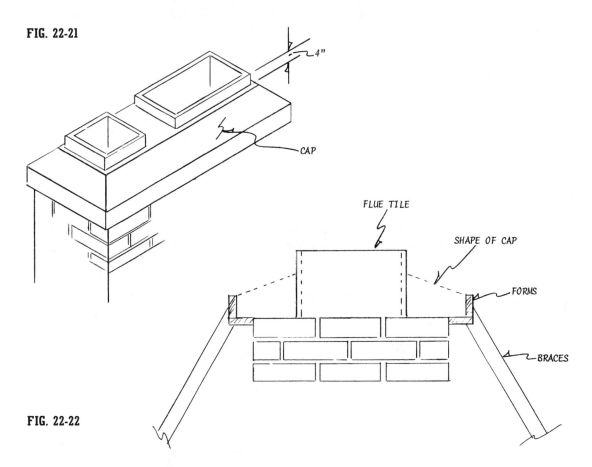

FIG. 22-22

THE COMPLETE PREFAB

This type of wood-burning fireplace has made a strong impact because it gives the builder complete latitude for individual styling and design without the considerable amount of work required by a conventional installation.

The products are really kits that include prefabricated fireboxes and interior and exterior chimney/flues. They come in many designs; a round-the-corner fireplace is shown in **Figure 22-23**. They do not require special foundations, and many are designed to permit zero clearance between them and wood framing members. The actual installation doesn't consist of much more than boxing in the units as shown in **Figure 22-24** and then covering the framing with materials of your choice.

FIG. 22-23

AROUND-THE-CORNER KIT

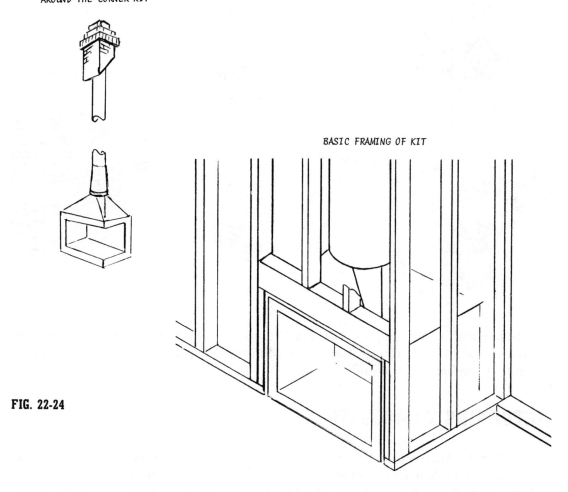

BASIC FRAMING OF KIT

FIG. 22-24

Some are made so multi-floor stacking installations are possible with venting occurring in a common chimney (**Figure 22-25**). A great variety of elbows, brackets, special flashings, and other components are available. There are even tops of simulated brick patterns in a choice of colors (**Figure 22-26**).

FIG. 22-25

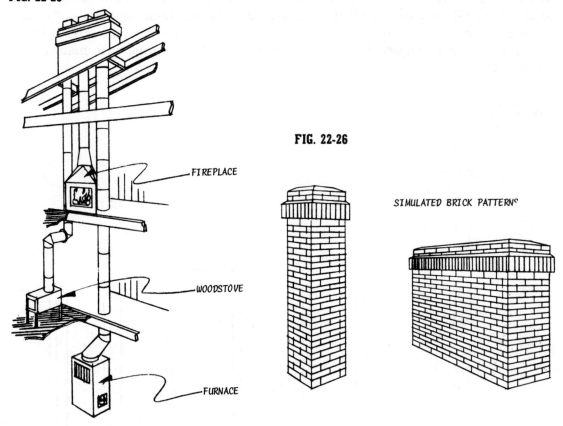

FIREPLACE

WOODSTOVE

FURNACE

FIG. 22-26

SIMULATED BRICK PATTERNS

Since such units come with complete and detailed installation instructions we won't use space to repeat them here. Manufacturers are listed in the reference section at the back of the book.

RETURNING TO WOODSTOVES

Woodstoves were once more prevalent than fireplaces but then lost ground to oil and gas heaters. Most fireplaces have been designed primarily to provide atmosphere in modern constructions. But the energy crisis of the 1970's made woodstoves popular home heaters once again. They are efficient heat-producers and, as the cracker-barrel group has always known, can be cheerful gathering centers where a coffee pot bubbles or where a tea kettle whistles.

I remember a friend's mountain cabin in which the only source of heat was a potbellied woodstove. That cabin was *warm*, and if the stove top wasn't being used to cook something, there was always a pot of water on it that did a good job of humidification.

I also recall an unheated two-car garage I used as a shop. At that time I bought a woodstove from a junk dealer for $15 and installed it with stovepipe for a flue. Although the garage was large, the woodstove heated it efficiently, and fuel consumption was less than the amounts of scrap wood that remained from projects.

Woodstoves can be nostalgic, yes, but in our time they have become good thinking.

WHAT TO CONSIDER

A woodstove, like a fireplace or a furnace, requires a flue; *its own flue.* You shouldn't plan to construct a fireplace chimney thinking you can cut into it someplace later and insert a stovepipe from a woodstove. For one thing, this would not be efficient. For another, codes may not permit it.

But you can construct a chimney that will take more than one flue by following the suggestions made earlier in this chapter. Codes will permit this so long as proper construction procedures are followed. There are many advantages here. It is easier and cheaper to build a multi-flue chimney than it is to install separate ones. A house with separate chimneys will probably be assessed more than one with a single chimney even though the latter might contain two or more flues. That adds up to a lower tax rate. It would seem that providing an extra flue makes some sense even if you don't plan to install a woodstove right off.

Of course you can also think about a woodstove instead of a fireplace, in which case you would construct a basic chimney/flue and provide for a smoke-pipe intake (**Figure 22-27**). Here, the lower part of the chimney should be solid masonry. The flue should start at a point not more than 8 inches below the centerline of the smoke-pipe intake. A clean-out door can be installed just below the smoke pipe but it doesn't seem necessary since removing the smoke pipe for cleaning — which you should do at least once a year — makes it easy to remove the soot that accumulates in the pocket merely by reaching through the pipe hole.

The end of the smoke pipe should be flush with the surface of the flue. A restricted draft would result if the pipe were allowed to project into the chamber.

The opening through which the smoke pipe passes should be airtight. Some old installations indicate that this was often done by packing around the pipe with cement mortar but this is not a good way to go today because many manufacturers produce easy-to-install fireproof components. This applies even if you are passing the smoke pipe through a wood-framed wall. Usually, the job consists of cutting an oversize opening through the wall for a special shield through which

FIG. 22-27

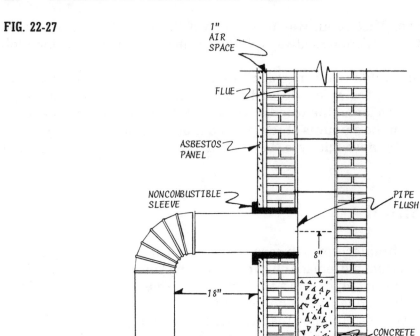

the smoke pipe will pass. There are many types and sizes of shields but going with the ones produced by the stove manufacturer will simplify matters.

It is a good idea to study manufacturer literature first. Armed with this, you can talk with the building inspector to see if the installation you plan is allowed by local codes.

SAFETY FACTORS

A minimum clearance is always required between a woodstove and a combustible wall. This will differ, depending on whether the stove is a radiating or circulating type. For the radiating type, which is simply a metal shell in which the fire is built, clearances may have to be as much as 36 inches. On the other hand, a circulating stove is a double-wall affair that creates an air chamber through which air circulates. It needs much less clearance for safe operation—in some cases as little as 12 inches. But clearances can often by reduced by attaching a noncombustible material to the nearby wall. A relatively easy solution is to install a sheet of at least 28-gauge metal to protect the wall. This should be mounted on fireproof spacers so air can circulate between the wall or ceiling and the shield. Protection is also required under a stove if it is mounted on a combustible floor. Here, sheet metal may be used, or you can construct a base of masonry materials.

Do check local codes before you start. As always, it's better to work *with* the inspector from the start than to risk having to make alterations later.

23

HEATING SYSTEMS

We are now faced with a situation that should have been anticipated by more of us than just a few realists and innovators. The fossil fuels on which we are almost totally dependent are limited in quantity and will inevitably be exhausted. If this fact alone doesn't make you pause and reflect, then also consider that as conventional energy sources diminish, the cost of using them will continue to go up.

Alternate sources of energy, namely solar, wind and nuclear, are in development stages. But even when and if they arrive for hookup to existing systems or new conventional installations, the heat sources should be designed for minimum consumption, if only to hold down maintenance costs. So the heating decision for the new-house builder is twofold: what to use and how to make the most of it.

If you go for a conventional fossil fuel system, will it last for your needs? Or will there soon be sufficient amounts of electricity from nuclear energy? Components of various types of conventional systems are readily available. Some are supplied almost in kit form with detailed instructions.

As this is written, going the new way — with sun and wind — is not yet practical in most parts of the country for total heat and power needs. But these new sources are practical *now* for supplemental use at least. Developments are coming fast. Certainly you ought to check on progress in your area. The use of solar energy for water and space heating is not new. The optimum in space heating is feasible only if the house is specially designed as a "solar-tempered" unit. We have already discussed the major factors: (1) house-size, (2) orientation, (3) minimum glass especially on cold walls, (4) more-than adequate insulation, (5) weathertightness. Add solar collector panels of sufficient area for the square footage of the building and you may have a house that is heated independently of commercial energy sources. If the solar-heated home is designed from scratch, the collector panels will be integral parts of the roof and maybe the wall systems, and you should anticipate this in your architectural plans.

One fact to remember is that the solar-tempered house, designed for minimum consumption, is the way to go regardless of the type of heating system you plan to install.

USING THE SUN

A solar heating system must include four components:

● *Collectors* to harvest sunlight and convert radiant energy to thermal energy.

● A *storage means* that can hold enough heat to last through the night or through periods of cloudy weather.

● A *distribution system* for delivery of the heat to points of use.

● A *backup system* that will contribute heat when stored solar heat is exhausted because the sun has not been cooperating.

This ends up as a combination of conventional energy and solar energy, with the conventional heating system designed to function as the backup.

The collector is the nucleus of the system since it must gather the sun's energy and make it suitable for storage and home heating. There are various types of collectors but the flat-plate design is the one most commonly used for both space heating and water heating.

The collector is a heat-absorbing metal plate with an array of tubing which may be an integral part of the plate or merely bonded to it. This assembly is, usually, in a black frame with a layer of insulation at the back and a glass or plastic covered air space on the surface that should face the sun.

The heat transfer can be liquid or air. It passes through the collector, picks up the heat, and transports it to storage. A southern exposure for the collectors is optimum but tests indicate that variations from due south do not cut efficiency much. A variance of from about 20 degrees east to about 45 degrees west of the ideal results in a less than 10 percent drop in performance. This is encouraging since it makes for more flexible design and placement—consideration of special interest to owners of existing homes if the house was not designed or oriented too well to begin with.

A point you should remember when buying or making collectors is that they will be most efficient when the heat transfer medium gathers the heat quickly. The higher the temperature at the collector itself, the more heat waste in that area and therefore the less efficiency.

Efficiency of the collectors becomes more important as the space available for them decreases. A "cheap" collector that is less efficient than a very expensive one might be the way to go if you judge in terms of cost per Btu.

There are two methods for heat storage that have proved successful. If water is the heat carrier, then a large tank of water may be used for storage. In essence, the water makes a constant circle through collectors and storage tank, becoming hotter with each circuit. Space heat can be extracted from the hot water by using a heat exchanger which may simply be a series of finned pipes through which the hot water passes. If you force air over the fins, you get space heat (**Figure 23-1**). For hot water, you pipe the storage tank contents to points of use.

FIG. 23-1

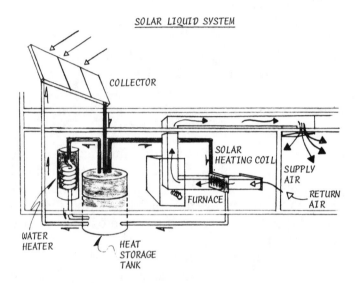

If air is the heat carrier, then the heat can be stored in a large insulated bin that is filled with rocks that range from about 3-inches to 1½-inches in diameter. The idea is for the rocks to provide area for heat exchange but to leave sufficient passageways for air movement. Air from the bin, directed to points of use, provides space heating (**Figure 23-2**).

If you are using an air system, the warmed air can be directed through conventional heating ducts. When liquid is the heat-transfer medium, the best bet is to work with the finned pipe arrangement. Running the water through a conventional hot water baseboard installation is not a good idea. The water from your storage will not be hot enough for the baseboard units to function efficiently.

A backup system comes into play during long periods of inclement weather. Usually, this backup is a conventional heating system that kicks in to provide warmth. But if you are interested in becoming as independent as possible, you might consider a woodburning stove, or stoves, as your backup system.

Solar water heaters (for tap use) are coming into wide use even at this writing

FIG. 23-2

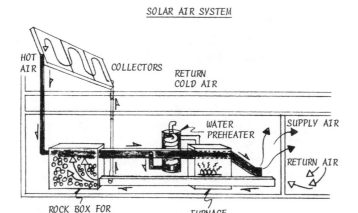

SOLAR AIR SYSTEM

simply because they are a fairly straightforward installation and are needed daily, year-round; whereas space heating is not.

Installations can be done by buying components (with instructions) or by getting complete kits. A recent study shows that costs will run between $300 and $500 if you do the job yourself and over $1200 if you let a contractor do it. Here, savings in energy costs can be impressive. A solar water heater will save you as much as $40 to $50 per year over a gas water heater and it will save you considerably more if you are thinking of heating water electrically. The systems, sizes, and prices do differ. Sources for further information are listed in the Appendix. Incidentally, solar units can replace or supplement an existing system.

On the whole, anyone building today is making a mistake by not checking out solar energy systems. There are tax incentive plans that will help offset installation costs. Resale value of the house will definitely increase. Your maintenance costs will definitely decrease, eventually to the point where the system will pay for itself.

The one factor that bothers me is the question of consumer protection. It is said that solar building codes and standards are a couple of years away. This means you should be extra careful when choosing a contractor or when buying components for a job you do yourself. When possible, check with people in the area who may have already installed solar systems. A call to the Better Business Bureau will let you know the reliability of the people you plan to deal with.

CONVENTIONAL SYSTEMS — WARM AIR

In central heating systems, heat is created at a central location and then delivered via ducts or pipes to points of use. The heat medium can be air or water or steam.

A *gravity warm-air system* is based on the fact that warm air, being lighter, tends to rise while colder air settles. Air heated by a furnace makes its way through various ducts which terminate at registers built into walls or floors. Return ducts bring cool air back to the furnace, where it is heated and recirculated.

For such gravity systems, the furnace must be lower than the registers, and it is usually located beneath a main floor. If the furnace is on the same floor as the rooms to be heated, then the inlet registers are usually located high on the walls. The gravity principle may also be used with space heaters. Here, small individual furnaces are located under floors or in the walls of rooms to be heated. The furnace-warmed air enters the room, rises, and circulates. Cool air is drawn back into the furnace for reheating. Each furnace serves an established amount of space, which may be one or more rooms.

A gravity system works quietly but has disadvantages. The air flow is gentle, so for efficiency the hot-air ducts must be as short as possible to minimize heat loss. This places hot-air registers on warmer interior walls and the returns along colder exterior walls. Thus the outer areas of a room might be cooler. The gentle air flow also makes the system less sensitive to thermostatic control.

Introduce an electrically powered blower and you have a *forced warm-air system* that eliminates the basic disadvantages of a gravity feed. The blower moves the heated air from the furnace to registers, so distribution is equalized and can be controlled. Here there is much less temperature difference between incoming warm air and exiting cold air. Since air movement is now mechanically controlled and since it can be directed anywhere, the location of the furnace becomes less critical. It can be installed in a garage or even in an attic (**Figure 23-3**).

A forced-warm-air system can be designed with more than one blower or with special devices so the heated air can be directed to particular points of use, with each having its own thermostat. Thus you can decide, for example, that bedrooms will be cooler than the kitchen or the living room and so help minimize maintenance costs and lessen demand on energy sources, whatever they might be.

Furnaces of various designs are available for different locations. A horizontal type which draws, heats, and then discharges air on a horizontal plane is especially good for an attic or a crawl space. A downflow unit for a main-floor installation is good, since air enters at the top and leaves from the bottom. An "upflow highboy" does well in a full basement. This unit draws return air into the back and sends warm air out the top. A similar unit, designed for basements with limited space, is called an "upflow lowboy." This lowboy circulates both incoming and outgoing air through the top.

FIG. 23-3

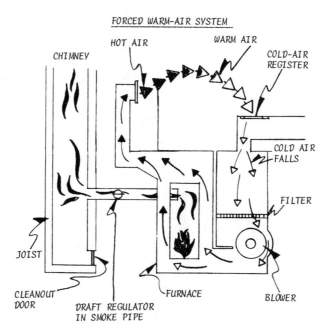

FORCED WARM-AIR SYSTEM

CONVENTIONAL HOT-WATER SYSTEMS

In principle, there is much similarity between air and water heating systems. For one thing, both can be gravity or forced types.

The gravity system works because water expands when it is heated. Hot water occupies more space than a similar amount—by weight—of cold water. Thus, in essence, hot water is lighter and rises. Furnace-heated water travels through riser pipes to radiators, where much of the heat is transferred to the metal in the radiators and then into the room. The water cools, becomes denser, and returns to the furnace via a separate piping system, where the circulatory action begins again.

Since water is not compressible and since it increases in volume when it is heated, the system must be designed with a safety factor that will work when needed to prevent components from rupturing. In an *open system* an expansion tank is included, located so it is higher than any part of the system (**Figure 23-4**). Its purpose is to maintain a consistent atmospheric pressure regardless of the temperature of the water. When heat increases to the point where expansion is excessive, the tank acts as a safety by taking in more water and exhausting it through an overflow pipe that is connected to a drain.

A *closed system* has a sealed expansion tank located near the furnace boiler.

FIG. 23-4

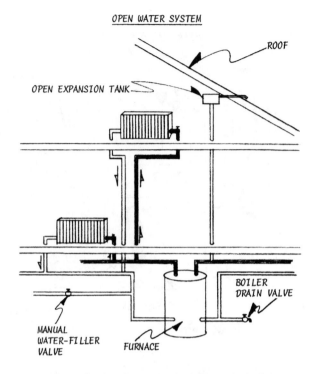

OPEN WATER SYSTEM

ROOF

OPEN EXPANSION TANK

BOILER DRAIN VALVE

MANUAL WATER-FILLER VALVE

FURNACE

At normal temperatures the air pressures inside and outside the tank are equal. When heated water enters the tank it compresses the captured air and puts the entire system under pressure, which, in effect, raises the boiling point of the water. In theory, such a system will regulate itself should temperatures get too high. But a relief valve is always included so that excess pressures will automatically be released.

The forced-hot-water (hydronic) system is like the closed-gravity design but includes a pump that circulates the water. If more than one pump is installed, with each controlled by a separate thermostat, various areas of the house can be heated to different temperatures even though the hot water is supplied by a single source.

SERIES-LOOP AND SINGLE-PIPE INSTALLATIONS

In a series-loop system the hot water that is pressure-fed from the furnace travels through pipes and through each of the radiators, or convectors, and then returns to the furnace (**Figure 23-5**). An obvious disadvantage is that you can't control or shut down individual radiators without affecting the total flow of water. With this system, rooms near the end of the run tend to receive less heat, since you can't

FIG. 23-5 CLOSED WATER SYSTEM (SERIES LOOP)

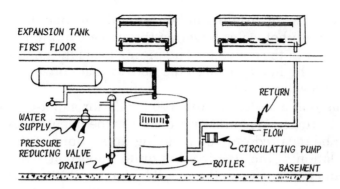

make any adjustment for the heat losses in rooms along the way. If you put your thermostat in the last room, then the rooms early in the run will be too warm. Such an installation is economical and easier to install than others but is practical only in very small houses.

The single-pipe system is better. Here, the hot-water carrier makes a continuous loop but has branches that lead to and come back from individual radiators, as shown in **Figure 23-6**. Thus the water flows through both the supply pipe and the radiator. The "used" water leaves the radiator cooler than when it entered, but then it merges with the water in the supply pipe, which is still hot, and continues on to the next radiator, where the action is repeated. (Double advantage here: When it reaches the last radiator, the water is much hotter than it would be in a series loop system. And because of valves in the system, you have individual room control without seriously affecting temperatures of other rooms on the run.)

FIG. 23-6 CLOSED WATER SYSTEM (SINGLE-PIPE)

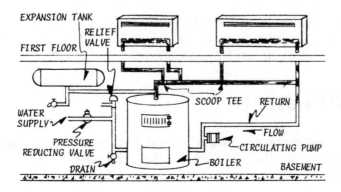

RADIANT HOT-WATER SYSTEMS

Most radiant systems are installed in houses built on concrete slabs. This is a circulatory system with hot water traveling through a continuous network of pipes or tubes embedded in the concrete. The piping heats the concrete, which in turn radiates heat into the house. A big plus here is that conventional radiators and registers are not needed.

It's obvious that a system of this type calls for super-careful workmanship and adequate testing under pressure before embedment. Leaks are difficult to locate and can be repaired only by breaking up sections of the slab.

The installation of the slab itself is quite important. You must provide for more than adequate drainage so there will be no possibility for water to get under the slab. Any water in contact with the concrete will greatly decrease the slab's effectiveness as a radiant panel.

Usually a thick, well-tamped, level layer of gravel is put down and then covered with waterproof insulation at least ¾ to 1 inch thick. Since much heat can escape at perimeter points, the thickness of the insulation placed there should be increased to 1½ inches.

It takes time for a large mass of concrete to heat up or to cool down, so some time elapses between a thermostat's signal and results in the slab. So special outdoor controls should be included. These can forecast temperature changes and initiate action long before an indoor thermostat would send out a signal.

Radiant heat can also be installed in walls or ceilings by embedding the water carriers in plaster. You might first guess that one installation could serve two rooms, but it doesn't work that way. In a ceiling, the installation should be designed for maximum effect on the room below. A room above will get minimal heat, especially if the ceiling is insulated. Radiant heat can be installed in exterior walls but will be efficient only if the walls are specially designed and insulated. Because of the problems, and additional costs, radiant heat is usually limited to interior walls.

Actually, a ceiling installation has much going for it. A floor might be covered with heavy rugs and many pieces of furniture, but there is nothing to interfere with heat radiating down from a ceiling.

HEATING WITH STEAM

This works like a hot-water system except that water is brought to the boiling point and becomes steam before it travels through pipes to radiators. Here, it gives its heat to the metal, condenses back to water, and returns to the boiler via another

FIG. 23-7

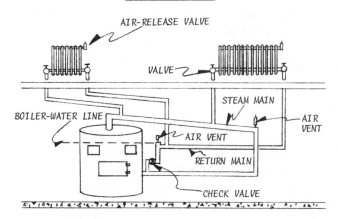

STEAM-HEAT SYSTEM

AIR-RELEASE VALVE

VALVE

STEAM MAIN

AIR VENT

BOILER-WATER LINE

AIR VENT

RETURN MAIN

CHECK VALVE

pipe, as shown in **Figure 23-7**. The radiators in a steam system can be smaller than those in a water system and still give off a comparable amount of heat, simply because steam leaves the boiler at about 212 degrees while a water system usually operates at about 180 degrees.

The safety measures in a steam system include pressure-relief valves and a boiler-water gauge. The valves work automatically when internal pressures approach the burst limits of the system. The gauge provides a visual check of the water level in the boiler. Also, there should be a steam-pressure gauge so you can read the amount of pressure in the system. Most residential systems are set up to operate at a maximum of 15 pounds pressure.

HEATING WITH ELECTRICITY

Electrical systems, in which the electricity is used directly as a heat source, can be the least troublesome to install and maintain. But operating costs can be very high and will probably continue to rise, especially in certain areas of the country.

In essence, electricity is converted to heat when you move it through special conductors that resist the flow of current. The conductors (heating elements) become hot and you get heat.

By substituting wires for pipes or tubes, you can turn walls or ceilings into radiant panels just like those in a hot water system. This can be done as an integral installation with heating cables embedded in plaster or set between layers of wallboard, or with ready-made panels that are designed into the structure.

Baseboard heaters are very popular and come in a wide range of sizes, with different output ratings so they can used for general heating or for supplemental purposes. It's also possible to install small panels in a wall. Since these too come in various heat-output ratings, you have choices in relation to the size and type of room to be heated. Often, areas such as bathrooms and utility rooms are excluded from the general house-heating system and are provided with manual-control electric units. Or these units can be controlled by individual thermostats, so you can have complete climate control over any room or area.

HEAT PUMPS

The concept of a heat pump — technically a reverse-flow air conditioner — is intriguing. If outside temperatures on a given day should range, for example, from 40 to 80 degrees, the unit would automatically maintain an inside temperature of your choice. It will heat or cool as required.

Heat pumps have had a bad reputation because of the bugs that exist in any new device and because of the current lack of knowledgeable service people and installers. A strong turnaround is happening, mostly because of continued research and development which has resulted in more reliable designs with less complicated mechanisms.

In nontechnical terms, a heat pump could be described as an air conditioner that runs two ways. In warm weather, it takes heat from the inside air and pumps it outside. In cold weather,the unit takes heat from outside air and brings it inside. (There is heat in winter air.) Heat-pump efficiency falls off as the temperature drops, so the most practical design includes conventional electrical-resistance elements that come into play automatically should temperatures fall to about 15 degrees.

Many experts are saying heat pumps are the way to go now and are predicting ever increasing enthusiasm. Heat pumps cost more to install than other heating/cooling systems but tend to consume less energy than conventional heating systems. Thus they may end up costing less in the long run.

Two important points: Don't install anything less than a high-quality unit with a good guarantee agreement. Heat pumps make the most sense economically when they are installed for year-round use to cool in the summer and to heat in the winter.

24

COVERING THE INSIDE WALLS

GYPSUM BOARD

Gypsum board is probably the most popular wall cover, especially when the final finish will be paint or wallpaper. Often, though, a waterproof type is used as a base for tile, or regular gypsum board is installed as a first cover under other materials. The latter idea contributes solidness to the wall if you are using thin plywood paneling as a finish cover, and it increases the R-value of the wall as a whole. There are also special insulating boards which have a bright aluminum foil laminated to the back surface, sound deadening boards, boards with increased firestopping values, and prefinished vinyl-surfaced products in a variety of colors and surface textures.

Regular gypsum board is available in ¼-inch, ⅜-inch, ½-inch, and ⅝-inch thicknesses. The ½-inch product is ususally selected for a normal installation. Regular panel size is 4×8 feet, but they can be purchased in lengths up to 16 feet. Longer panels cover greater areas with fewer joints, but their size and weight make them difficult to handle, especially if you are working alone. A factor that can help determine the length of boards to work with is whether you do a vertical or a horizontal installation. If horizontal, use longer boards; if vertical, work with 8-foot sheets, since in rooms of normal height a single panel will run floor-to-ceiling without a joint. The long dimension of panels placed on a ceiling should run at right angles to the joists. All joints must fall on the centerline of a framing member.

MEASURE AND CUT

Do all measuring very carefully, especially when openings are required for electrical boxes. Since such openings will probably have to be made quite often, it pays to make a cardboard template you can use whenever the openings are required.

FIG. 24-1

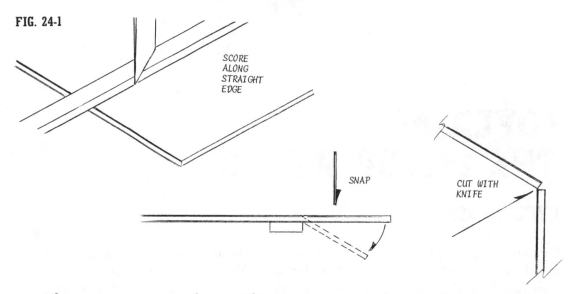

SCORE
ALONG
STRAIGHT
EDGE

SNAP

CUT WITH
KNIFE

The easiest way to make straight cuts is to use a sharp knife with a straight piece of wood as a guide. Score deeply enough with the knife to sever the surface paper. Don't, as some do, separate the boards by pulling them apart. As shown in **Figure 24-1**, take the time to cut through the back paper after the board is snap-broken.

Openings will be neater if you drill holes at opposite corners and then make saw-cuts in the directions shown by the arrows in **Figure 24-2**. You can use a keyhole saw or equip yourself with special saws made for the purpose such as

FIG. 24-2

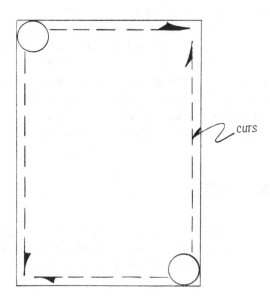

CUTS

FIG. 24-3

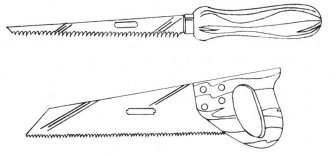

the two in **Figure 24-3.** It's also possible to do a good job with a saber saw. Cuts can be smoothed with sandpaper, but since it clogs quickly you're better off making a special sander with metal screen; two versions are shown in **Figure 24-4.** The screen is a good abrasive for the material and can be cleared easily simply by knocking it against a hard surface.

FIG. 24-4

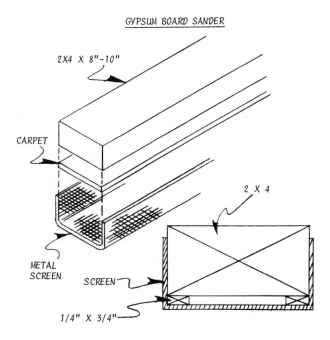

GYPSUM BOARD SANDER

2X4 X 8"-10"

CARPET

2 X 4

METAL SCREEN   SCREEN

1/4" X 3/4"

INSTALLING GYPSUM BOARD

First check all studs to be sure none has developed a bow. Bowing is more likely in partition walls than outside walls, where the framing is secured with sheathing, siding, and bracing. Bowed studs can be replaced, or sawn partly through and then reinforced in a straight position.

The boards can be attached with nails, screws, or adhesive. Many professionals use screws, but if you go that way be sure to rent an electric screwdriver with a built-in slip clutch that disengages automatically as soon as the screwhead is seated, and you must work with special wallboard screws. Nails also are of special design, the most common being a coated type of drywall nail with a straight shank or one of annular-ring design. A drywall screw and both types of nail are shown in **Figure 24-5**.

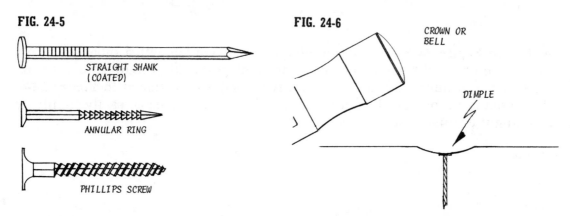

FIG. 24-5

STRAIGHT SHANK
(COATED)

ANNULAR RING

PHILLIPS SCREW

FIG. 24-6

CROWN OR BELL

DIMPLE

Drive nails with a hammer that has a crowned head so the last blow will form a shallow dimple (**Figure 24-6**). Do not dimple so deeply that you break the paper. Hold the board firmly against the framing; don't depend on the nail to bring the board up tight. Start nailing at the center of panels and work out toward edges, using the nailing patterns shown in **Figure 24-7**.

FIG. 24-7

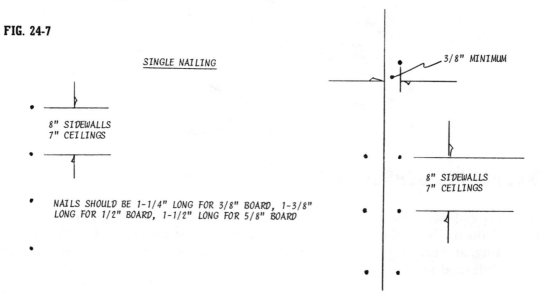

SINGLE NAILING

8" SIDEWALLS
7" CEILINGS

NAILS SHOULD BE 1-1/4" LONG FOR 3/8" BOARD, 1-3/8"
LONG FOR 1/2" BOARD, 1-1/2" LONG FOR 5/8" BOARD

3/8" MINIMUM

8" SIDEWALLS
7" CEILINGS

Double nailing is a newer technique, designed to provide firmer contact between panels and framing and additional insurance against nails working out (popping). Best procedure is to apply the boards conventionally but using a 12-inch nail spacing on interior areas. Then, return and drive a second nail not more than 2 inches from the first one (**Figure 24-8**).

FIG. 24-8   **FIG. 24-9**

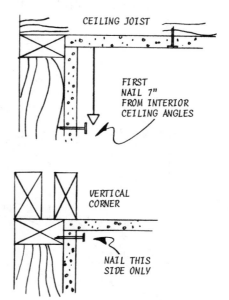

DOUBLE NAILING

SIDEWALLS
AND
CEILINGS

12"

ABOUT 12"

12"

CEILING JOIST

FIRST
NAIL 7"
FROM INTERIOR
CEILING ANGLES

VERTICAL
CORNER

NAIL THIS
SIDE ONLY

Some professionals prefer not to do complete nailing along corners where wallboards meet, feeling that this permits some movement, which will prevent cracking that can be caused by normal structural stresses. In a ceiling corner, the nails are kept about 7 inches back from panel edges, and in a vertical corner, the edge nails are omitted on one of the panels (**Figure 24-9**).

Wallboard can be stuck on with a special adhesive that you squeeze out of a caulking gun. The method eliminates much nailing and is said to provide a sturdier wall, since the bond between panel and framing is continuous. Apply the adhesive in a bead that is about ⅜ inch in diameter to all surfaces the panel will contact. A zigzag pattern is best, especially on studs that will back up a joint. Press the panels firmly in place and use nails only along edges. The adhesive is tacky enough to hold the panels, but if necessary you can set up some temporary bracing or even drive a nail here and there to hold the contact.

COVERING JOINTS IN GYPSUM BOARD

Joints are covered with a special joint compound (often called joint cement) and paper tape in a series of steps that should produce a practically invisible cover (**Figure 24-10**). How it turns out has everything to do with how careful you are. What makes the technique possible without resulting in an unsightly bulge is the recess that occurs where panels join (**Figure 24-11**). The joint compound is available as a powder that you mix with water, but it's better to work with a ready-mix that you can buy in 1-gallon and 5-gallon cans. It saves time and work and assures you of a proper mix.

FIG. 24-10

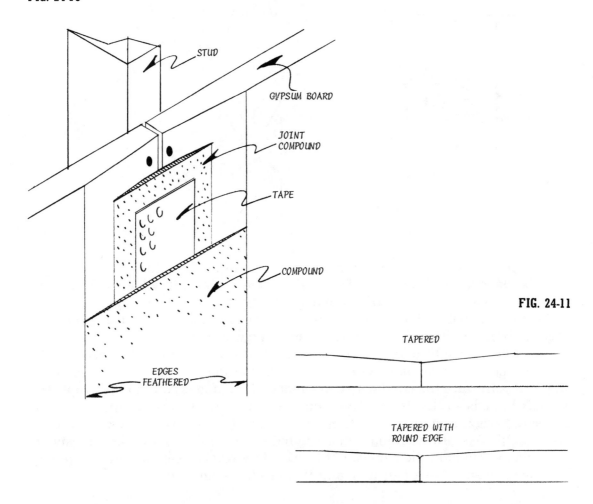

FIG. 24-11

The first step is to use a wide-blade knife to place a layer of compound over the joint between two sheets (**Figure 24-12**). Many amateurs attempt this, and other steps, with an ordinary putty knife, but it doesn't work too well and imposes a handicap you really don't need. It's best to get a set of broad-bladed tools made for the purpose (**Figure 24-13**). They don't cost too much and will set you up right to begin with.

FIG. 24-12

FIG. 24-13

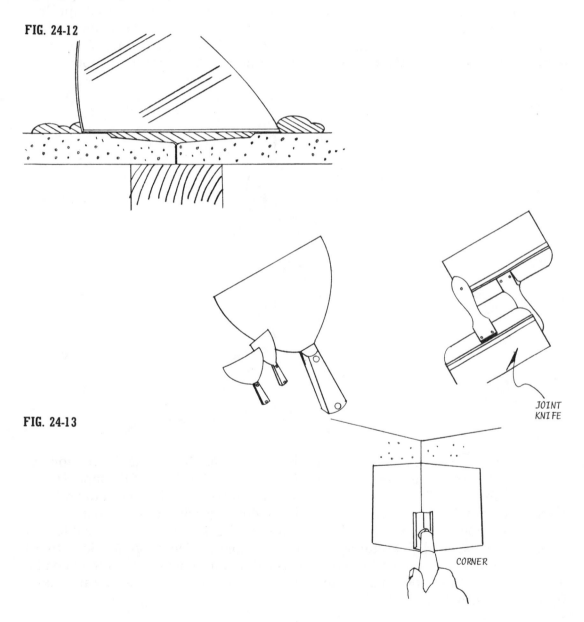

JOINT
KNIFE

CORNER

Place the tape as shown in **Figure 24-14** as soon as the first layer of compound is down. Many workers use the tape dry, but I have found that if it is dampened about ten minutes or so before use it is easier to handle and will not suck moisture from the compound. Work with the knife to smooth the paper. It is not necessary to embed the tape in the compound, but you should have total, firm contact. Scrape off compound that squeezes out and smooth it down on top of the tape. Don't allow high spots or ridges to form anywhere. Let this initial application dry thoroughly. You'll know it is dry when the compound turns pure white and is hard to the touch. Sanding, at this stage, will be necessary only if you have done the first step poorly and have left roughness, ridges, and high spots.

FIG. 24-14

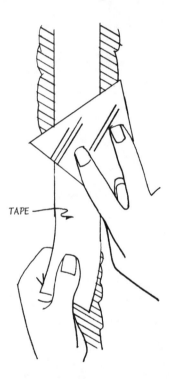

TAPE

Apply a second coat of compound, being sure that it fills the depression between sheets and that it extends a few inches beyond the edges of the tape. Use as wide a knife as possible and work so the compound will feather out into nothingness. When you have to make a joint between cut edges of the board that have no depression, you will have to make all compound coats as thin as possible and feather them especially carefully to minimize bulges (**Figure 24-15**). Sometimes this last application is done in two stages, with much depending on how good a job you did the first time. At any rate, the final compound width can be anywhere from 12 to 14 inches.

FIG. 24-15

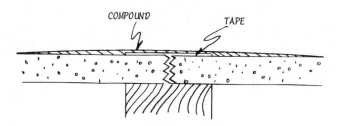

COMPOUND

TAPE

When this is dry, do a minimal amount of sanding, using a 220- or 320-grit paper wrapped around a block of wood (**Figure 24-16**). The best abrasive to use is the open-mesh type, since it has much less tendency to clog than conventional papers. A special sanding tool is available, and it's not a bad idea to have one, since it will let you reach any area of a wall or ceiling without standing on a stool or ladder (**Figure 24-17**). No matter what sanding you do, be careful not to rough up the paper surface of the wallboard.

FIG. 24-16

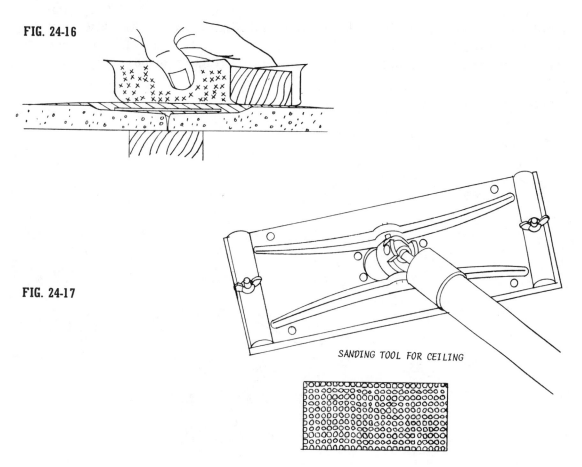

FIG. 24-17

SANDING TOOL FOR CEILING

Vertical inside corners are also covered with tape. Fold it down the middle before placing it against the compound (**Figure 24-18**). Follow the procedure outlined for the edge-to-edge joint, but work even more carefully, since here you don't have a depression to receive the compound. Use as little compound as possible and don't try to feather out more than 2 or 3 inches beyond the tape.

FIG. 24-18

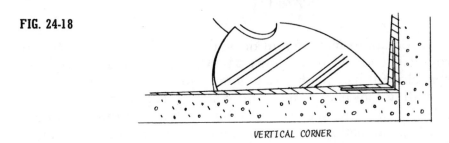

VERTICAL CORNER

Outside corners should be protected with a special metal corner that is nailed in place this way before the finishing is done (**Figure 24-19**). Work with compound only (no tape), applying a minimum of two coats with total coverage extending about 4 or 5 inches away from the corner (**Figure 24-20**).

FIG. 24-19

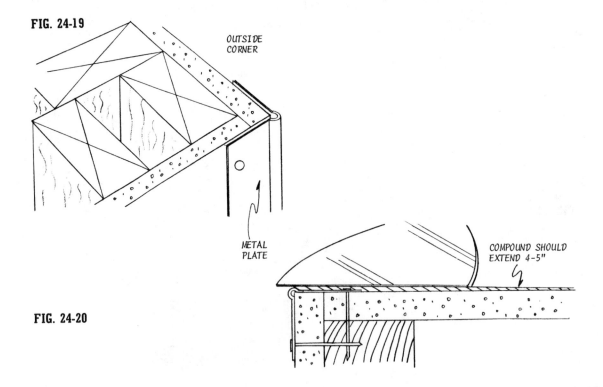

OUTSIDE CORNER

METAL PLATE

COMPOUND SHOULD EXTEND 4-5"

FIG. 24-20

You don't have to use tape and compound on wall-to-ceiling joints if you would rather use molding (**Figure 24-21**).

FIG. 24-21

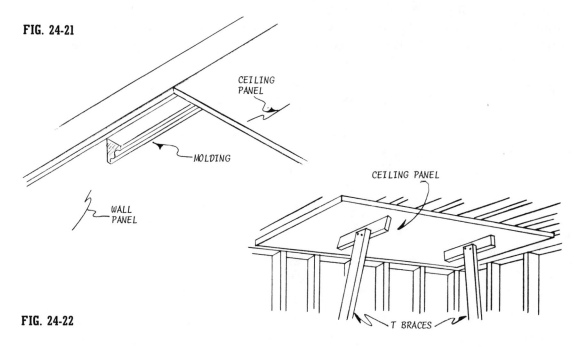

CEILING PANEL

MOLDING

WALL PANEL

CEILING PANEL

FIG. 24-22

T BRACES

Incidentally, the easiest way to hold ceiling panels in place while you are nailing them is to make a couple of T-braces and use them as shown in **Figure 24-22**. A 1×4 nailed across the end of a 2×4 that is about 1 inch longer than the floor to ceiling height will do.

The nails are easy to cover, since so little area is involved. Work with a 4-inch knife and apply the compound with only as much pressure as you need to fill the dimple (**Figure 24-23**). Apply a second coat after the first one dries and then do a light sanding job to finish up.

FIG. 24-23

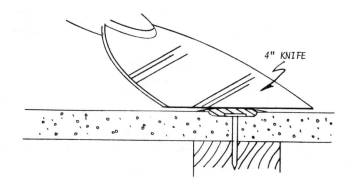

4" KNIFE

Prefinished gypsum wallboard does not require the joint treatments described above. Instead, the panels are attached with special color-matched nails. Various types of panels with edge-to-edge joints which may be left as is are available (**Figure 24-24**). Inside and outside corners and wall-to-ceiling joints are usually covered with special moldings.

FIG. 24-24

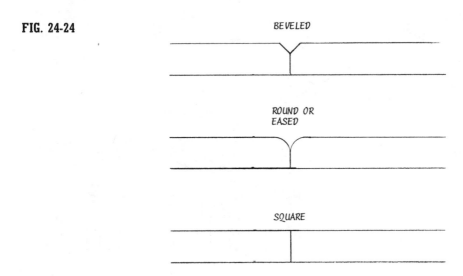

BEVELED

ROUND OR EASED

SQUARE

PREFINISHED PLYWOOD PANELS

Plywood panels are popular because the panels are light and easy to handle, they come in 4×8-foot sheets that cover a lot of area quickly, and once up, they require no further attention beyond an occasional damp-dusting. Designs and patterns can be listed by the hundreds and run the full gamut of softwoods, hardwoods, and exotic hardwoods. Some are sleek and very modern; others like Georgia-Pacific's "Ol'Savannah" in **Figure 24-25** are hard to tell from used boards that have been restored and refinished. Others, although plywood like Georgia-Pacific's "Corkblock" in **Figure 24-26**, look more authentic than the material they imitate.

The thickness usually used is ¼-inch, but you can get thinner or thicker panels. Anything under ¼ inch makes for a rather flimsy-feeling wall, so when attachment is directly to studs it's recommended that ⁵/₁₆-inch plywood sheathing or ³/₈-inch gypsum board be used as an underlayment.

Some panels are edge-matched to come together without a visible joint, or the mating might form a groove which is compatible with other grooves in the panel. Square-edge panels may be butted, or the joints may be finished in a variety of

FIG. 24-25

FIG. 24-26

ways that can add some distinction to the installation. You can work with ready-made moldings or design dividers of your own (**Figure 24-27**). It's possible too to find panels for which ready-made matching moldings are supplied by the manufacturer (**Figure 24-28**).

FIG. 24-27

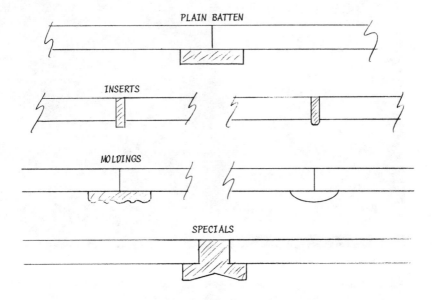

FIG. 24-28

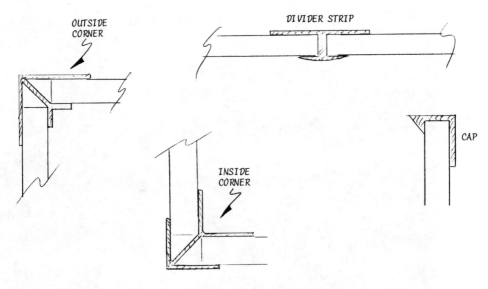

INSTALLING PLYWOOD PANELS

First check all studs to be sure none has developed a bow since the time of installation, and repair any bowing walls, by replacing the stud or cutting and reinforcing it (**Figure 24-29**).

FIG. 24-29

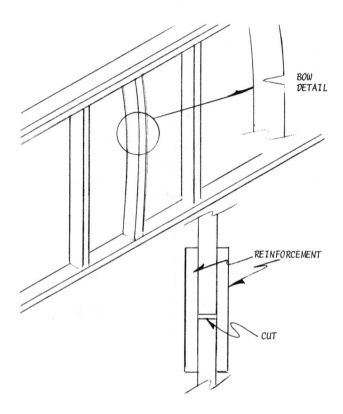

BOW
DETAIL

REINFORCEMENT

CUT

Buy your paneling a few days before installation time and stand the individual pieces about the room so they will become acclimated. This is also the right time to plan the placement of panels to match tone and grain patterns. Plan the placement so the end panels on each wall will be about equal in width and not less than the space across three studs.

It's usually best to start the installation at a corner, cutting the panel, if necessary, so the outboard edge is plumb and falls on the centerline of a stud (**Figure 24-30**). Do the cutting on the inboard edge of the panel, making sure that it conforms to any irregularities in the wall it abuts. The best way to do this is to hold the panel in place and use a pair of dividers or a compass as shown in **Figure**

FIG. 24-30

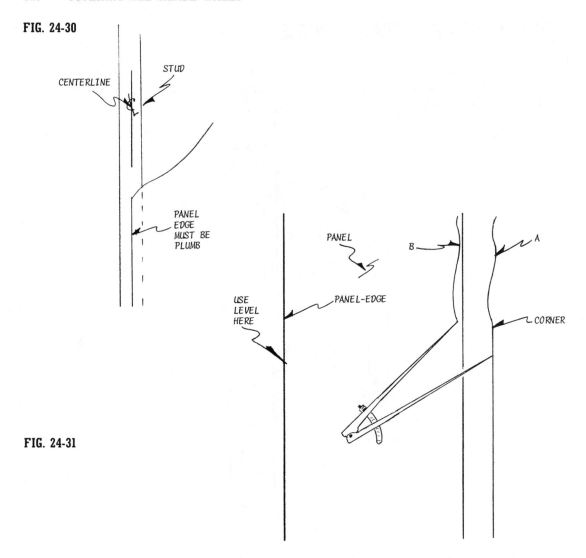

FIG. 24-31

24-31 to transfer the irregularity to the panel. Use a level at the opposite edge to be sure of plumbness there. Get this first piece placed correctly and all others will fall into place as they should.

Measure carefully when you need cutouts, but don't be so precise that the panels must be forced into place since this might create stresses that will cause the panels to bow away from the wall.

If you cut by hand, use a fine-tooth crosscut saw and keep the face of the panel up. If you work with an electric saw, either cutoff or saber, keep the face of the panel down.

Panels are usually secured with either color-matched nails or adhesives. In either case follow the recommendations of the manufacturer as to nailing schedules or adhesive placement. With adhesives, it's usually suggested that a zigzag line of the material be applied to all contact points on the framing. The panel is pressed into place, pulled away from the wall to break the contact, and then returned to its original position. To firm the bond, go over the entire panel, hitting down on a length of 2×4 wrapped in some soft carpeting (**Figure 24-32**).

FIG. 24-32

At corners and openings you can use moldings of your choice, or those that are specially made by the manufacturer to match the paneling (**Figure 24-33**).

FIG. 24-33

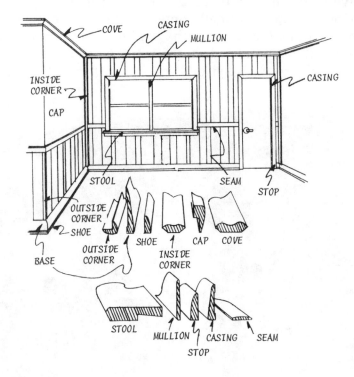

HARDBOARD PANELS

What we said about plywood in relation to the wide range of designs, textures, and patterns that are available applies to hardboard paneling as well. In addition to realistic wood finishes like the Royalcote walnut-grained effect shown in **Figure 24-34**, you can get panels that look and feel like marble, stucco, brick, stone, and other materials.

Basic installation of hardboard paneling doesn't differ materially from plywood products unless you choose a type that is attached with special metal clips. If so, follow the instructions that are supplied with the product. Generally, panels are installed along the lines shown here: nailing directly to studs, **Figure 24-35**; nailing over an underlayment, **Figure 24-36**; adhesive attachment directly to studs, **Figure 24-37**; and adhesive attachment over a solid backing, **Figure 24-38**.

Prefinished accessories in matching colors and wood grains are available for the final touches (**Figure 24-39**).

FIG. 24-34

FIG. 24-35

NAILS 8" O.C. AT
INTERMEDIATE SUPPORTS

FIG. 24-36

NAILS 8" O.C. AT
INTERMEDIATE SOLID
BACKING SUPPORTS

STUDS 16" O.C.

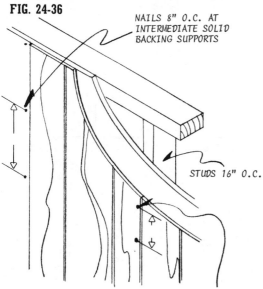

STUDS 16" O.C.

NAILS 4" O.C. AT JOINT
AND ALONG ALL EDGES

NAILS 4" O.C. AT JOINT
AND ALONG ALL EDGES

NOTE: FOLLOW PROCEDURE FOR
NAILING OVER OPEN FRAMING
BUT USE SPECIAL 1-5/8" NAILS
TO PENETRATE AT LEAST 3/4"
INTO STUDS

CONTINUOUS
ADHESIVE BEAD
1/2" FROM ALL
EDGES OF PANEL

16" O.C.
MAXIMUM

INTERMITTENT
3" ADHESIVE BEAD
6" SPACE ON
INTERMEDIATE
STUDS

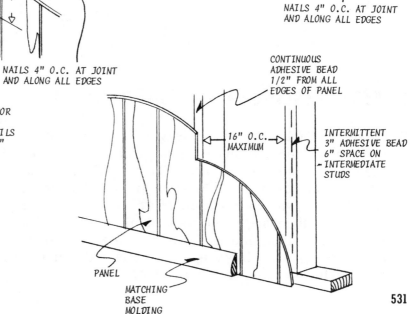

FIG. 24-37

PANEL

MATCHING
BASE
MOLDING

531

FIG. 24-38

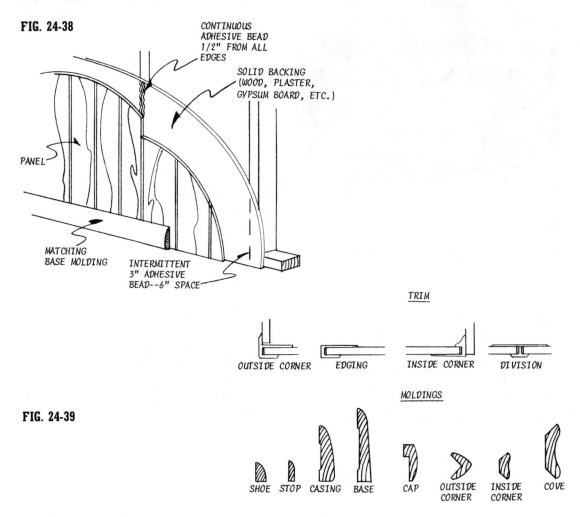

CONTINUOUS
ADHESIVE BEAD
1/2" FROM ALL
EDGES

SOLID BACKING
(WOOD, PLASTER,
GYPSUM BOARD, ETC.)

PANEL

MATCHING
BASE MOLDING

INTERMITTENT
3" ADHESIVE
BEAD--6" SPACE

TRIM

OUTSIDE CORNER   EDGING   INSIDE CORNER   DIVISION

MOLDINGS

FIG. 24-39

SHOE   STOP   CASING   BASE   CAP   OUTSIDE CORNER   INSIDE CORNER   COVE

PLASTIC-FINISHED HARDBOARD

These are special hardboard panels with high-heat-baked modified-melamine finishes. Here too, choices run a wide range of simulated materials and textures that include rough-wood appearances like the "Country House Plank" by Marlite shown in **Figure 24-40**, and high-gloss finishes that are suitable for kitchens and even bathrooms, shown in **Figure 24-41**.

A new wall-cover product is a plank (Marlite) which measures 16 inches wide by 8 feet long and has tongue-and-groove edges that simplify fitting. Adhesive and special clips are used to attach the planks so there are no visible nails when the job is done (**Figure 24-42**).

FIG. 24-40

FIG. 24-41

FIG. 24-42

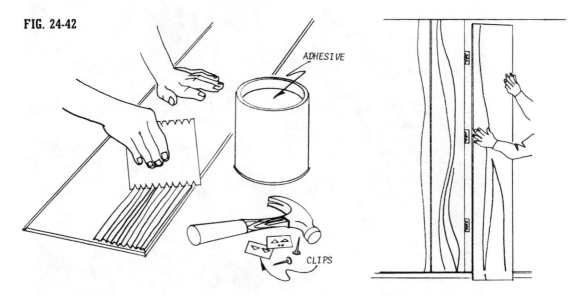

These products are often supplied in kit form for particular installations. One is a tub-recess kit which includes the necessary panels, moldings, adhesive, and caulking and step-by-step instructions for doing the job (**Figure 24-43**).

Another kit includes all the materials you need to wainscot a 12-foot wall. In **Figure 24-44** both the wainscot and the top of the wall are Marlite plastic-coated hardboards.

FIG. 24-43

FIG. 24-44

SOLID WOOD

Wall coverings of solid wood have been with us for centuries and are still popular with people who like the feel and look of the material. You can be pretty creative with solid lumber paneling, or end up with something distinctive merely by choosing boards which have been shaped and machine-molded to provide shadow lines and decorative edges. Those most generally available are of the interlocking tongue-and-groove types (**Figure 24-45**). If you use boards that are not more than 6 inches wide, you can do blind nailing (**Figure 24-46**). Wider boards will probably require a nail or two between joints, and these will have to be set and concealed with a wood dough.

FIG. 24-45

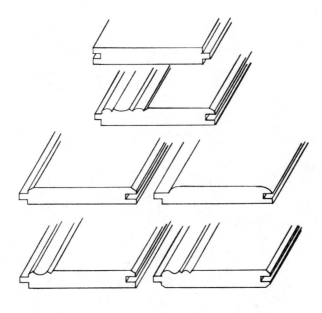

FIG. 24-46

BLIND NAILING

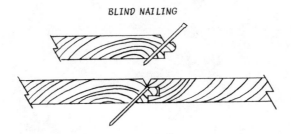

Plain, square-edge boards—as long as they are kiln-dried—are also usable, and you can do a board-and-batten or a board-on-board installation as you would for an exterior wall. Ponderosa pine, sugar pine, and redwood are popular species, but there's no reason why you can't work with western red cedar, white fir, Engelmann spruce, and others. Much will depend on the availability of species in your area, but choices will never be limited even if the material must be ordered for you.

Boards can be smooth or rough-sawn, clear or knotty, or even contain surface defects like those in pecky cypress and wormy chestnut. It all depends on how you want the wall to look. Whatever you choose, be sure to store the boards in the room for as much as a week so they will become acclimated before you start to saw and nail. Use spacers between the boards so air can circulate (**Figure 24-47**).

The easiest but not the most popular installation is with the boards placed horizontally so they can be secured to studs without the need for additional

FIG. 24-47

FIG. 24-48

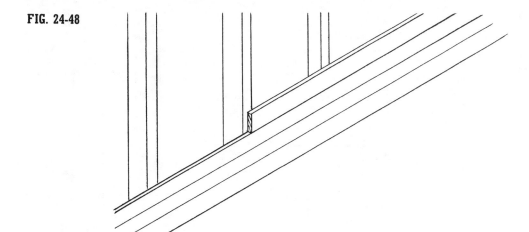

nailing surfaces (**Figure 24-48**). The job goes quickly, since a minimum number of cuts are required. A horizontal installation can make a room look longer, but will look its best only if the panel pieces are of uniform width.

A vertical installation calls for additional blocking between the studs (**Figure 24-49**). An alternative is to make a backing frame *over* the studs (**Figure 24-50**).

FIG. 24-49

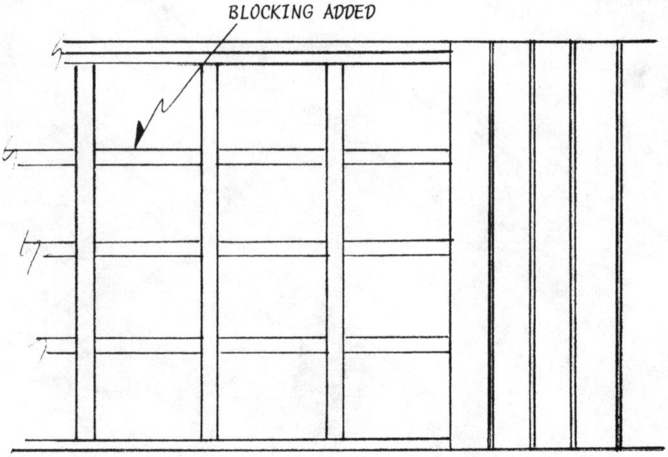

FIG. 24-50

FIG. 24-51

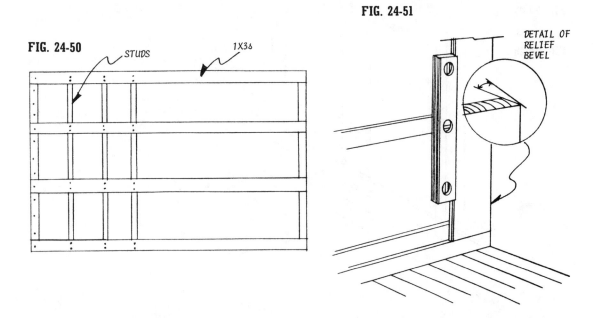

Cut a slight relief bevel on the edge of the first board and install it so it will be perfectly plumb (**Figure 24-51**). If necessary, shape the inboard edge so it will conform to any irregularity in the wall it abuts.

Continue to add boards, but check occasionally with a level to be sure of alignment. Work carefully so board tongues will seat completely in mating grooves.

The last board should also have a relief bevel on the edge abutting the wall. The fit of the boards at the beginning and the end of the wall is not so critical if you plan to add molding to conceal the joint.

The bottom of the wall can be done in either of two ways, as shown in **Figure 24-52**. You can install a baseboard first and then cut the panels to form a tight butt joint, or cut the boards to extend almost to the floor and add a baseboard on the outside. The joint at the ceiling is easy to finish with molding of your choice. An example is shown in **Figure 24-53**.

FIG. 24-52

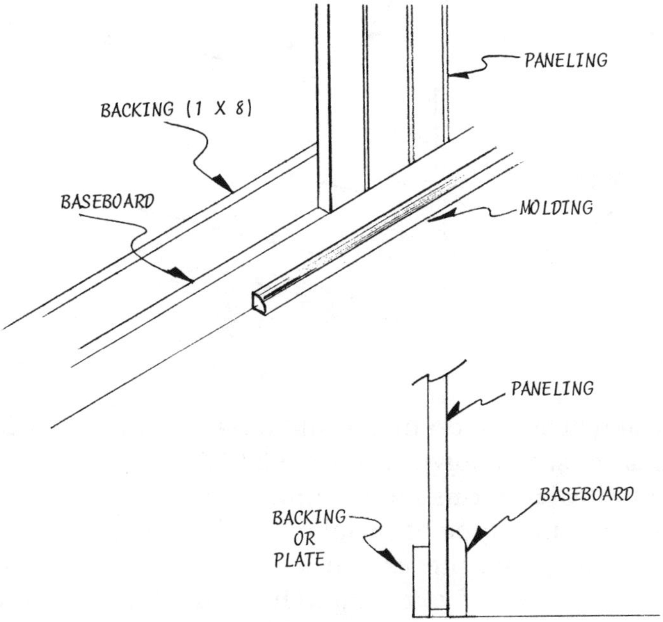

PANELING

BACKING (1 X 8)

BASEBOARD

MOLDING

PANELING

BACKING
OR
PLATE

BASEBOARD

FIG. 24-53

CEILING
MOLDING

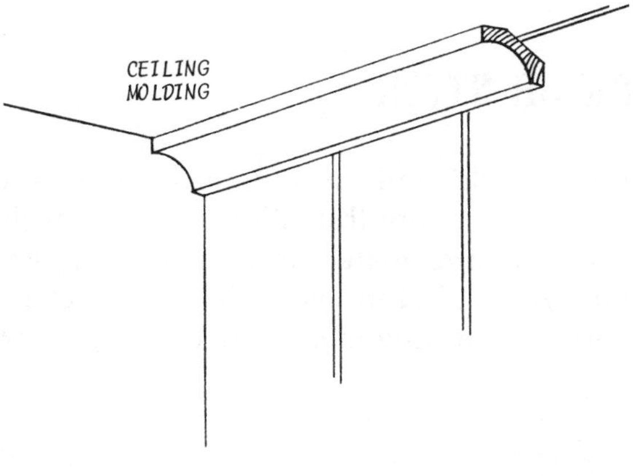

If the paneling is horizontal and makes a turn, you can do the corners with miters. Careful work is called for, but you will get a tighter visible edge on outside corners if you overcut the miter just a bit (**Figure 24-54**).

FIG. 24-54

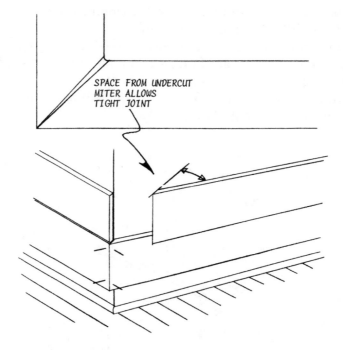

SPACE FROM UNDERCUT MITER ALLOWS TIGHT JOINT

There are various ways to do inside and outside corners for both vertical and horizontal applications, as shown in **Figure 24-55**.

A herringbone design forms an exciting pattern, as shown in **Figure 24-56**, but it is a very demanding type of installation. Well done, it is a mark of super craftsmanship. The diagonal pieces can be butted against vertical dividers, or board ends can simply meet. If you want to fudge a bit, place the vertical dividers *over* the joints; then the angle cuts will not be so critical.

IMITATION BRICK OR STONE

There are many materials in this area on the market now and more are coming. Some are made of plastic, others are thin slices of the real thing, still others are combinations of crushed natural materials and bonding agents. They may be available as individual units or, in the case of bricks, for example, as panels; you install a dozen or so bricks at a time merely by driving a few nails.

FIG. 24-55

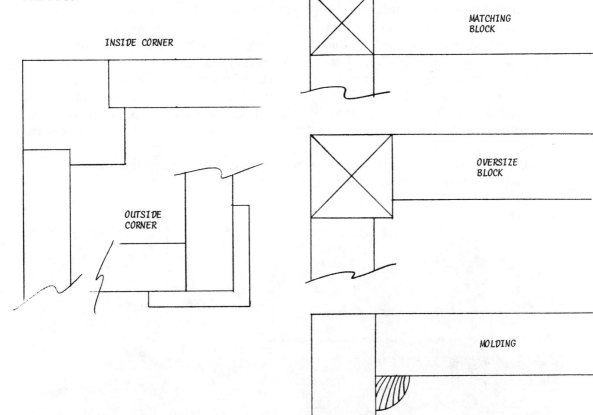

INSIDE CORNER

OUTSIDE CORNER

MATCHING BLOCK

OVERSIZE BLOCK

MOLDING

FIG. 24-56

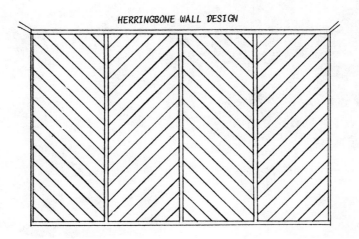

HERRINGBONE WALL DESIGN

Application methods differ depending on the product. Some require a wire-mesh installation over a solid backing to hold a special "grout" which in turn receives the brick or stone. Others will stick to a mastic that is applied directly to the backing. Where they may be used — indoors, outdoors, moisture areas, and so on — is also variable. Best bet is to check with manufacturers before making a choice. The following are just a few examples of what is available.

"Bricover," **Figure 24-57**, plastic brick, is hard to tell from the real thing even when you feel it. Roxite, produced by Masonite, comes with the appearance of brick too, or of stone as in **Figure 24-58**, and can even be used outdoors (**Figure 24-59**).

One thing all these materials have in common: They are wall *covers;* they are not intended to be structural, load-bearing components.

FIG. 24-57

FIG. 24-58

FIG. 24-59

MODERN TILE

Now you can buy genuine ceramic tile in pregrouted sheets that you stick to a wall with a mastic. One such product is American Olean's "Easy-Set," which you can buy in sheets of nine tiles or in tub-surround kits that include larger sheets of 49 tiles each plus special cove molding. In new contructions the installation should be done over waterproof gypsum board.

Spread the mastic, or adhesive, with a $3/16$-inch V-notch trowel (**Figure 24-60**). Don't cover more than a few square feet at a time until you gain some experience in placing the sheets of tile without having to fuss too much. Place the tile sheets in position and press firmly to get good contact with the adhesive. Butt adjoining sheets tightly together, since the tiles are designed to provide correct spacing automatically. Use a damp cloth as you go to remove any adhesive that squeezes out of the joint or sticks to the surface of the tiles.

FIG. 24-60

Cuts will have to be made for fitting tiles and for providing openings for pipes (**Figure 24-61**). If the cut is irregular, you can work with tile nippers, breaking off very small pieces of tile until you achieve the shape you want (**Figure 24-62**). The secret here is not to break off large pieces that can cause cracks where you don't want them. The parts you break off can be quite sharp, so watch your hands, and wear safety goggles.

FIG. 24-61

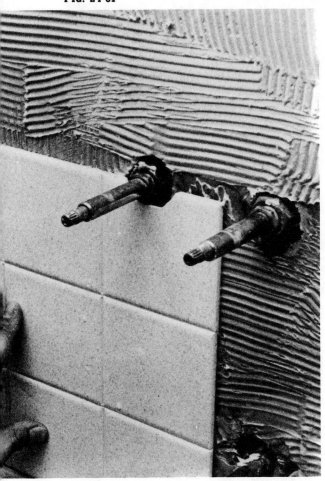

FIG. 24-62

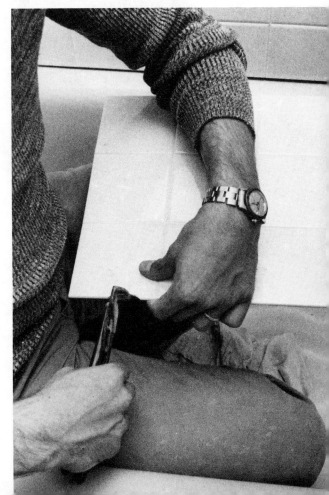

Straight cuts can be done with a regular tungsten-carbide-tipped tile cutter by scoring the glazed side of the tile where you wish to make a break (**Figure 24-63**). Position the scored line over the edge of a board or a length of heavy wire and apply firm pressure with your hands on both sides of the score line.

FIG. 24-63

FIG. 24-64

FIG. 24-65

The joints between sheets are filled with a special material, as suggested by the manufacturer, which is applied with a caulking gun (**Figure 24-64**).

Smooth out the joint lines with your thumb or the eraser end of a pencil and then clean off excess material with a soft cloth dampened with denatured alcohol (**Figure 24-65**). Don't wipe across the joint, since any grout material left to dry on the face of the tiles will be difficult to remove later.

25

COVERING
THE CEILINGS

GYPSUM BOARD

Gypsum boards cover ceilings as they do walls. The first challenge is to hold the heavy sheets in place until they are nailed. A solution is to have helpers, but even so, the job can be done more conveniently by using temporary supports. For this, nail a 2×4 support across the wall, keeping it ⅝ inch or so lower than the top surface of the plate. Place the sheet as shown in **Figure 25-1** and lift up the free end while bearing forward so the inboard edge will remain on the support. A helper puts the T-brace in position, unless you are dexterous enough to handle both chores.

FIG. 25-1

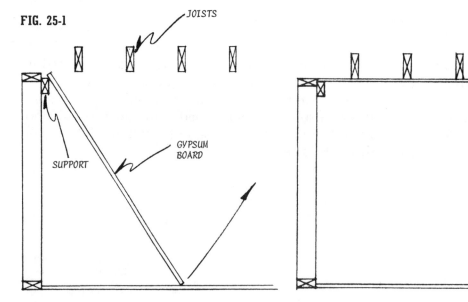

Place sheets so they run at right angles to the joists. Joints should be staggered, and all those "in the field" must fall on the centerline of a framing member. Plan the installation so that no piece will be shorter than the span across three joists. Let cut edges occur at the perimeter of the room so that most joints in the field will be done along the tapered edges of the sheets. Nailing patterns and joint treatments are just as explained in Chapter 24 for walls.

Gypsum-board ceilings are usually installed for a paint or paper finish, but there is no reason why you can't include one to provide additional tightness if you plan, for example, an acoustical-tile ceiling, regardless of whether the tiles will be stapled to furring strips or stuck on with an adhesive. Obviously in such a situation you don't have to be persnickety about the appearance of joints.

CEILING TILE

Ceiling tiles, whether acoustical or purely decorative, may be designed for either staple or adhesive application. Whichever you use, some planning is in order so you don't end up with one of the two situations shown in **Figure 25-2**.

FIG. 25-2

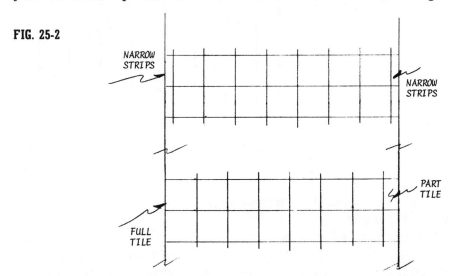

Narrow strips at perimeters will be unsightly and difficult to install. Having uneven borders is as objectionable. The solution is to mark intersecting centerlines of the room on the floor. Starting at the intersection, place a line of tiles on the floor and run them along both centerlines so you can see how the end tiles will fit. If the actual centerlines do not result in end tiles that are at least 6 inches wide, then shift the tiles so you will have a *working* centerline that will result in a more acceptable layout (**Figure 25-3**).

FIG. 25-3

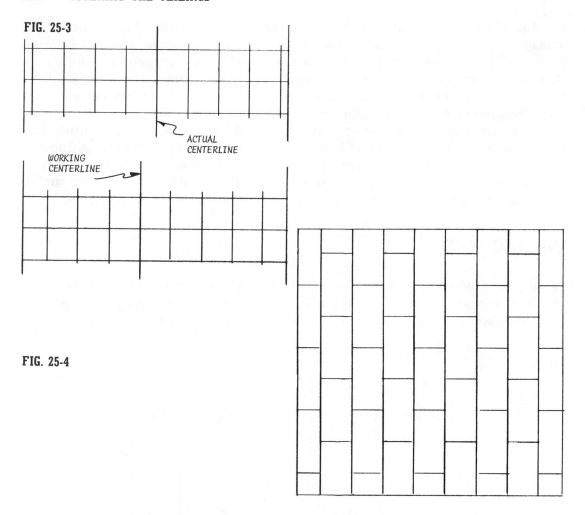

ACTUAL
CENTERLINE

WORKING
CENTERLINE

FIG. 25-4

The technique will work regardless of the size or shape of the tiles; rectangular tiles are shown in **Figure 25-4**.

Mechanical attachment of tiles is done against 1×3 or 1×4 strips of sound wood (furring strips) that you nail directly to joists (**Figure 25-5**). The spacing of the strips depends on tile size, but measurements are always center-to-center. If the tiles are 12 inches wide, then the furring is spaced 12 inches O.C. Best bet is to start the first strip on the working centerline, snapping a chalk line as a guide and being sure the line is 90 degrees to the joist run. With the first strip up, you can cut a few pieces of wood to correct length so they may be used as gauges to position the other strips correctly. Use two 7d or 8d box or common nails at each bearing. Let end joints occur at a joist; fill in around the perimeter of the room as shown in **Figure 25-6**.

FIG. 25-5

FIG. 25-6

Check the furring for levelness, regardless of whether you are installing over a backing or directly to joists. When necessary, use thin pieces of wood as shims to make adjustments (**Figure 25-7**). When you are nailing through a backing, increase the length of the nails by ½ inch.

Start the installation at the wall, being sure to cut the border tiles to correct width (**Figure 25-8**). It's not a bad idea to snap a chalk line on the centerline of the first furring strip to get you started accurately. Other tiles will simply fall into place. How careful you are with the tile-to-wall joint will depend on whether you intend to install molding or will just let the tiles abut. Soft tiles can be cut easily if you work with a straightedge and a very sharp knife. Some workers score the tiles and then snap them, but better results are obtained if you score several times to cut right through the material.

FIG. 25-7

FIG. 25-8

FIG. 25-9

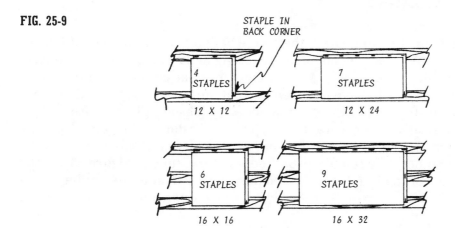

STAPLE IN
BACK CORNER

4 STAPLES
12 X 12

7 STAPLES
12 X 24

6 STAPLES
16 X 16

9 STAPLES
16 X 32

The tiles may be nailed, but the job is much faster and easier if you use a stapling gun. The patterns shown in **Figure 25-9** are pretty common, but check the literature that comes with the product you buy to be sure no special treatment is in order.

Tiles, usually butt-edged, may be secured with adhesive against a solid backing. Usually instructions say to use a small putty knife to place blobs of the adhesives on the back of the tiles (**Figure 25-10**). Place the tile a bit out of position, then press against the ceiling and slide the tile to where it should be. Don't force against adjacent tiles; moderate contact is all you want.

Other types of adhesives have appeared on the market, some that you can apply by brush—so be sure to check the instructions that are printed on labels.

FIG. 25-10

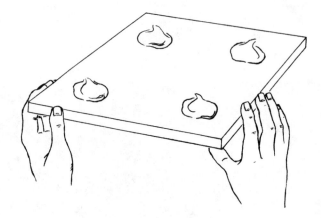

SUSPENDED CEILINGS

When systems like this first appeared, they were hailed because they could be installed with minimum fuss to cover an existing ugly ceiling or to conceal ductwork or plumbing more easily than a conventional ceiling would permit. New concepts and materials have changed all that, and now suspended ceilings are used on new constructions as well as remodeling jobs simply because of their attractiveness. Basically, there are two types of systems. In one, like the Armstrong product in **Figure 25-11**, the ceiling material conceals the means of suspension. In the other, like the Simpson Redwood product in **Figure 25-12**, the grid supporting the ceiling is a decorative element and is available in either metal or wood. In either case, the job can be done under open joists or a closed-in ceiling.

The best way to start, after you have established the working centerline as we have already discussed, is to mark the ceiling height you want on the four walls of the room by using a level and snapping chalk lines. The perimeter molding, which is part of the materials you buy, is nailed up on the lines you have es-

FIG. 25-11

FIG. 25-12

FIG. 25-13

tablished. In **Figure 25-13** the job is shown with a true suspended ceiling. The next step is to stretch lines across the room from molding to molding so you will know the correct position of the hangers that will support the gridwork. Some systems use special clips as in **Figure 25-14**, but lengths of wire may be used, and in fact may be called for. With the hangers up, the main grid runners are put in place (**Figure 25-15**). Then you can start placing tiles, using short, interlocking crossrunners as dividers (**Figure 25-16**).

FIG. 25-14

FIG. 25-15

FIG. 25-16

One of the advantages of doing a ceiling this way is that each tile is an independent element which can be removed at will for access to what might be above or for replacement. Also, translucent plastic panels may be substituted for as many regular tiles as you wish if you plan to install indirect lighting.

The Armstrong system we mentioned may be installed against a covered ceiling or directly to exposed joists. A basic requirement is that the ceiling be dropped 2 inches. Again, the job is started by installing perimeter molding, which may be metal angle or a wood of your choice (**Figure 25-17**). Next, special metal furring channels are nailed across the joists (**Figure 25-18**). Since these are quite sturdy, they can easily span across low spots without the need of shims. Also, a nail into every other joist is sufficient to hold them. The cross tees are designed to interlock with the channels, but they are still moveable so they can slide along the channel and be concealed in the slot on the leading edge of the tile (**Figure 25-19**).

The system is organized so it can be used with various types and sizes of tiles.

FIG. 25-17

FIG. 25-18

FIG. 25-19

EXPOSED BEAMS

Like the idea of heavy, handsome beams running across a ceiling? It's possible to enjoy them without their being integral parts of the structure if you buy or make nonstructural units that are as visually effective as the real thing. The idea has caught on, and manufacturers are offering ready-made, lightweight beams of wood, hardboard, or plastic. Most are available in channel form, since this reduces weight and provides a passageway in which you can install electrical wiring. Those of wood or hardboard are nailed to 2×4s or 2×6s which are first attached to joists, preferably with lag bolts; those of plastic are secured with zigzag lines of adhesive that you squeeze from a tube.

To do your own thing, you can attach 2×4 or 2×6 "beams" across joists before placing the ceiling (**Figure 25-20**). This requires pretty accurate fitting and additional nailing surfaces so if you wish to work this way it will be better to form rabbets in the beam before you attach it (**Figure 25-21**). Another, easier way is to attach the beams after the ceiling is installed (**Figure 25-22**).

FIG. 25-20

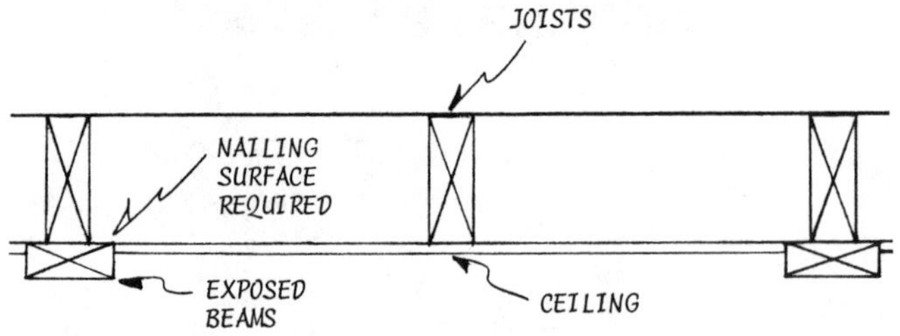

FIG. 25-21

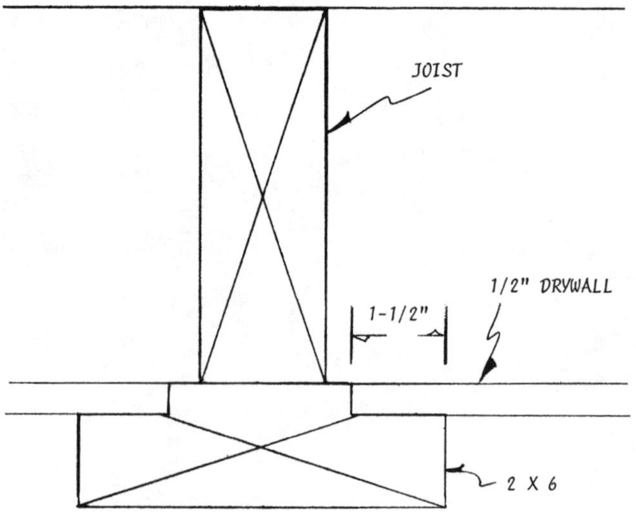

FIG. 25-22

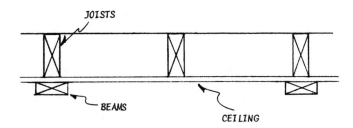

If you do this with solid stock, limit the thickness to 2 inches and make the attachment with lag bolts driven into every other joist. Sink the lag bolts in counterbored holes that you then plug with dowel to get a pegged effect. You can also make the beams like long U-channels, using ½-inch or ¾-inch stock to reduce weight. In **Figure 25-23** a ¾-inch-thick nailer is attached across the joists first and the box beam is secured to the nailer.

FIG. 25-23

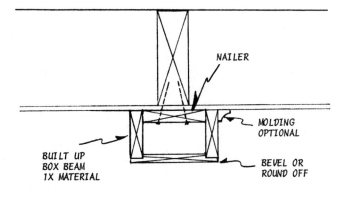

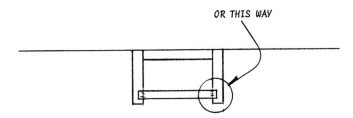

One idea you might check out is using baseboard or casing molding as the sides of the beams (**Figure 25-24**). You can get some interesting architectural effects without the need for elaborate equipment.

Other ideas are to work with rough wood or material that you deliberately distress with a rasp or even an ax.

Don't be too generous with the number or the size of the beams you install; they can be overpowering, especially in a small room.

FIG. 25-24

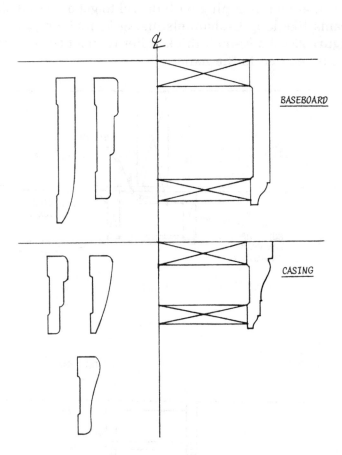

26

INSTALLING FINISH FLOORS

WOOD

The durability and the appearance of wood amply justify its timeless popularity as a flooring material. Today it is available in many forms, but all of them can be classified as being strip, plank, or block. Modern versions of strip and plank may be installed with either nails or adhesives; blocks, which really produce a parquet floor, are usually placed with adhesive. All three are available prefinished so that the job is complete as soon as the material is down. Woods, patterns, and finishes are so numerous that the only way to make a proper choice is to visit a flooring-supply establishment and actually see and feel the selections.

Strip flooring is basically of two types. The more sophisticated version has an interlocking tongue-and-groove arrangement on all edges and ends (**Figure 26-1**).

FIG. 26-1

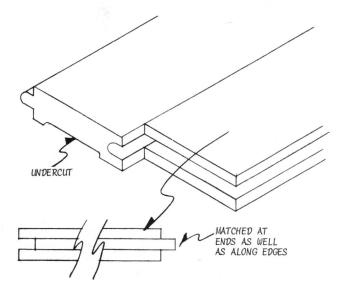

UNDERCUT

MATCHED AT
ENDS AS WELL
AS ALONG EDGES

Technically, it is described as being side- and end-matched. The undercut on modern products is provided so the pieces are more likely to stay flat and remain stable even if the subfloor is less than perfect. Nails, if used, are driven at an angle through edges so they are concealed.

Square-edge strip flooring, as the name suggests, is simply narrow lengths of wood which are butted edge to edge and end to end (**Figure 26-2**). Here, surface nailing is required and the flooring chore is extended, since you must set and then conceal the nails with a wood dough. It can be economical, especially if you buy thin material, but you can't rely too much on its stability unless you have a substantial subfloor and include an underlayment.

FIG. 26-2   SQUARE-EDGED STRIP FLOORING

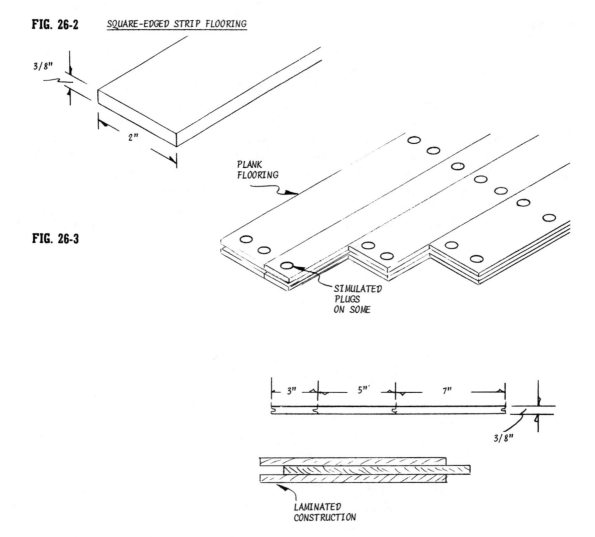

3/8"

2"

PLANK
FLOORING

FIG. 26-3

SIMULATED
PLUGS
ON SOME

3" · 5" · 7"

3/8"

LAMINATED
CONSTRUCTION

Plank flooring comes in random widths and can be either solid boards or one of the more modern laminated constructions (**Figure 26-3**). The material is available with false pegs already in place or is designed for surface attachment with screws or nails driven through counterbored holes which are then filled with plugs. One type is installed with real wrought-iron nails driven through surfaces. While some of these products merely simulate the kind of floors you would expect to find in old colonial homes, others are the real thing—for example, solid oak planks with authentic walnut pegs.

There are many choices in block flooring, both in patterns and in sizes. Some are assemblies of small, solid pieces of wood. Others are laminations, and still others resemble short pieces of strip flooring that are edge-joined into square blocks (**Figure 26-4**). An interesting innovation is a laminated-oak block (Bruce Flooring) which has a polyethylene foam pad on the back to soak up noises. It may be used anywhere but is especially suitable for upper floors in, for example, a two-level home.

FIG. 26-4

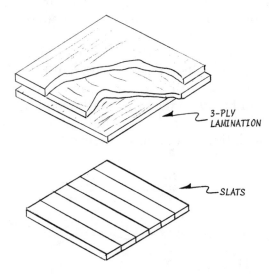

Many of the modern versions of both plank and block flooring are designed for adhesive application on wood or directly to concrete.

UNDERLAYMENTS

An underlayment is a ¼- to ½-inch thickness of plywood, hardboard, or particleboard that you may have to place over the subfloor, depending on the type of material used for the finish floor (**Figure 26-5**). It is usually required under resil-

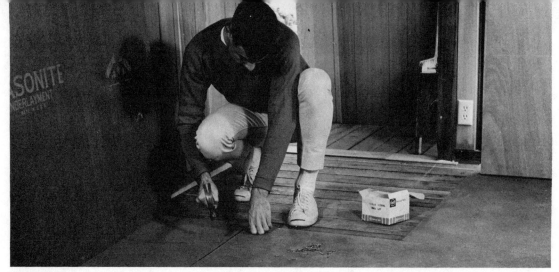

FIG. 26-5

ient materials like linoleum and vinyl tiles and under carpeting. It may be necessary to add it under thin wood flooring, but it is not mandatory under thick strip or plank flooring. There is nothing wrong with adding it under any circumstances if you like the idea of increased rigidity, more insulation value, and guaranteed smoothness for the final cover.

A point to remember if you use different-thickness floor coverings in various rooms is that you might be able to end up with a uniformly even floor throughout the house by compensating with underlayments of various thicknesses.

Recommended underlayment grades of plywood have a solid surface backed with inner ply construction that resists deformations caused by concentrated loads. A typical installation is shown in **Figure 26-6**.

FIG. 26-6

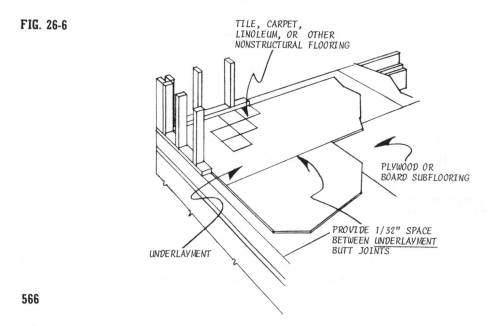

TILE, CARPET, LINOLEUM, OR OTHER NONSTRUCTURAL FLOORING

PLYWOOD OR BOARD SUBFLOORING

PROVIDE 1/32" SPACE BETWEEN UNDERLAYMENT BUTT JOINTS

UNDERLAYMENT

Place panels so that the surface grain will run across joists and end joints will occur over a framing member. Stagger end joints in relation to each other and plan for all joints to be staggered in relation to those in the subfloor. Use 3d ring-grooved or cement-coated nails to secure plywood that is ½ inch or less in thickness, or use 16-gauge staples that you can drive with a rentable tool. The staples should be long enough to penetrate the subfloor by at least ⅝ inch.

Space nails 6 inches apart along edges, 8 inches apart in the field. Check the installation carefully to be sure all nailheads are flush and no unevenness has occurred at joints. Use a few extra nails wherever you think they are needed; do some sanding if nails have raised splinters or if sawed edges are rough. There's no taboo against filling all joints with a wood dough.

Hardboard underlayment is manufactured from wood fibers that are bonded under heat and pressure to form panels which are dense, durable, and grainless. Most have one coarse-sanded surface designed to increase adhesive bonding qualities. Installation doesn't differ from that outlined for plywood except for the specifics shown in **Figure 26-7**.

FIG. 26-7.   FASTENING HARDBOARD

FASTENER SIZE AND TYPE	FASTENER SPACING	
4d annular groove or ring groove cement-coated sinker nails	Panel edges 3″ (⅜″ from edge)	Intermediate 6″
18 ga. staples	4″ to 6″ (⅜″ from edge)	4″ to 6″

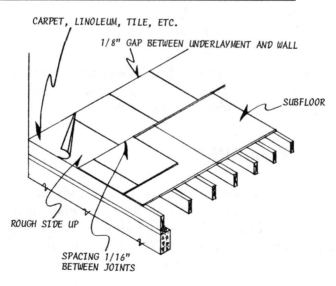

CARPET, LINOLEUM, TILE, ETC.

1/8″ GAP BETWEEN UNDERLAYMENT AND WALL

SUBFLOOR

ROUGH SIDE UP

SPACING 1/16″ BETWEEN JOINTS

Particleboard is a panel of wood particles bonded with special additives under heat and compressed to a uniform density. It is flat and grainless, and while it is strong, it should be carefully handled to avoid damage to edges. Usually it is recommended that a vapor barrier be placed between it and the subfloor. A typical installation is shown in **Figure 26-8**.

FIG. 26-8.   FASTENING PARTICLEBOARD

FASTENER SIZE AND TYPE	FASTENER SPACING	
	Panel edges	Intermediate
Ring-groove or coated box nails (length should be sufficient to enter sub-floors 1″ to 1½″)	6″ (½″ to ¾″ from edge)	10″
16 ga. staples	3″ (½″ to ¾″ from edge)	6″

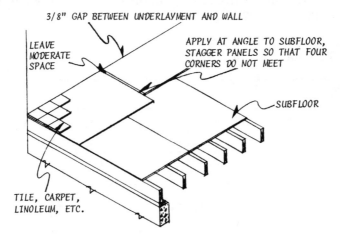

All underlayment material should be placed immediately before the installation of the final cover. If there must be a time lapse, take steps to protect the underlayment from physical damage and moisture.

PLACING STRIP FLOORING

The complexity of the layout will relate to whether you use strip flooring throughout the house or in just one room. If the coverage is overall, then strips must run parallel and continuously through all areas with lengths following the long dimension of hallways and rooms. You are likely to have alignment problems when you encounter projections into a room, doorways, openings into closets, and so on.

A good starting point is to figure that you are going to run the pieces the long dimension of the house as a whole, since this would, normally, place them at right angles to joists, which is advisable. Do a test run in a critical area (a hallway is good) by actually placing pieces of flooring loosely to see how they might lay out. By some trial and error you can find where to make a start so strips can run as they should with minimum fuss on your part. In the very basic example in **Figure 26-9**, it is easy to see that strips started lengthwise in the hallway could easily move into rooms while maintaining necessary parallelism. Of course all installations are not so simple, but the idea is there. Also, no law says you can't change the direction of the strips if you choose to, but it is wise to do that only as a last recourse.

FIG. 26-9

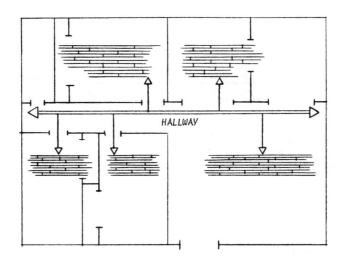

Make a last inspection of the subfloor to be sure all nails are down and correctly seated. Sweep away dust and dirt or, preferably, do the job with a vacuum cleaner. Place a layer of double-weight building paper over the subfloor, lapping joints at least 4 inches, and then snap chalk lines to indicate the joist runs. Note that the paper, like the joists, should run at right angles to the strips (**Figure 26-10**).

The type, size, and number of nails you use to secure the strips is critical if the floor is to feel solid and be without squeaks. General recommendations are given in the table in **Figure 26-11**.

Actually, wood floors will be more solid if nails are long enough to pass through the subfloor and enter the joists. Some of the nail spacing specified in Figure 26-11 will not permit this at every joist, so it seems logical to adopt an 8-inch-O.C. spacing generally. With joists 16 inches O.C. you would have a nail at each joist crossing and one between.

FIG. 26-10

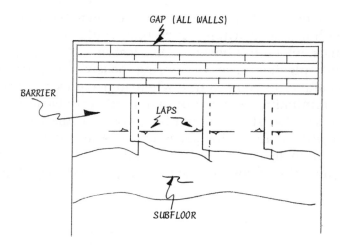

FIG. 26-11.   BLIND-NAILING OF TONGUE & GROOVE FLOORING

FLOORING NAILS

SIZE	LENGTH
4d	1-1/2"
5d	1-3/4"
6d	2"
7d	2-1/4"
8d	2-1/2"

SQUARE-EDGE TYPE

SCREW TYPE

FLOORING SIZE		NAILING REQUIREMENTS		
Thick.	Width	Size	Spacing	Type
$^{25}/_{32}$	3¼	7d	10 to 12	
$^{25}/_{32}$	2¼	or	O.C.	
$^{25}/_{32}$	1½	8d		cut flooring
½	2	5d	10	or
½	1½		O.C.	screw type
⅜	2	4d	8 O.C.	
⅜	1½			
SURFACE-NAILING SQUARE-EDGE FLOORING				
$^{5}/_{16}$	2	15 gauge	7 O.C.	special
$^{5}/_{16}$	1½	X 1"	(2 nails)	cement-coated, fully barbed

Start the first line of strips parallel to a wall, placing them as shown in **Figure 26-12**. Be careful with this beginning, since it will affect the placement of all pieces that follow. The surface nails you use to start with can be placed so they will be hidden later by the baseboard, or you can set them and conceal them with a wood dough. Work with the longest pieces of flooring, saving shorter ones to end runs. Cutoffs can be utilized in closets and similar areas. After the first strip is down, place the next five or six lines loosely on the floor so you can judge positions of end joints, which should be staggered and not closer than 6 to 8 inches in succeeding courses. This is also the time to judge how adjacent strips (or courses) will blend in terms of grain patterns and color tones.

FIG. 26-12

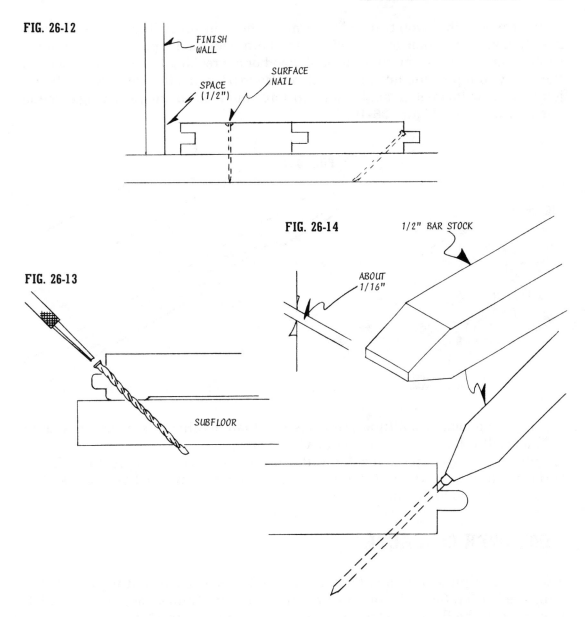

FIG. 26-14

FIG. 26-13

Courses that follow the starter strips are blind-nailed (**Figure 26-13**). Drive the nail so it bisects the angle formed by the tongue and shoulder on the wood. To avoid damage to the flooring, switch to a nail set while the nail still projects about $1/8$ inch. A homemade nail set like that shown in **Figure 26-14** might make the job easier. If you try the idea, be sure you don't drive the nail set so far that you split the tongue. Work with a scrap piece of flooring and a hammer when you encounter pieces that won't fit tightly against those in preceding courses (**Figure 26-15**). If you find some are especially stubborn, tack-nail a piece of wood to the subfloor and then use another piece of wood or a wrecking bar as a lever to force the baddy into position until it is nailed. Use extra nails, if necessary, at all end joints, driving them at an angle that will force the piece being placed against the one already there (**Figure 26-16**).

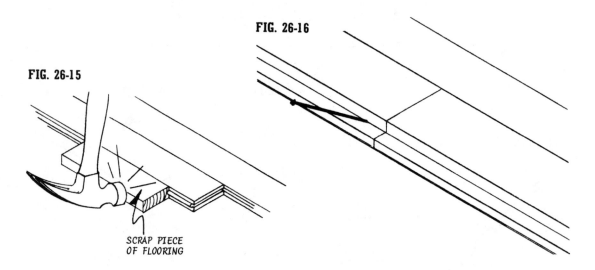

FIG. 26-16

FIG. 26-15

SCRAP PIECE
OF FLOORING

If you encounter splitting problems in situations like this, take the time to drill a small pilot hole before you drive the nail.

When you reach the opposite wall, rip the final strips to correct width and surface-nail them as you did the starters. Remember, there must be a space of ½ inch between flooring and all walls.

WOOD OVER CONCRETE

There are two problems here. First, the concrete must be covered with a reliable moisture barrier. Second, you must provide joist substitutes (sleepers) to which you can nail the flooring. One solution is shown in **Figure 26-17**.

The bottom sleepers are placed directly in the asphaltic mastic, which must cover the slab completely. Spread the mastic uniformly with a toothed trowel. The top sleepers are nailed in place after the polyethylene film is laid down.

Another system is to place the polyethylene film between two layers of mastic, and put 2×4 sleepers in the top layer of mastic as shown in **Figure 26-18**.

FIG. 26-17

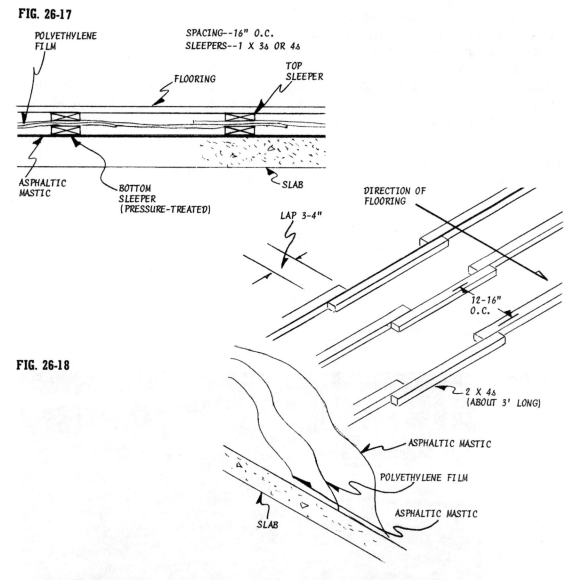

POLYETHYLENE FILM

SPACING--16" O.C.
SLEEPERS--1 X 3s OR 4s

FLOORING

TOP SLEEPER

ASPHALTIC MASTIC

BOTTOM SLEEPER (PRESSURE-TREATED)

SLAB

FIG. 26-18

LAP 3-4"

DIRECTION OF FLOORING

12-16" O.C.

2 X 4s (ABOUT 3' LONG)

ASPHALTIC MASTIC

POLYETHYLENE FILM

ASPHALTIC MASTIC

SLAB

In all cases, the sleepers adjacent to the concrete should be secured with special hardened concrete nails spaced about 20 inches apart. These can be driven with a heavy hammer, or you can rent special equipment that sinks the nails with explosive power.

It's obvious that sleepers over concrete won't supply adequate support for thin strip flooring; consider $25/32$-inch flooring a minimum thickness. However, you can use an alternate method that will permit the use of any type of flooring.

The slab is waterproofed and sleepers are placed as described above, but then a substantial subfloor of plywood is placed just as if you were working over conventional joists (**Figure 26-19**).

Some types of modern wood flooring materials, like the plank design by Bruce Flooring in **Figure 26-20**, are designed to be installed directly on concrete. Procedures for such products vary; you must be sure to understand the manufacturer's instructions and follow them to the letter.

FIG. 26-19

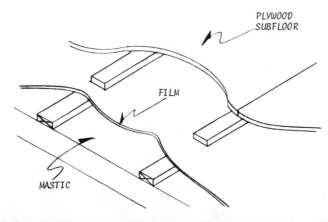

FIG. 26-20

BLOCK FLOORING

An "easy" and quite adequate way to install block flooring is to use a modern type that is installable with a mastic. There are dozens of types and patterns available — two are shown in **figures 26-21** and **26-22** — so your choice is far from limited. Do check installation specifics, especially in regard to the adhesive recommended and to whether a vapor barrier is required; these may vary in different products. A good layout procedure is shown on the next page.

FIG. 26-21

FIG. 26-22

Use a chalk line to mark the intersecting centerlines of the area (**Figure 26-23**). Then, working from the intersection, place loose blocks in both directions to see how they will end up at the walls. Make adjustments until, as shown in **Figure 26-24**, A and B are equal and at least one-half the width of a block. At this point you can snap additional lines to show the position of the starters (**Figure 26-25**). Spread mastic from the centerpoint toward the area you will start to cover. It will probably be necessary to wait a bit for the adhesive to become tacky before you can place the blocks, but instructions on the container will explain this.

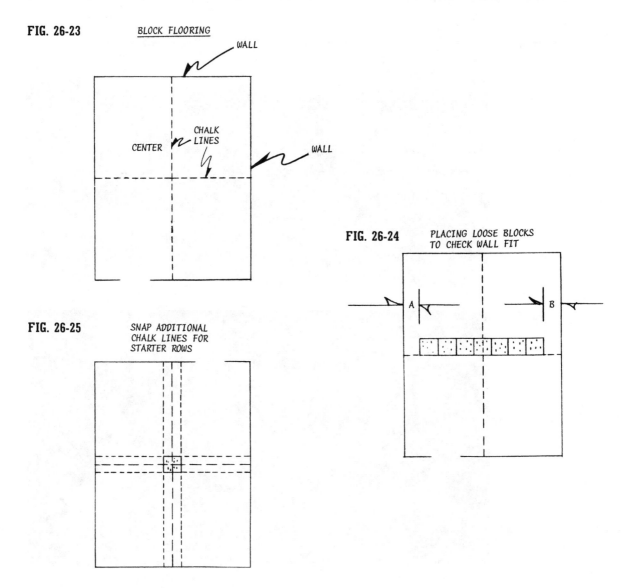

FIG. 26-23   BLOCK FLOORING

WALL

CENTER   CHALK LINES

WALL

FIG. 26-24   PLACING LOOSE BLOCKS TO CHECK WALL FIT

A   B

FIG. 26-25   SNAP ADDITIONAL CHALK LINES FOR STARTER ROWS

Start placing blocks in pyramid fashion, as in **Figure 26-26** until you have covered the spread of mastic. Then continue in similar fashion until all but the blocks adjacent to the walls have been placed. At this point, work as shown in **Figure 26-27** so you can mark the border pieces correctly for cutting to width.

FIG. 26-26

FIG. 26-27

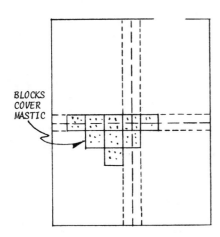

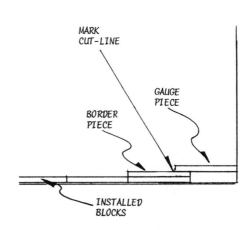

Design variations are possible when you plan placement of pieces so grain directions or designs are opposed in adjacent pieces (**Figure 26-28**). More impressive, and more complicated, departures from straightforward layouts are possible when you do a diagonal pattern (**Figure 26-29**). The actual installation of the blocks doesn't differ but more planning, more cutting, and often more material are required.

FIG. 26-28

FIG. 26-29

CHALK LINES FOR DIAGONAL LAYOUT

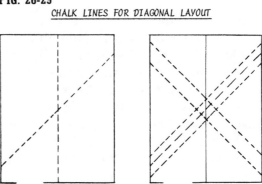

Figures 26-30 and 26-31 show two methods of making an attractive and unobtrusive joint when block flooring (or any wood flooring) abuts a carpet.

FIG. 26-30

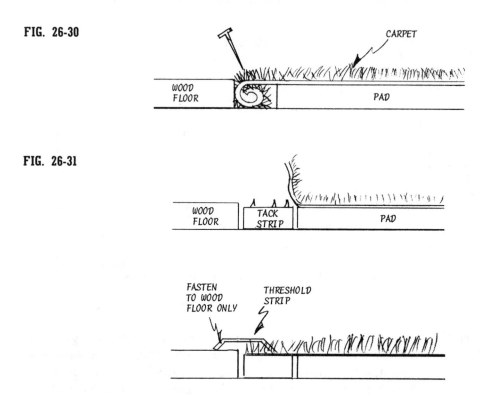

FIG. 26-31

RESILIENT FLOOR TILES

This type of floor cover, except in the case of some remodeling work, must always be installed over a sound, smooth underlayment. A typical job schedule is shown in the following illustrations supplied by the Armstrong Cork Company.

Be sure the underlayment is secured with coated or ring-grooved nails. The type shown in **Figure 26-32** is called Temboard and is marked with green dots to show where the nails should be placed. The worker is using a special nailing machine — which you can rent — to facilitate the operation.

Establish intersecting centerlines for the area as shown in **Figure 26-33**, and then work as we described for block flooring (and for ceilings) to find and mark the *working* centerlines.

FIG. 26-32

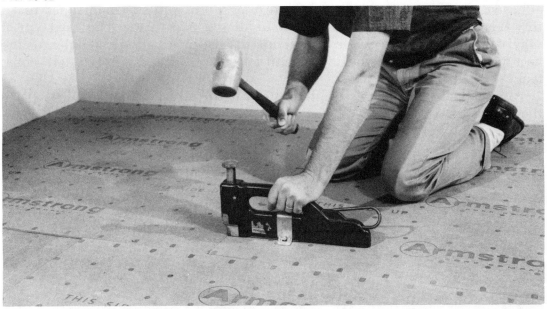

FIG. 26-33

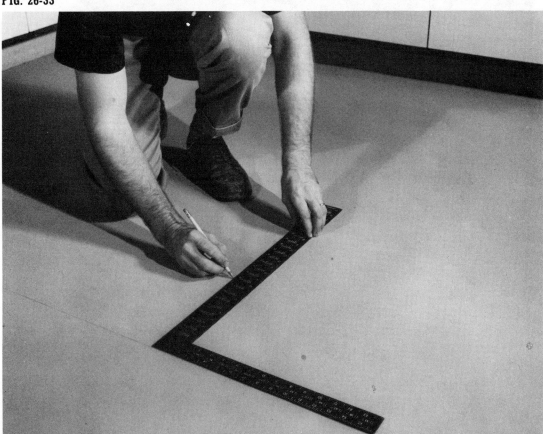

Spread a coat of adhesive over about one-quarter of the area. Some new adhesives may be applied by brush for regular tiles, but it's recommended that a trowel be used as in **Figure 26-34** if you are going to place a new embossed type of vinyl-asbestos tile. At any rate, follow the instructions on the container for the best method. Allow some drying time, if necessary, for the adhesive to become tacky.

Place the first tiles very carefully, being sure to follow the marked lines. Place the tiles exactly where they must go; do not slide them into place. Butt the edges firmly, but do not force them (**Figure 26-35**).

FIG. 26-34

FIG. 26-35

If you want to add some decorative detail, you can use ready-made feature strips that are placed along with the regular tiles and with the same adhesive (**Figure 26-36**). These are available in different colors so you can do combinations that will complement other colors in the room.

Finish the installation at walls or under cabinets with a ready-to-use cove base that is applied with the same adhesive used to stick down the tiles (**Figure 26-37**). This too is available in different colors, so you can work along with various decorating schemes.

FIG. 26-36

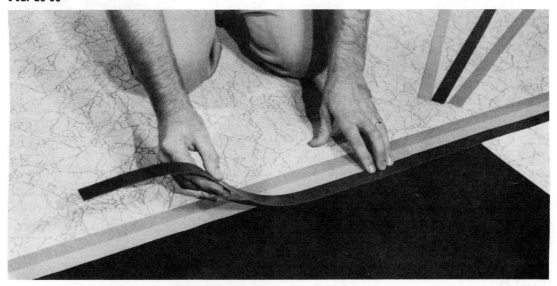

FIG. 26-37

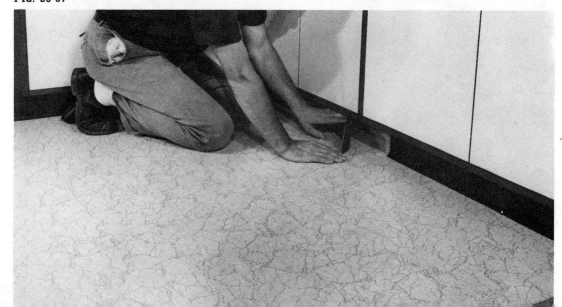

FIG. 26-38

Tiles can be cut with snips when it is necessary to fit them around obstructions (**Figure 26-38**). The best way is to make a paper pattern and test it before tracing the outline on the tile. Tiles will be easier to cut if they are warm.

Clean over the area by working with a manufacturer-recommended solvent and a soft cloth. The cloth should be damp only; you don't want excess solvent to be sucked down between the joints in the tile.

You might want to check out self-stick types of floor tiles; more and more of them are appearing on the market every day. With these, the installation is a simple matter of removing the release paper from the back of the tile and then pressing the tile firmly into place (**Figure 26-39**). The floor is ready to use as soon as the last tile is down.

FIG. 26-39

HARD TILE

In the old days generally, and still today for commercial applications and for heavy-duty usage, ceramic tiles were placed over a thick reinforced bed of mortar supported by a solidly nailed subfloor. Because of the installation's weight, extra-heavy joists or special bracing members may be required as safety factors in the framing. The basics of the design, when done on top of joists, are shown in **Figure 26-40**.

FIG. 26-40

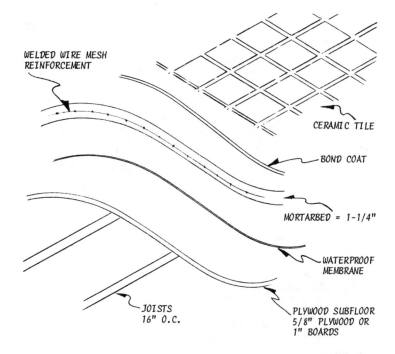

WELDED WIRE MESH
REINFORCEMENT

CERAMIC TILE

BOND COAT

MORTARBED = 1-1/4"

WATERPROOF
MEMBRANE

JOISTS
16" O.C.

PLYWOOD SUBFLOOR
5/8" PLYWOOD OR
1" BOARDS

Since ceramic-tile floors are often limited to small areas such as bathrooms and utility rooms and perhaps kitchens, the specific areas are often depressed as shown in **Figure 26-41** so finish floor heights can be uniform. The ledgers that support the subfloor must also reinforce the cut joists, so pieces that are substantially longer than the cuts should be used.

Another method, designed for a better-than-average installation in a residence, is shown in **Figure 26-42**. Note that a thinner epoxy mortarbed is substituted for the heavy mortar shown in the other illustrations. The advantage is less weight and a thinner total cross section.

Figure 26-43 shows the still easier method that is being used more and more today when ceramic tiles are placed over a wood floor and will be exposed only to light residential traffic.

FIG. 26-41

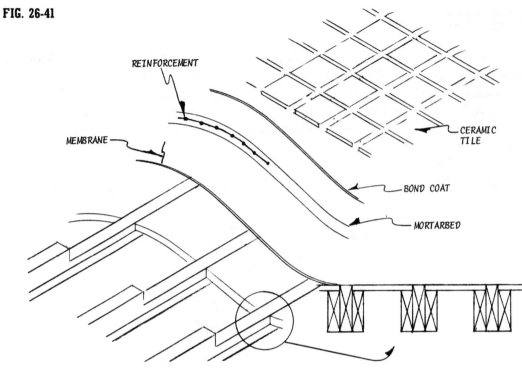

REINFORCEMENT

MEMBRANE

CERAMIC TILE

BOND COAT

MORTARBED

FIG. 26-42

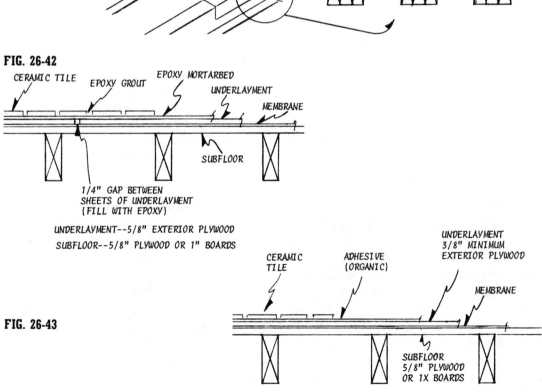

CERAMIC TILE

EPOXY GROUT

EPOXY MORTARBED

UNDERLAYMENT

MEMBRANE

SUBFLOOR

1/4" GAP BETWEEN
SHEETS OF UNDERLAYMENT
(FILL WITH EPOXY)

UNDERLAYMENT--5/8" EXTERIOR PLYWOOD

SUBFLOOR--5/8" PLYWOOD OR 1" BOARDS

UNDERLAYMENT
3/8" MINIMUM
EXTERIOR PLYWOOD

CERAMIC
TILE

ADHESIVE
(ORGANIC)

MEMBRANE

FIG. 26-43

SUBFLOOR
5/8" PLYWOOD
OR 1X BOARDS

The most modern methods of installing ceramic tiles over wood subfloors are about as simple as placing resilient tiles. The tiles, like Redi-Set by American Olean, come in pregrouted sheets that are, and remain, flexible, so the substrate requires minimum attention. In fact, the material may be placed directly on plywood subfloors if they are ⅝-inch five-ply or heavier. Board subfloors should be covered with an underlayment.

The layout is done as we have already described for wood blocks or resilient tile. Spread the adhesive with a notched trowel, starting at the intersection of the adjusted centerlines and covering about one-quarter of the area (**Figure 26-44**). Do not spread more adhesive than you can cover with tile in an hour or less.

FIG. 26-44

Place the sheets of tile as shown in **Figure 26-45**, making contact along one edge and then rolling the rest of the sheet down into position. Press edges tightly together as you go; do not slide tiles into position. Cuts that are required along a grout line are easily done with a sharp knife (**Figure 26-46**).

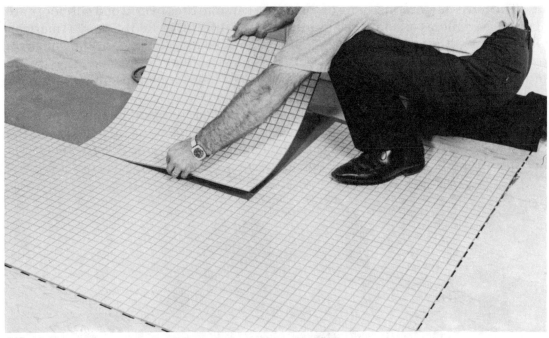

FIG. 26-45
FIG. 26-46

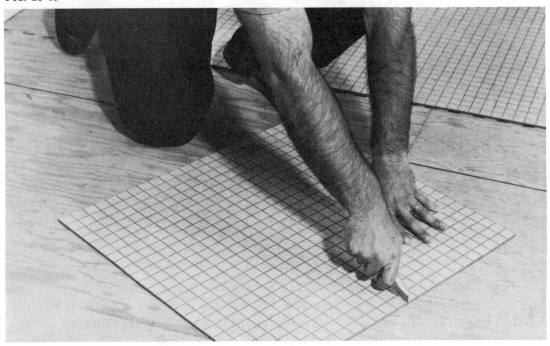

When the tile itself must be cut, use a regular tile cutter for straight cuts (**Figure 26-47**). Use tile nippers for irregular cuts (**Figure 26-48**).

FIG. 26-47

FIG. 26-48

Edges, and openings around plumbing fittings if any are present, are sealed with a special caulking that comes in colors to match the grout color in the material you have selected and is applied with a caulking gun (**Figure 26-49**). The manufacturer recommends that the floor be rolled in both directions with a 150-pound carpet-covered roller before edges are caulked. If you choose not to rent a roller, you can do the job by hammering down on a piece of carpet-covered 2×4.

FIG. 26-49

OTHER TYPES OF FLOORING

Tredway is a new type of sheet cushioned-vinyl flooring, made by Armstrong, that is quicker and easier to install because it is fastened by stapling it around the edges of the room (**Figure 26-50**). Because it is flexible and can be stretched it has, within limits, a built-in margin for error. Trimming and cutting are easy to do if you work with a straightedge and a sharp knife (**Figure 26-51**).

The new material can be applied to concrete with cement; cementing may also be done along edges that can't be secured with staples.

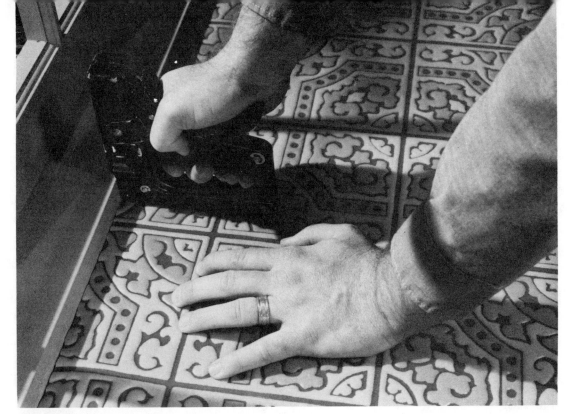

FIG. 26-50

FIG. 26-51

FIG. 26-52

FIG. 26-53

How about a flagstone floor? **Figure 26-52** shows what I did in my studio over a concrete slab, but don't attempt it unless you are prepared for a good deal of labor. We purchased the material in bulk, which meant pieces had to be cut up with a hammer and chisel so they would fit together reasonably well (**Figure 26-53**). "Reasonably well" allows room for some errors, since joints that are not perfectly uniform have an attractive rustic effect. One of the installation problems is caused by the unequal thickness of the bulk sheets. This can range from not much more than ¼ inch to over 1 inch, so the mortarbed must be carefully adjusted if the floor surface is to be smooth and even. The rewards are a distinctive floor that is very easy to maintain and will last indefinitely.

How about cut-and-loop sculptured shag that you can put down in the form of self-adhesive 12-inch squares? The material has a foam backing so you don't need the conventional padding. The Armstrong example shown in **Figure 26-54** is called Bandwagon and is one of a line which includes eight different colors.

To cover a porch floor or a completely outdoors area like a patio or the deck around a swimming pool, consider indoor-outdoor carpeting that can take abuse and is easy to clean with a stream of water from a hose (**Figure 26-55**). Such materials come in sheet form in 6-foot and 12-foot widths. Usual colors are either green or brown, but others are appearing, since the concept is beginning to become popular for indoor as well as outdoor areas. One type is available in self-adhesive 9-inch squares. We tried it in a small utility area and it's working fine.

FIG. 26-54

FIG. 26-55

27

INSTALLING TRIM

Trim work is done with ready-made moldings that are available in many shapes and sizes. Generally, they can be classified as structural or decorative, but the terms describe applications more than they do the molding itself. From a practical point of view, moldings used to cover and to seal gaps between door frames and walls and those used around windows like those in **Figure 27-1** are structural pieces. Yet the same moldings, or similar ones, can be used merely as decoration. **Figure 27-2** shows how moldings can transform the blank look of flush doors.

At any rate, moldings add the final touches to house constructions and contribute to appearance while doing necessary chores. A little-appreciated fact is that moldings provide clean lines to cover necessary spaces between construction elements, as baseboard does over a wood floor (**Figure 27-3**). The baseboard also serves to protect the wall from damage when furniture is placed and when floors are cleaned.

FIG. 27-1

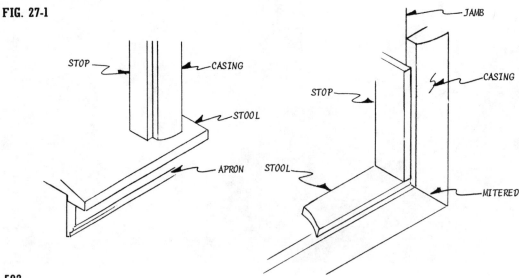

FIG. 27-2

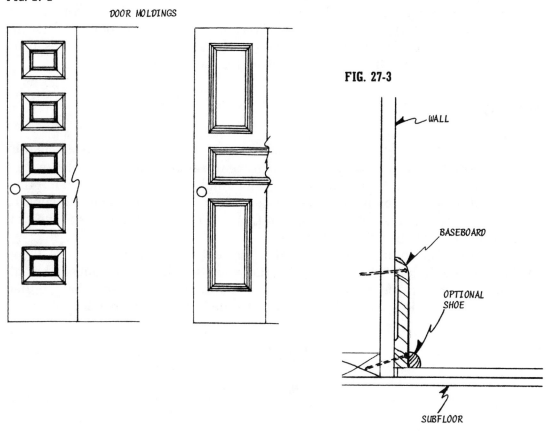

DOOR MOLDINGS

FIG. 27-3

Planning for the use of moldings in particular areas can relieve you of the chore of having to do a joint that is good-looking in itself, for example in wall-to-ceiling corners. Finally, moldings when used correctly will act as covers over unsightly cracks that might appear because of normal expansion and contraction of framing members and the settling of the structure as a whole.

BUYING MOLDING

Any building-supply company or lumber yard worthy of the name will stock a good supply of various types of molding with lengths running from 6 to 16 feet. When you determine molding lengths for your needs, always round off to the next highest foot. It's much better to end up with a short cutoff than have to splice lengths to complete a run. Always add the amount of material required for miters.

For example, if you are framing around a window, each piece will have a miter at each end. If the molding is 3 inches wide, you would add a minimum of 6 inches and then round off to the next even foot. This applies, of course, if you are getting moldings cut to specific lengths. When you use bulk lengths you cut pieces to accommodate only the extra material required for the miters.

NAILING

Moldings are attached with either finishing or casing nails so that heads can be driven below the surface of the wood with a nail set and then concealed with a wood dough. The size of nails depends on the thickness of the molding, but they must always be long enough to penetrate well into framing members such as studs, plates, trimmers, and headers. Don't forget to add the thickness of the wall cover when you choose a nail length.

Nails driven into profile corners will be easier to conceal (**Figure 27-4**). If you pre-tint the molding before you install it, the wood dough used to conceal the nail can be colored with the same stain. Prefinished moldings that may be supplied along with a paneling material should be attached with color-matched nails.

FIG. 27-4

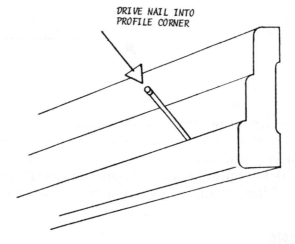

DRIVE NAIL INTO PROFILE CORNER

CUTTING AND JOINING

All cuts on moldings should be made with a fine-tooth crosscut saw. The best way to get accurate cuts is to use a miter box. Commercial ones such as the sophisticated version in **Figure 27-5** are available, or you can make a simple one as

FIG. 27-5

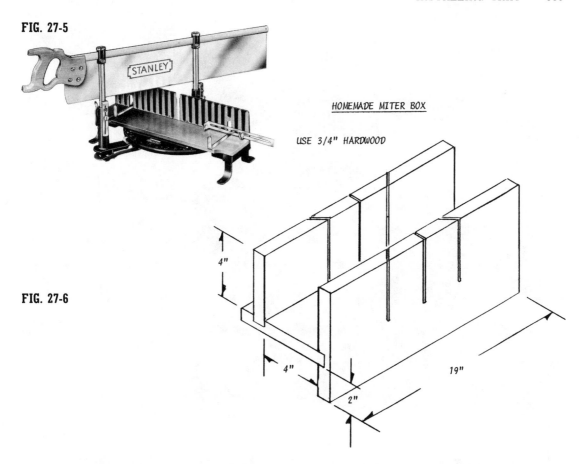

HOMEMADE MITER BOX

USE 3/4" HARDWOOD

4"

FIG. 27-6

4"

19"

2"

shown in **Figure 27-6**. Actually, even the simple type can be purchased ready-made, but may lack the lip that permits securing the unit with a clamp so it will hold steady as you saw (**Figure 27-7**).

In addition to straight cuts and simple miters, the jig can be used to saw compound angles if you set it up properly (**Figure 27-8**). The width of the strip at the bottom determines the slope angle of the work.

General molding rules are to miter all outside corners but to *cope* inside ones. The coped joint takes a little more doing than a simple miter, but it's wise to do, for example, where baseboards meet at an inside corner, since nailing won't cause the joint to open and shrinkage will not result in an obvious crack. Coping a joint merely means that you shape the end of one piece so it will conform to the profile of the piece it abuts (**Figure 27-9**).

The job can be done in one of two ways. You can transfer the profile of one piece to the end of the mating piece by using a compass (**Figure 27-10**). Then make the cut with a coping saw. With the second method, you cut a routine inside

FIG. 27-7

FIG. 27-8

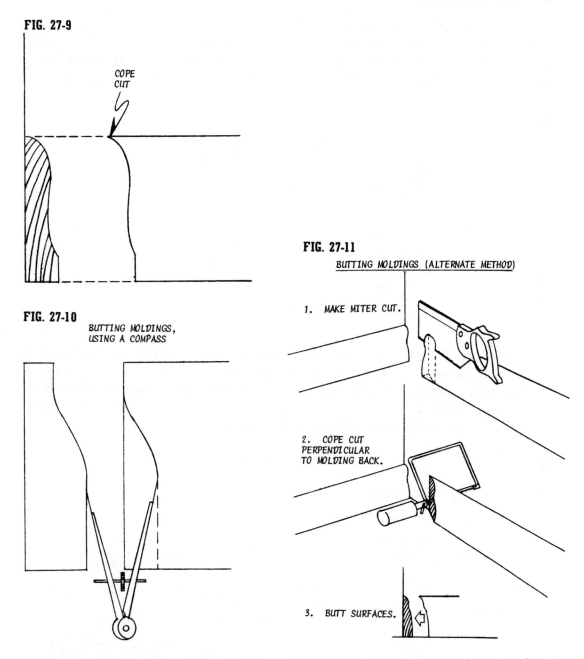

FIG. 27-9

COPE CUT

FIG. 27-10

BUTTING MOLDINGS, USING A COMPASS

FIG. 27-11

BUTTING MOLDINGS (ALTERNATE METHOD)

1. MAKE MITER CUT.

2. COPE CUT PERPENDICULAR TO MOLDING BACK.

3. BUTT SURFACES.

miter first and then follow the line of the miter with a coping saw, keeping the cut perpendicular to the back surface of the molding (**Figure 27-11**). It's a good idea to undercut a bit so you will be assured of a tight fit at the front edge of the moldings where they join.

Runs of molding are best done with single pieces, but if you must extend lengths it's better to scarf the joint, as shown in **Figure 27-12**, instead of butting the ends. There is less likelihood that a scarfed joint will separate, especially if you apply some glue and drive a nail at an angle through the mating ends.

FIG. 27-12

SCARF JOINT

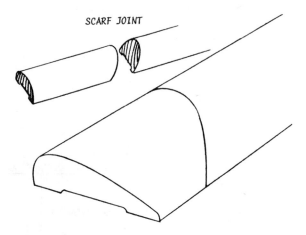

If moldings intersect, you should let at least one piece run continuously. In the example shown in **Figure 27-13** the cut pieces would be easy to shape by working with a coping saw and then smoothing with a drum sander. A butt block as shown in **Figure 27-14** is often used to provide a decorative detail or a transition point for intersecting dissimilar moldings.

FIG. 27-13   **FIG. 27-14**

ONE PIECE CONTINUOUS

WITH BUTT BLOCK

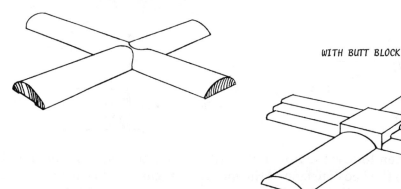

TYPES OF MOLDING

The drawings in this section show standard ready-made moldings. All are available in different sizes, and some in different materials. While all were designed for a specific purpose, most are flexible enough to be used for decorative purposes as well as the intended structural applications. Many craftspeople use leftovers to do decorative things on cabinets and furniture and to make picture frames.

Casing is used to trim around doors and windows; *baseboard* is used at the bottom of walls (**Figure 27-15**). While each classification includes particular designs, there is enough overlap so the types are practically interchangeable.

FIG. 27-15   CASING DESIGNS

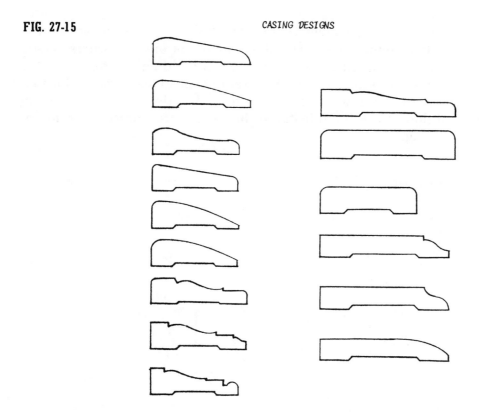

Base shoe, shown in **Figure 27-16**, is designed for use where the baseboard and the floor meet. *Base caps*, shown in **Figure 27-17**, may be used to finish off the top of plain flat baseboards. It ends the job nicely and also serves to close any gap that might exist between the wall and the baseboard.

FIG. 27-16   BASE SHOE

FIG. 27-17

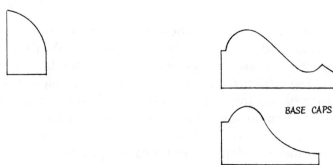

BASE CAPS

Crown-and-bed moldings, shown in **Figure 27-18**, are used to soften the sharp lines where two planes meet. Usual applications are at corners where ceilings and walls meet, or outside under eaves. Such moldings are very suitable for decorative trim work, around a mantle for example, or to make frames for pictures. The back areas are "hollow," because the moldings thus require less material to produce and they are easier to install if the corners they must cover do not meet at a true right angle.

FIG. 27-18   CROWN-AND-BED MOLDINGS

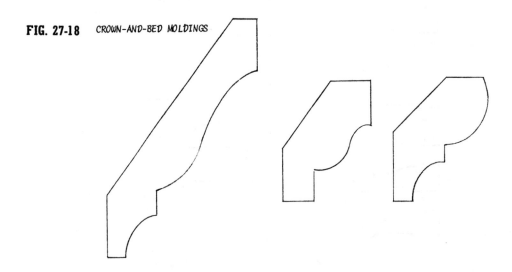

Stop molding, shown in **Figure 27-19**, is nailed to jambs to stop the door when it is closed. It is also used on windows with sliding sash, and frequently selected for surface-mounting on cabinet doors when a raised-panel effect is desired.

FIG. 27-19   *STOP MOLDINGS*

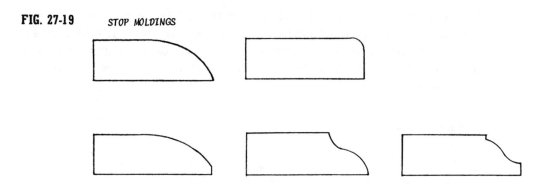

Stools, shown in **Figure 27-20**, are used at the bottom of windows to provide a snug joint with the lowered sash. A saw cut to remove the bevel on the molding's undercut would make this an excellent material to finish off the edge of a table or counter.

FIG. 27-20   *STOOLS*

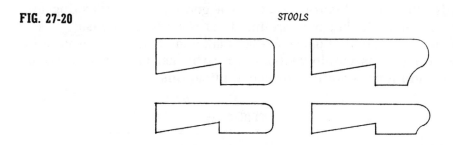

Picture molding, shown in **Figure 27-21**, got its name because it was originally designed as perimeter trim from which pictures could be hung. It may still be used that way, but the modern application is as a substitute for crown molding, since it is less conspicuous.

FIG. 27-21   *PICTURE MOLDING*

Shelf edge or *screen* molding, shown in **Figure 27-22**, was originally de-signed to cover the raw edges of screening on doors or windows. It may also be used to decorate the edges of wood members like shelves, or to conceal exposed plywood edges. Since the moldings are thin, they may also be used as canvas-backed slats for tambour doors.

FIG. 27-22 *SHELF-EDGE OR SCREEN MOLDINGS*   **FIG. 27-23**

CORNER GUARDS

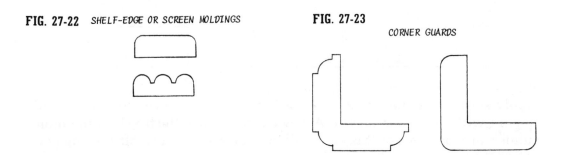

Corner guards, shown in **Figure 27-23**, do a good job of protecting and finishing outside corners, whether they are inside or outside the house.

Shingle molding, shown in **Figure 27-24**, makes a neat, decorative joint where the house siding abuts window sills and overhangs. The strips are often used as shelf cleats and for decorative surface applications.

FIG. 27-24   **FIG. 27-25** *BRICK MOLDING*

SHINGLE MOLDING

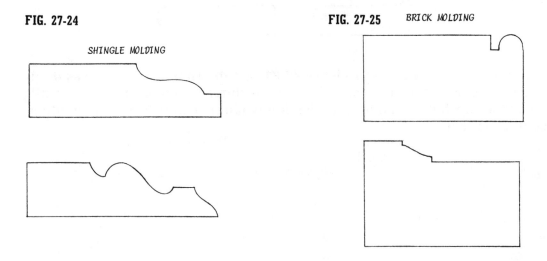

Brick molding, shown in **Figure 27-25**, is the trim used at the joint that occurs when an exterior wall is done partly in brick and partly in wood. It is also used if a wall is a stucco-wood combination.

Drip caps, shown in **Figure 27-26**, are specially designed for use at top edges on the exterior side of doors and windows to prevent moisture from getting inside the walls.

FIG. 27-26

FIG. 27-27

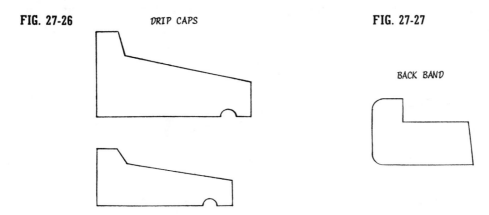

Back bands, shown in **Figure 27-27**, are meant to be caps for baseboards and casings, but they may also be used as corner guards when only one edge of the material turning the corner is exposed.

Ply caps, shown in **Figure 27-28**, may be used at the top of wainscoting to supply a smooth finish. They are also effective for edging plywood and for framing any panel, especially if the panel will be used as a slab for, say, a table top.

FIG. 27-29

FIG. 27-28

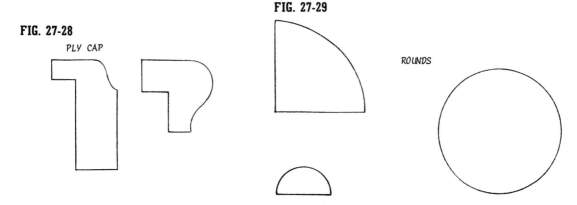

Rounds are available as quarter-rounds, half-rounds, and full-rounds, as shown in **Figure 27-29**. Typical applications for full rounds are as closet poles, curtain rods, and banisters; for quarter-rounds, as decorative trim for inside corners and as shelf cleats; for half-rounds, as decorative surface trim or as seam covers.

Chair rail, shown in **Figure 27-30**, is applied to walls at particular heights to protect the wall from chair backs. It is not used so much today, but might still be a good idea in areas like playrooms.

Hand rails, shown in **Figure 27-31**, are specially made for stairway applications.

FIG. 27-30   **FIG. 27-31**

CHAIR RAIL   HAND RAILS

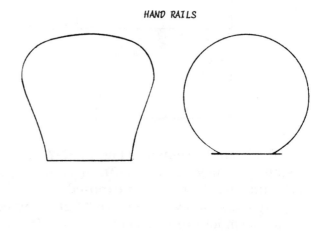

28

SPECIAL CONSTRUCTION METHODS

The word "special" here applies to construction techniques, not to particular architectural designs, although the very fact that an unusual method is involved can lead to distinctive results. All the concepts have been field-tested by federal or independent agencies, yet you may encounter resistance because of conflicts with local codes. In most cases, the new idea doesn't drastically affect the *details* of housebuilding; it may be a different way to do a foundation, or a framing method that can save material. At any rate, any departure from the norm should be checked out with local building authorities before you adopt it.

A HOUSE ON POLES

Sounds like a child's dream of a backyard retreat, but it's a way to hang a house on poles on otherwise unusable sites or ones that would require such extensive grading that the house would appear to sit on a scar in the earth. The idea can be used on level ground, but it is especially appealing for steep sites, since you get a "foundation" merely by digging holes and without changing the nature of the land. Of course you don't use any old telephone poles; only those that have been produced for the purpose by being pressure-treated with a preservative that might be creosote, waterborne salts, or pentachlorophenol. Of the three, creosote is the least desirable, especially if the poles will be exposed inside the house. Poles treated with waterborne-salts preservatives are comparatively dry to the touch, do not have an odor, and are paintable. Pentachlorophenol applications may be in a light petroleum solvent or in heavy oils. The light petroleum results in a fairly dry surface which can usually be painted or stained. The heavy oil leaves an oily surface which is especially apparent when temperatures are high. People ordering poles should specify whether they want a paintable surface.

Principal methods of construction are pole-frame or platform. In a pole-frame structure the principal vertical members are pressure-treated wood poles that run continuously from below the ground to the roof (**Figure 28-1**). They support the entire vertical weight of the building and also provide considerable lateral resistance to forces caused by winds and even earthquakes.

FIG. 28-1

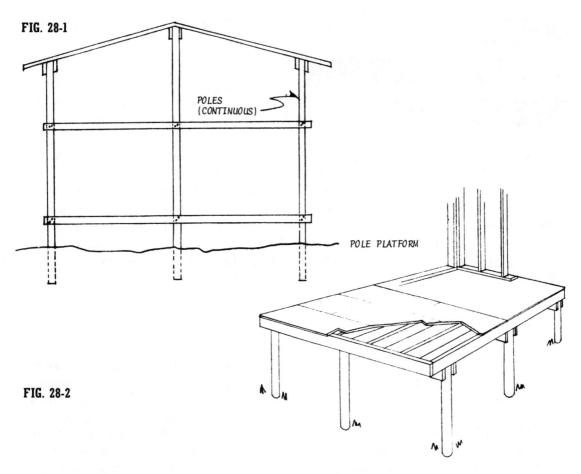

POLES
(CONTINUOUS)

POLE PLATFORM

FIG. 28-2

The pole platform is an easy way to convert a steep lot to a "building site" that then permits conventional construction methods (**Figure 28-2**).

Proper embedment of the poles is, of course, crucial, with vital factors being the number and size of the poles, climate and soil conditions, height of poles from floor to plate, the tributary area per pole, and the method of embedment. While tables like that in **Figure 28-3** relate embedment to soil conditions and the degree of the slope, you would do well to consult an engineer here so that professional specifications can be established. This is advisable not only so that the supports

FIG. 28-3.   EMBEDMENT DEPTHS FOR POLES IN POLE-FRAME BUILDINGS ON SITES WITH SLOPES LESS THAN 1:10

H, feet	Pole spacing, feet	GOOD SOIL				AVERAGE SOIL				BELOW-AVERAGE SOIL			
		Embedment depth, feet		D, inches	Tip size, inches	Embedment depth, feet		D, inches	Tip Size, inches	Embedment depth, feet		D, inches	Tip size, inches
		(A)	(B)			(A)	(B)			(A)	(B)		
1½ to 3	8	5.0	4.0	18	6	6.5	5.0	24	6	*	6.0	36	6
	10	5.5	4.0	21	7	7.0	5.0	30	7	*	6.5	42	7
	12	6.0	4.5	24	7	7.5	5.5	36	7	*	7.0	48	7
3 to 8	8	6.0	4.0	18	7	7.5	5.5	24	7	*	7.0	36	7
	10	6.0	4.5	21	8	8.0	6.0	30	8	*	7.5	42	8
	12	6.5	5.0	24	8	*	6.0	36	8	*	8.0	48	8

* Embedment depth is greater than eight feet and is considered excessively expensive.
Note: Where a concrete floor slab is used at grade, embedment depths may be reduced to 70 percent of those shown.

EMBEDMENT OF POLES

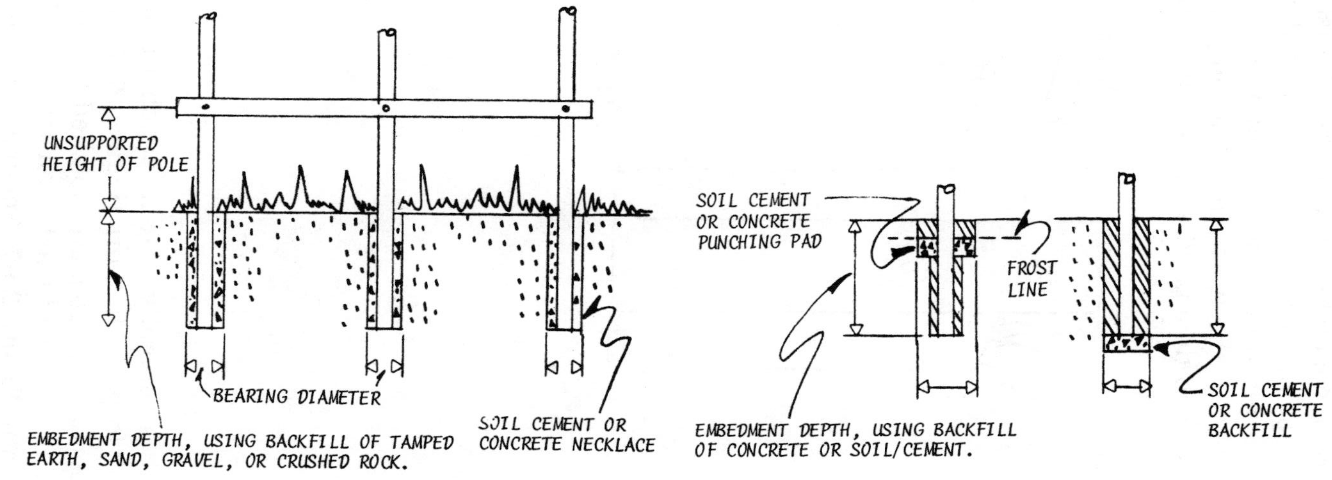

UNSUPPORTED HEIGHT OF POLE

BEARING DIAMETER

SOIL CEMENT OR CONCRETE NECKLACE

EMBEDMENT DEPTH, USING BACKFILL OF TAMPED EARTH, SAND, GRAVEL, OR CRUSHED ROCK.

SOIL CEMENT OR CONCRETE PUNCHING PAD

FROST LINE

SOIL CEMENT OR CONCRETE BACKFILL

EMBEDMENT DEPTH, USING BACKFILL OF CONCRETE OR SOIL/CEMENT.

will be strong enough but so that the job won't be overdone. Overembedment and overuse of concrete are two typical mistakes that undermine the system's inherent economies. You can see the big difference, as shown in **Figure 28-4**, between being able to backfill with sand and having to do a concrete job that includes a collar below the frost line. The only way to determine the least troublesome method is to do a good study of site conditions.

FIG. 28-4

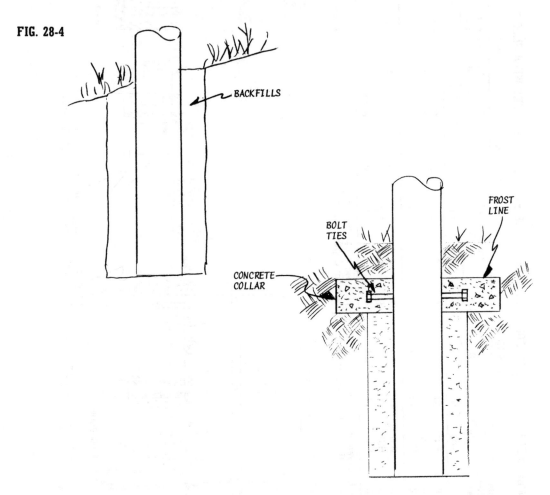

It is generally best to attach framing members such as beams, girders, or rafters to poles in pairs (**Figure 28-5**). Horizontal beams will receive additional support if the poles are notched to form seats (**Figure 28-6**). Top-seated beams, such as you might use to construct a pole-supported platform, can be attached as shown in **Figure 28-7**. Special spike grids are available (**Figure 28-8**). When inserted between the pole and the beam, they substantially increase the strength of the bolted connection. These, however, require special installation techniques.

FIG. 28-5

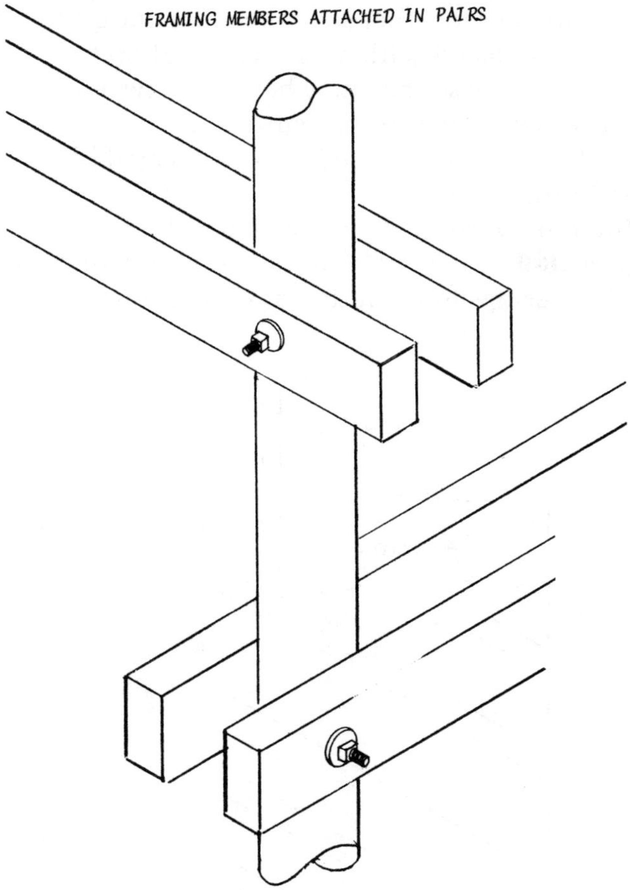

FRAMING MEMBERS ATTACHED IN PAIRS

FIG. 28-6

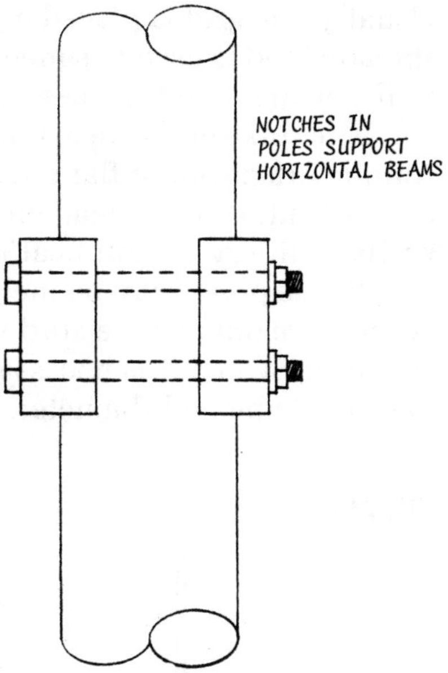

NOTCHES IN
POLES SUPPORT
HORIZONTAL BEAMS

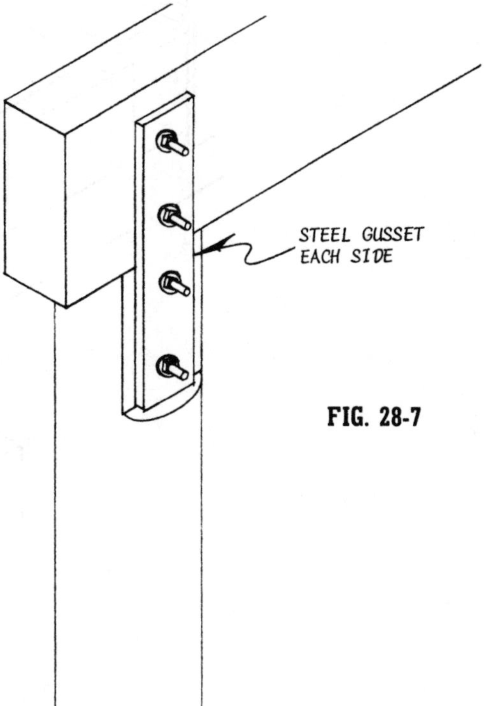

STEEL GUSSET
EACH SIDE

FIG. 28-7

FIG. 28-8

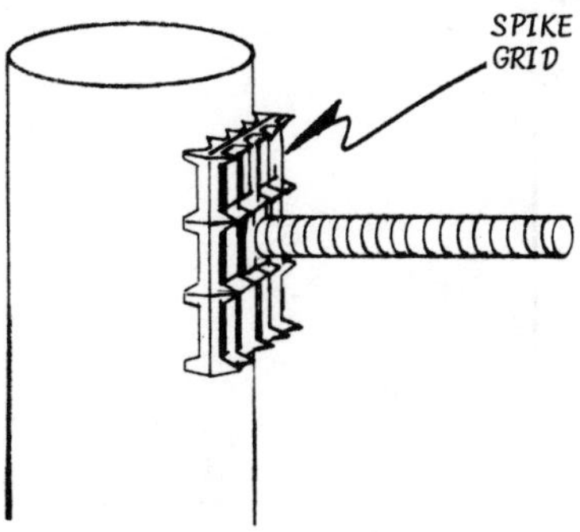

SPIKE
GRID

Usually, the grid is placed against the pole over drilled holes and a high-strength threaded rod is used to squeeze the wood surfaces together so the teeth on the grid will penetrate both pieces. Then the rod is replaced with a conventional bolt.

The spike grids are available in several shapes so they may be used between a round surface and a flat one, two flat surfaces, or two round surfaces.

All cuts or holes made in the poles after they are received must be treated liberally with several applications of wood preservative.

While poles may be included inside the house, it is generally best to design so they are outside the stud walls (**Figure 28-9**). This can eliminate the problems that occur when it is necessary for wall coverings to be butted against poles that vary in shape and diameter.

FIG. 28-9

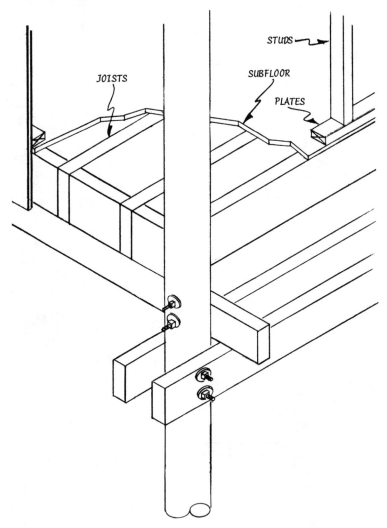

A good way to look at pole construction overall is as a means of supplying a nonmasonry, nonconcrete support for a house that will be built in fairly conventional ways. Framing members, as always, should be spaced so that standard lengths and widths of lumber, plywood, and other building materials can be used.

THE MOD 24 BUILDING SYSTEM

This system is not brand-new but is becoming increasingly popular because of wider acceptance by general and local codes. In essence, it is based on the use of plywood installed over lumber framing spaced on a 24-inch module (**Figure 28-10**). Cost savings are obvious if you only consider that each 4-foot length of wall will require one less stud than a similar length of wall that is framed with studs 16 inches O.C. It also means less time and labor, since you are installing fewer pieces. The Mod 24 system works best when trusses, joists, studs, and the like are aligned. Structurally, the idea is to produce a series of in-line frames that will use both lumber and plywood to best advantage.

FIG. 28-10

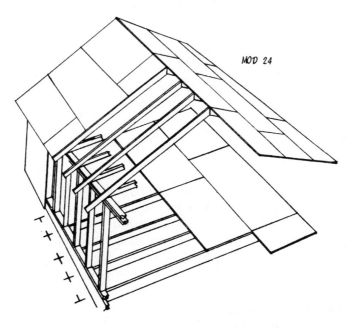

MOD 24

Recommended floor framing consists of in-line joists spaced 24 inches O.C. (**Figure 28-11**). This can be covered with a conventional two-layer floor (subfloor and underlayment) done as shown in **Figure 28-12**, or, for the best utilization of materials, with a combination subfloor-underlayment plywood installed as

FIG. 28-11

FIG. 28-12

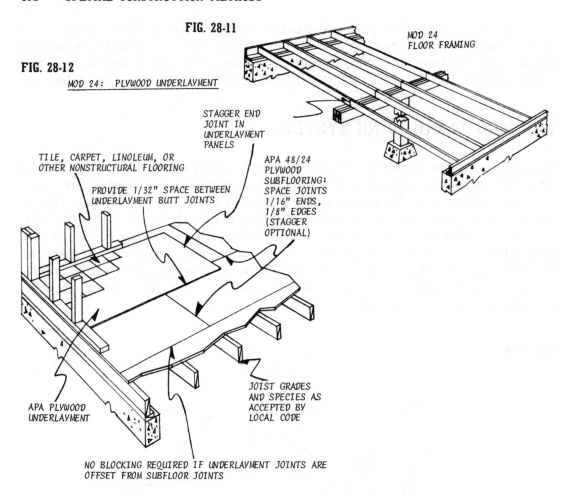

MOD 24: PLYWOOD UNDERLAYMENT

MOD 24
FLOOR FRAMING

STAGGER END
JOINT IN
UNDERLAYMENT
PANELS

TILE, CARPET, LINOLEUM, OR
OTHER NONSTRUCTURAL FLOORING

APA 48/24
PLYWOOD
SUBFLOORING:
SPACE JOINTS
1/16" ENDS,
1/8" EDGES
(STAGGER
OPTIONAL)

PROVIDE 1/32" SPACE BETWEEN
UNDERLAYMENT BUTT JOINTS

APA PLYWOOD
UNDERLAYMENT

JOIST GRADES
AND SPECIES AS
ACCEPTED BY
LOCAL CODE

NO BLOCKING REQUIRED IF UNDERLAYMENT JOINTS ARE
OFFSET FROM SUBFLOOR JOINTS

shown in **Figure 28-13**. If you do the single-layer installation the plywood should be at least ¾ inch thick and should, preferably, have tongue-and-groove edges. In both cases, the designers of the concept recommend that the American Plywood Association's glued-floor method of installation be used. A special adhesive, applied with a caulking gun, is put down over joists and between panels. This way, there's no chance of nails loosening or of squeaking floors. The bond between panel surfaces and between joist edges and panels is continuous, and this promotes a stiffer construction.

The best way to utilize the modular spacing is to plan for openings through the walls to be compatible. Compare the off-module rough-window opening in **Figure 28-14** with the one that works along with 24-inch stud spacing and you will see the savings in framing materials and labor. The off-module location demands two extra studs.

FIG. 28-13

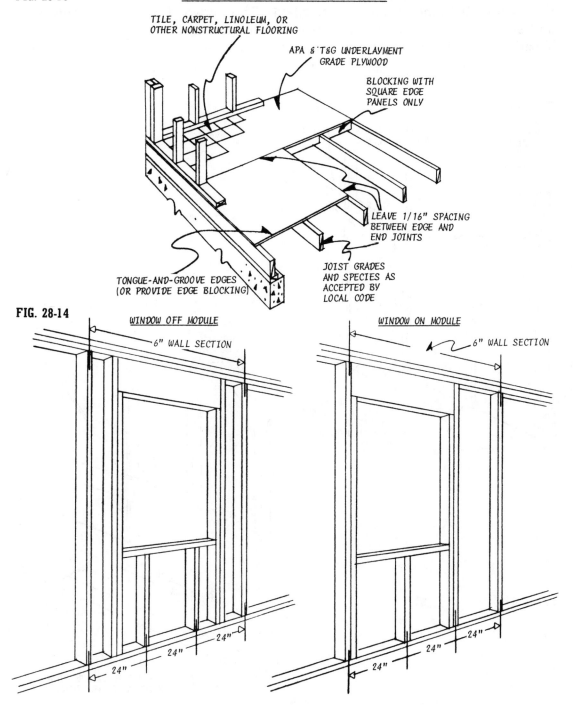

MOD 24: COMBINED SUBFLOOR UNDERLAYMENT

TILE, CARPET, LINOLEUM, OR
OTHER NONSTRUCTURAL FLOORING

APA & T&G UNDERLAYMENT
GRADE PLYWOOD

BLOCKING WITH
SQUARE EDGE
PANELS ONLY

LEAVE 1/16" SPACING
BETWEEN EDGE AND
END JOINTS

JOIST GRADES
AND SPECIES AS
ACCEPTED BY
LOCAL CODE

TONGUE-AND-GROOVE EDGES
(OR PROVIDE EDGE BLOCKING)

FIG. 28-14

WINDOW OFF MODULE

6" WALL SECTION

24"   24"   24"

WINDOW ON MODULE

6" WALL SECTION

24"   24"   24"

Outside walls may be covered with a single layer of exterior-grade plywood that is at least ½ inch thick (**Figure 28-15**). The plywood acts as both sheathing and siding. However, there is no reason why you can't do a two-layer wall— sheathing *and* siding—in order to increase stiffness and improve insulative qualities (**Figure 28-16**).

FIG. 28-15

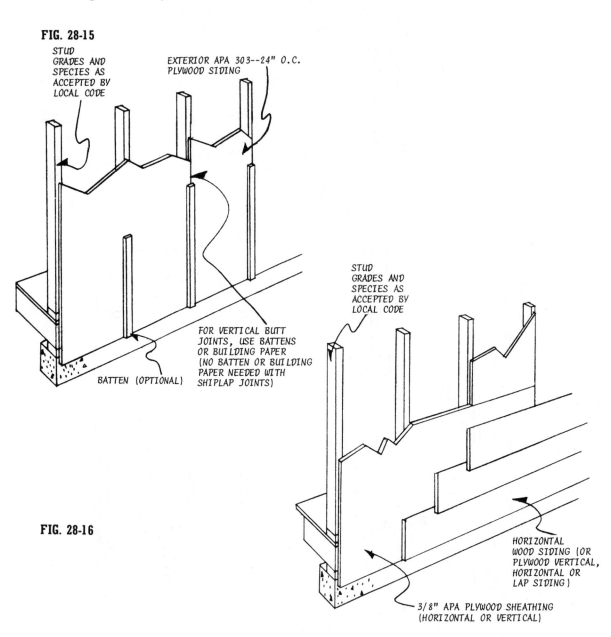

STUD GRADES AND SPECIES AS ACCEPTED BY LOCAL CODE

EXTERIOR APA 303--24" O.C. PLYWOOD SIDING

FOR VERTICAL BUTT JOINTS, USE BATTENS OR BUILDING PAPER (NO BATTEN OR BUILDING PAPER NEEDED WITH SHIPLAP JOINTS)

BATTEN (OPTIONAL)

STUD GRADES AND SPECIES AS ACCEPTED BY LOCAL CODE

FIG. 28-16

HORIZONTAL WOOD SIDING (OR PLYWOOD VERTICAL, HORIZONTAL OR LAP SIDING)

3/8" APA PLYWOOD SHEATHING (HORIZONTAL OR VERTICAL)

Roof framing can be done in conventional fashion following the modular spacing, but usually, prefab or made-on-the-site trusses that meet the requirements of local codes are used. Overall, it would seem that the 24-inch spacing is critical to local building authorities. If they buy that, then the application of inside and outside covers shouldn't differ radically from the techniques we have been discussing.

The system is not limited to single-story homes, although stud lengths may be affected if you plan to build up. For example, for a single-story home, the wall studs can be up to 10 feet long. If you add a second floor, then the wall studs in the bottom wall must not be more than 8 feet, owing to code concern for structural strength.

THE PLANK-AND-BEAM SYSTEM

This type of residential construction is a spinoff from the heavy timber designs often encountered in commercial buildings with large areas of open floor space. The concept is characterized, as shown in **Figure 28-17**, by the use of fewer but larger-size bearing pieces that are able to support concentrated structural loads.

FIG. 28-17

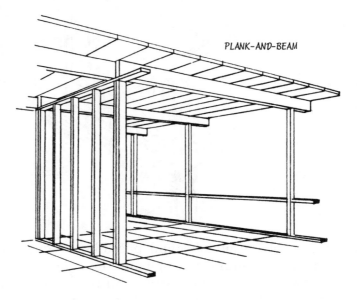

PLANK-AND-BEAM

To really understand the idea it is wise to make some comparisons. In **Figure 28-18** the conventional frame with studs 16 inches O.C. and the plank-and-beam frame with studs 24 inches O.C. are of similar length and have the same openings for doors and windows. It's obvious which design requires the least number of

FIG. 28-18

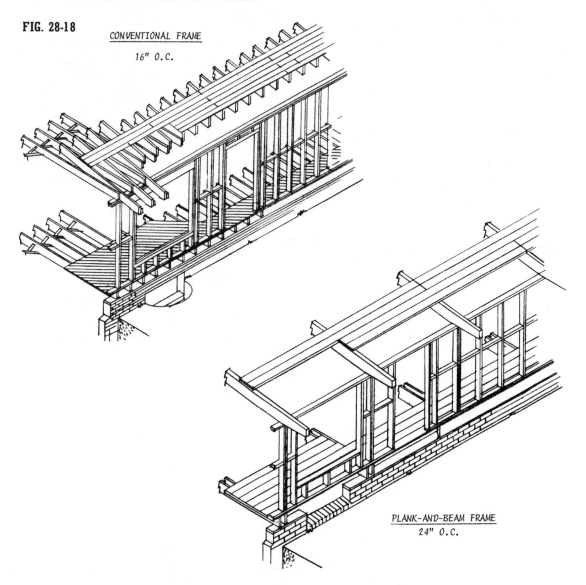

CONVENTIONAL FRAME
16" O.C.

PLANK-AND-BEAM FRAME
24" O.C.

pieces, so you can see a savings in labor and, possibly, out-of-pocket costs. We say possibly, because much depends on design and the materials that are used. For example, if a roof span requires 4×12 beams and they cost as much as or more than the number of smaller pieces you would use for a conventional frame, then obviously you are not saving money. But the thought must go a bit further. The 2× planks over the beams can provide both roof sheathing and finished inside ceiling; the extension of beams and planks beyond walls is certainly a simple way to provide wide, structurally acceptable overhangs without the cost and labor required by the features of a regular cornice (**Figure 28-19**). Overall, time and money have much to do with the materials you work with and your design. An important factor is to plan the framing on a modular basis so standard-size building materials may be used with a minimum of cutting and waste. For example, roof beams that are 5 feet O.C. will take 10-foot planks very nicely. Joint bearings 4 feet O.C. will be compatible with plywood and drywall sheets that are 4 feet wide, and so on.

FIG. 28-19

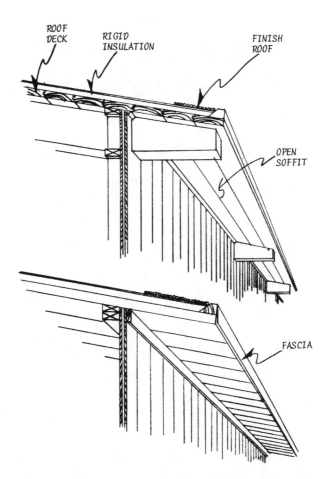

Of course a choice in this area of construction can also be influenced by architectural effects. An exposed plank-and-beam ceiling is quite distinctive and makes a considerable contribution to the spaciousness of a room.

Often, plank-and-beam is combined with conventional framing—for example, a floor and roof done with plank-and-beam but with perimeter walls of regular studs and headers. This is what I did when putting up my studio. It has standard walls on a concrete slab, but the roof is 4×12 beams spanning 20 feet and covered with 2×8 tongue-and-groove boards. The result was a higher-than-average ceiling and uncluttered floor space.

Post-and-beam floor framing can be done over continuous foundation walls, but the idea lends itself very well to pier foundations. The floor joists in a conventional design are supported by girders which rest on piers, but in a plank-and-beam system, the flooring is nailed directly to the pier-supported beams (**Figure 28-20**).

FIG. 28-20

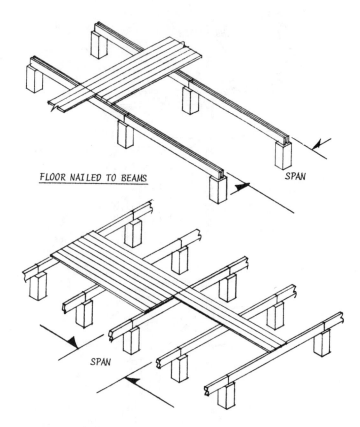

FLOOR NAILED TO BEAMS

SPAN

SPAN

Posts must be strong enough to support concentrated loads and large enough to provide adequate bearing surfaces. The material used is never less than 4×4, but if appearance is not a factor you can work by nailing together pieces of 2× stock. Beam ends must always fall over a post, and it's a good idea to increase the bearing surface by adding additional blocks (**Figure 28-21**).

The spacing of wall posts is determined by considering both design and allowable spans; the thickness of floor and roof planks is a major contributing factor. Generally, it is probably wise to reduce the maximum allowable span to a

FIG. 28-21

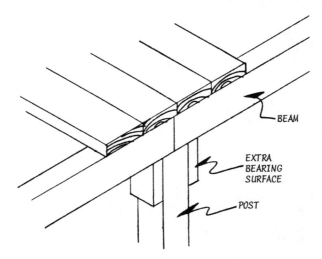

BEAM

EXTRA
BEARING
SURFACE

POST

dimension that permits a modular system. Once the dimension is established, the posts can be spaced evenly around the perimeter. Top plates are installed conventionally and roof beams are positioned so each will bear directly over a post (**Figure 28-22**).

FIG. 28-22

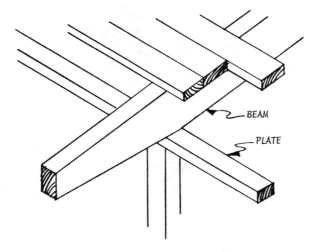

BEAM

PLATE

At the bottom end, posts sit on plates just as studs would (**Figure 28-23**). In situations where the post is not located over a floor beam, additional blocking can be added under the floor to provide extra support.

The run of the roof beams does not have to be from plates to ridge but can be longitudinal as shown in **Figure 28-24**. The purlin beams (longitudinal ones) are often called on to span greater distances and carry heavier loads than conven-

FIG. 28-23

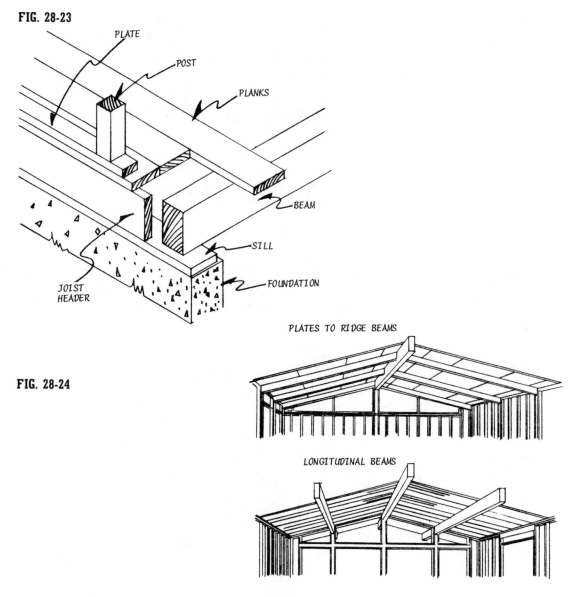

PLATE

POST

PLANKS

BEAM

SILL

FOUNDATION

JOIST HEADER

FIG. 28-24

PLATES TO RIDGE BEAMS

LONGITUDINAL BEAMS

tionally placed beams and, so, are usually larger in cross section. Advantages are that you can be more flexible when designing end walls. And end-wall overhangs are easy to provide merely by extending the beams.

All roof beams must receive sufficient support either from individual posts or a conventional stud wall with adequate top plates. The beam-to-post connection is often reinforced with inside framing that is done as shown in **Figure 28-25** for a ridge and as shown in **Figure 28-26** for a beam.

FIG. 28-25

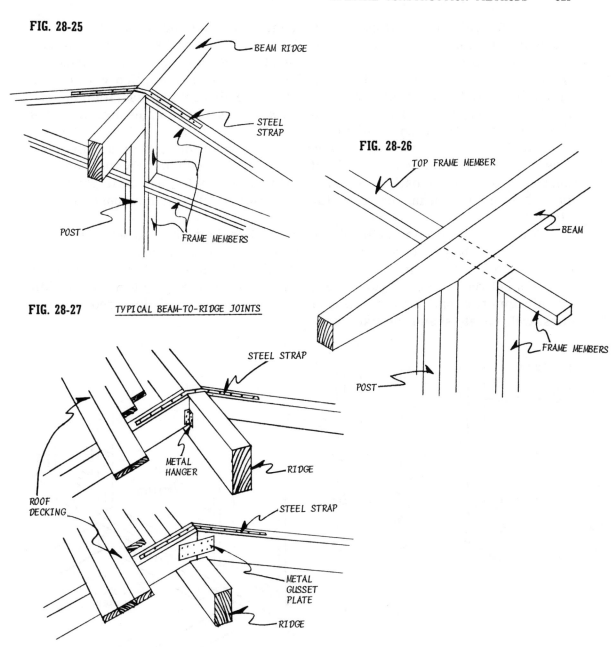

BEAM RIDGE

STEEL STRAP

POST

FRAME MEMBERS

FIG. 28-26

TOP FRAME MEMBER

BEAM

FRAME MEMBERS

POST

FIG. 28-27   TYPICAL BEAM-TO-RIDGE JOINTS

STEEL STRAP

METAL HANGER

RIDGE

ROOF DECKING

STEEL STRAP

METAL GUSSET PLATE

RIDGE

Typical beam-to-ridge joints are shown in **Figure 28-27**.

Plank-and beam construction is not all peaches and cream; there are limitations and problems, and while they can be solved, they should be anticipated and understood before construction starts.

A 2-inch-thick roof/ceiling provides some insulation, but it may be far from adequate for some areas. Insulation over the roof must be of the rigid type that can stand abuse inside insulation is not subjected to. There are insulations that can do the job, but some checking is in order since many must be applied with mastic and only on roofs with minimum pitch. There are also ceiling/insulation materials that can be fastened to the *underside* of roof decks, but here, the appearance of the material is a major consideration.

You have less freedom in locating partitions when they rest on a plank-and-beam floor. The planking may not be stiff or strong enough to carry the load. The answer is to design so partitions, load-bearing or otherwise, will be directly supported by a floor beam, or to include additional framing members to boost strength.

A reasonable solution is to substitute a 4×4 or doubled 2×4s set on edge for the customary bottom partition plate. These must span across the floor beams. Where there is an opening through the partition, the 4×4 can be fastened under the floor between beams (**Figure 28-28**). It will also be necessary to provide additional support in specific areas to take the concentrated loads imposed by bathtubs, water heaters, and the like.

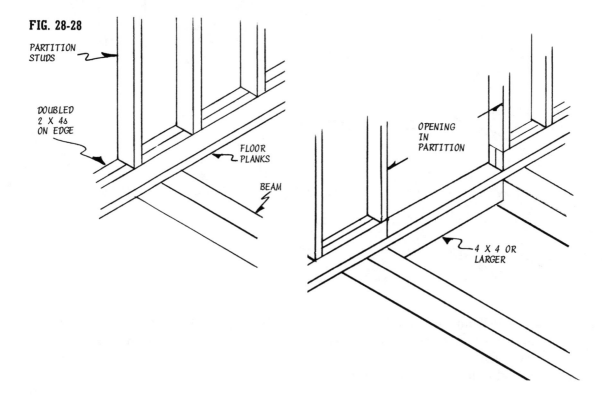

FIG. 28-28

PARTITION STUDS

DOUBLED 2 X 4s ON EDGE

FLOOR PLANKS

BEAM

OPENING IN PARTITION

4 X 4 OR LARGER

A plank-and-beam roof doesn't provide space in which you can hide wiring systems. Many designers substitute wall fixtures for ceiling fixtures as much as possible or use trim raceways so wires can be surface-mounted. Another out is to construct "hollow" beams as shown in **Figure 28-29**.

FIG. 28-29

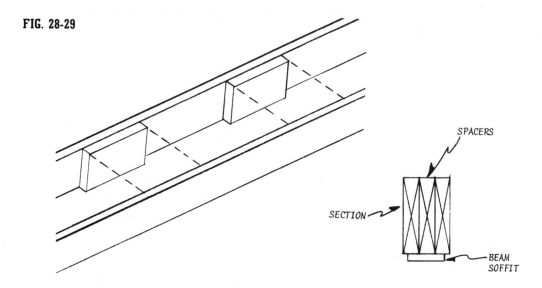

Fixtures can be attached to the soffit; wires can run through the open spaces of the beams. Another way is to pre-rout the beams to provide grooves for wires.

All-in-all, and as with any construction system, you must do a good deal of planning before you drive the first nail.

WOOD FOUNDATIONS

The wood foundation is not only possible, it's here and well beyond the testing stage. The system has been accepted by the model building codes and is becoming more evident in state and local building codes throughout the country. The fact that the All-Weather Wood Foundation system (AWWF) works is not so strange if you view it as an extension of the time-proven idea of providing support for a building by using specially treated wood that resists damage in soil-contact situations—for example, the pole house. If the foundation is designed and installed so that it provides structural strength and if the wood used can't be damaged by decay or moisture or termites, the concept should, and does, make sense.

Basically, the AWWF system is a set of prefabricated, pressure-treated, plywood-sheathed stud walls that are set below grade. A typical wall panel is made with 2×4 or 2×6 studs placed 16 inches or less O.C., sheathed with plywood (**Figure 28-30**). The panels can be any convenient width and as high as 8 feet when you measure from the footing plate to the foundation sill. All the materials used must be pressure-preservative treated. All fasteners must be silicon bronze or copper, or hot-dipped zinc-coated steel. Electrogalvanized or mechanically plated nails or staples and hot-dipped zinc-coated staples are not permitted. Any framing anchors that are used must be zinc-coated sheet steel.

FIG. 28-30

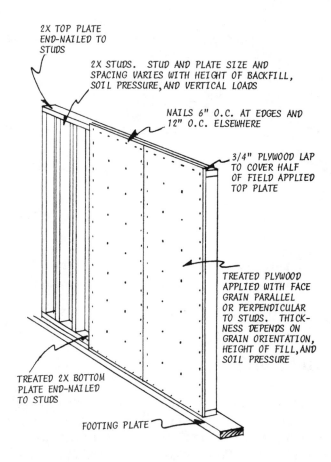

2X TOP PLATE
END-NAILED TO
STUDS

2X STUDS. STUD AND PLATE SIZE AND
SPACING VARIES WITH HEIGHT OF BACKFILL,
SOIL PRESSURE, AND VERTICAL LOADS

NAILS 6" O.C. AT EDGES AND
12" O.C. ELSEWHERE

3/4" PLYWOOD LAP
TO COVER HALF
OF FIELD APPLIED
TOP PLATE

TREATED PLYWOOD
APPLIED WITH FACE
GRAIN PARALLEL
OR PERPENDICULAR
TO STUDS. THICK-
NESS DEPENDS ON
GRAIN ORIENTATION,
HEIGHT OF FILL, AND
SOIL PRESSURE

TREATED 2X BOTTOM
PLATE END-NAILED
TO STUDS

FOOTING PLATE

The footing plate on the panels is set in at one end but extends out at the other end so adjoining panels will overlap. The design permits an interlock at corners, as shown in **Figure 28-31**, which is similar to the organization of top plates in a conventional wall frame. Panel end-to-end joints are done as shown in **Figure 28-32**.

FIG. 28-31

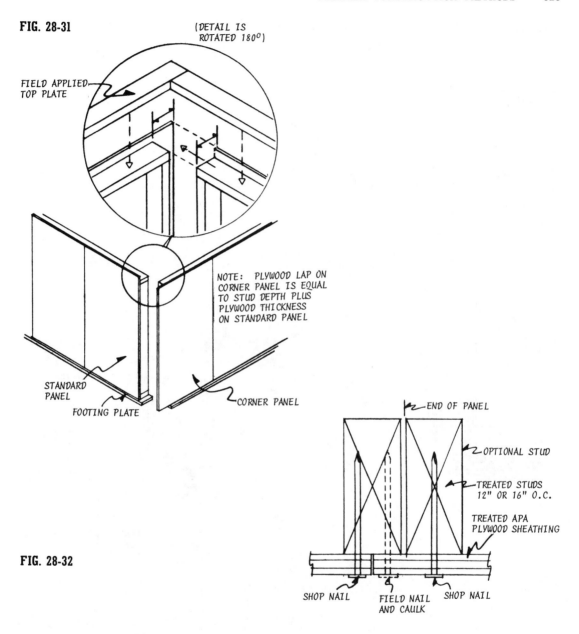

(DETAIL IS ROTATED 180°)

FIELD APPLIED TOP PLATE

NOTE: PLYWOOD LAP ON CORNER PANEL IS EQUAL TO STUD DEPTH PLUS PLYWOOD THICKNESS ON STANDARD PANEL

STANDARD PANEL

FOOTING PLATE

CORNER PANEL

END OF PANEL

OPTIONAL STUD

TREATED STUDS 12" OR 16" O.C.

TREATED APA PLYWOOD SHEATHING

FIG. 28-32

SHOP NAIL   FIELD NAIL AND CAULK   SHOP NAIL

An intriguing feature is that panels may be placed on footings of gravel, coarse sand, or crushed stone. The levelness of the footing and the levelness and plumbness of the panels are very critical. Great care should be exercised in plumbing (**Figure 28-33**).

FIG. 28-33

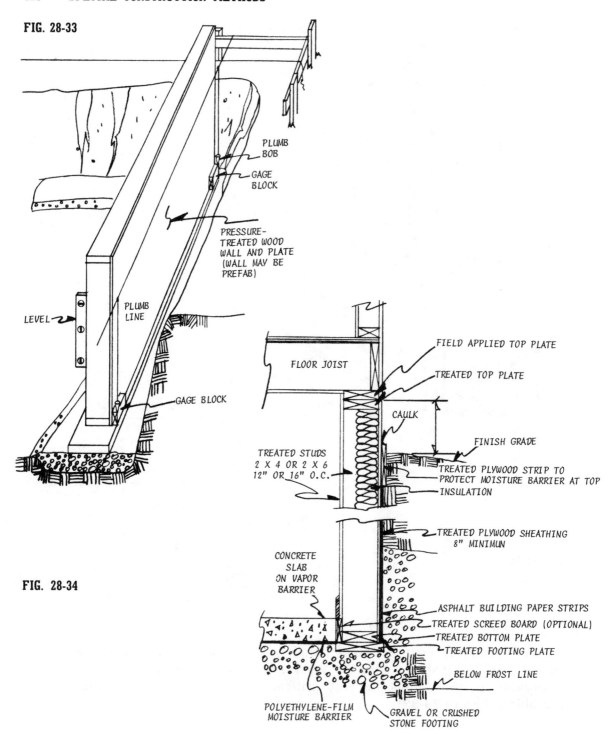

PLUMB BOB

GAGE BLOCK

PRESSURE-TREATED WOOD WALL AND PLATE (WALL MAY BE PREFAB)

LEVEL

PLUMB LINE

GAGE BLOCK

FLOOR JOIST

FIELD APPLIED TOP PLATE

TREATED TOP PLATE

CAULK

FINISH GRADE

TREATED STUDS 2 X 4 OR 2 X 6 12" OR 16" O.C.

TREATED PLYWOOD STRIP TO PROTECT MOISTURE BARRIER AT TOP

INSULATION

TREATED PLYWOOD SHEATHING 8" MINIMUN

CONCRETE SLAB ON VAPOR BARRIER

ASPHALT BUILDING PAPER STRIPS

TREATED SCREED BOARD (OPTIONAL)

TREATED BOTTOM PLATE

TREATED FOOTING PLATE

BELOW FROST LINE

FIG. 28-34

POLYETHYLENE-FILM MOISTURE BARRIER

GRAVEL OR CRUSHED STONE FOOTING

FIG. 28-35

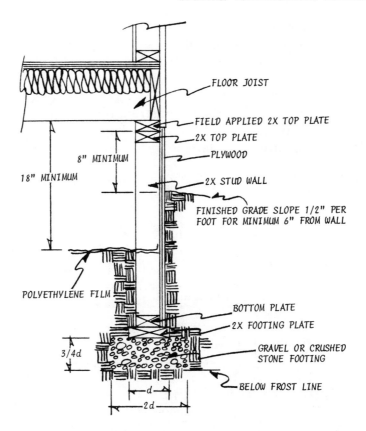

FLOOR JOIST

FIELD APPLIED 2X TOP PLATE

2X TOP PLATE

PLYWOOD

2X STUD WALL

FINISHED GRADE SLOPE 1/2" PER FOOT FOR MINIMUM 6" FROM WALL

8" MINIMUM

18" MINIMUM

POLYETHYLENE FILM

BOTTOM PLATE

2X FOOTING PLATE

GRAVEL OR CRUSHED STONE FOOTING

BELOW FROST LINE

3/4 d

d

2 d

General AWWF construction features for a house with a cellar are shown in **Figure 28-34**, and for a house with a crawl space in **Figure 28-35**.

While AWWF is basically a simple system, its design and application must not be approached casually. An in-depth examination of soil and drainage conditions at the site is a first must. While on-the-job construction of foundation panels is possible, they must be engineered to comply with acceptable standards. This is not, and doesn't have to be, a chore for the individual builder. Tested and recommended sizes and details, in addition to fabrication specifications and construction methods, are available in a DFI manual published by the National Forest Products Association.

THE PLEN-WOOD SYSTEM

The concept here is to utilize the entire area under the floor of a single-story residence as a plenum chamber that will move warm or cool air to floor registers in all rooms. While the idea can be incorporated in a concrete or masonry foundation, it was specifically designed for an AWWF installation topped with a lumber-and-plywood floor. You can picture it if you view a crawl space as a sealed, moisture-proofed, insulated chamber that will receive hot air from a downflow furnace. This provides a warm floor, and since hot air rises, it will enter any area where you provide a floor register.

Benefits are said to be easier construction and lower building costs, and lower energy costs for heating and cooling. The system has already been approved by FHA Minimum Property Standards, the Uniform Building Code, and other model codes.

FIG. 28-36

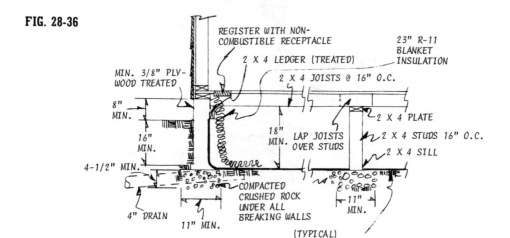

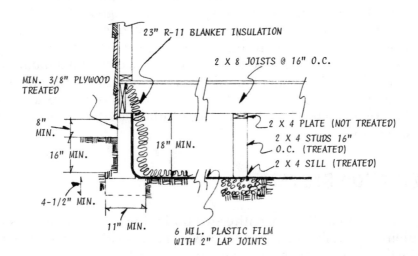

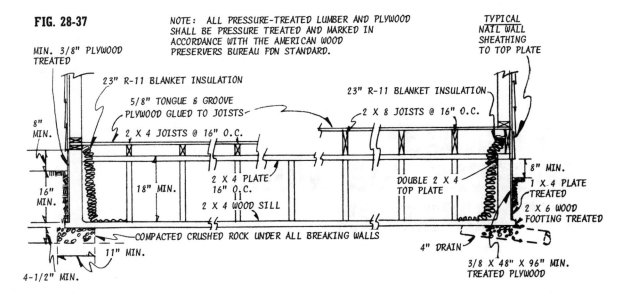

FIG. 28-37

NOTE: ALL PRESSURE-TREATED LUMBER AND PLYWOOD SHALL BE PRESSURE TREATED AND MARKED IN ACCORDANCE WITH THE AMERICAN WOOD PRESERVERS BUREAU FDN STANDARD.

TYPICAL NAIL WALL SHEATHING TO TOP PLATE

MIN. 3/8" PLYWOOD TREATED

23" R-11 BLANKET INSULATION

5/8" TONGUE & GROOVE PLYWOOD GLUED TO JOISTS

23" R-11 BLANKET INSULATION

2 X 8 JOISTS @ 16" O.C.

2 X 4 JOISTS @ 16" O.C.

8" MIN.

16" MIN.

18" MIN.

2 X 4 PLATE 16" O.C.

2 X 4 WOOD SILL

DOUBLE 2 X 4 TOP PLATE

8" MIN.

1 X 4 PLATE TREATED

2 X 6 WOOD FOOTING TREATED

COMPACTED CRUSHED ROCK UNDER ALL BREAKING WALLS

11" MIN.

4" DRAIN

3/8 X 48" X 96" MIN. TREATED PLYWOOD

4-1/2" MIN.

The excavation for the crawl space is done in a normal manner but with trenches added for the gravel footings on which the foundation panels will rest. Intermediate foundation supports are also treated stud walls but without sheathing. Insulation batts and vapor barriers are attached to the inside of the foundation walls before joists and the plywood flooring are added. A downflow furnace is installed after the house is closed in and then openings are cut through the floor where they are needed to accommodate adjustable heat registers.

It's recommended that the plywood-to-joist bond be achieved with an adhesive and not nails. Sections through typical installations are shown in **figures 28-36** and **28-37**.

APPENDIX: SOURCES FOR MORE INFORMATION

For addresses of sources for more information on systems, products, or materials, match key letters accompanying the subject list below with letters after the names and addresses that follow.

SUBJECTS

Siding	**A**	All-weather wood foundation	**O**
Paneling	**B**	Pressure-treated lumber	**P**
Gypsum board	**C**	Mod 24 system	**Q**
Plank-and-beam system	**D**	Pseudo brick	**R**
Plen-wood system	**E**	Solar water heaters	**S**
Pole constructions	**F**	(or collectors)	
Construction lumber	**G**	Heat pumps	**T**
Truss designs	**H**	Stairways	**U**
Hardboard products	**I**	House plans	**V**
Plastic-coated hardboards	**J**	Doors and windows	**W**
Ceramic tile systems	**K**	Roofing products	**X**
Ceiling systems	**L**	Garage doors	**Y**
Floor systems (or covers)	**M**	Concrete forms	**Z**
Fireplaces	**N**	Framing anchors	**AA**

SOURCES

WESTERN WOOD PRODUCTS ASSOCIATION   **A,B,E,G,P,Q,V**
1500 Yeon Building
Portland, OR 97204

AMERICAN PLYWOOD ASSOCIATION   **A,B,H,L,M,O,Q**
1119 A Street
Tacoma, WA 98401

GEORGIA-PACIFIC CORPORATION   **A,B,C**
900 S.W. Fifth Avenue
Portland, OR 97204

LOUISIANA-PACIFIC CORPORATION   **A,G,O,Q,W**
1300 S.W. Fifth Avenue
Portland, OR 97201

ROLSCREEN COMPANY   **W**
Pella, IA 50219

ANDERSEN CORPORATION   **W**
Bayport, MN 55003

AMERICAN WOOD PRESERVERS INSTITUTE   **F,O,P**
1651 Old Meadow Road
McLean, VA 22101

McCORMICK & BAXTER   **P**
300 Montgomery Street
San Francisco, CA 94104

NATIONAL FOREST PRODUCTS ASSOCIATION   **E,O (fee)**
1619 Massachusetts Avenue N.W.
Washington, DC 20036

SUPERINTENDENT OF DOCUMENTS   **D (fee)**
U.S. Government Printing Office
Washington, DC 20402

CON-FORM EQUIPMENT CORPORATION   **Z**
225 N. Arlington Heights Road
Elk Grove Village, IL 60007

Continued next page

RED CEDAR SHINGLE & HANDSPLIT SHAKE
  BUREAU   **X**
5510 White Building
Seattle, WA 98101

SHAKERTOWN CORPORATION   **X**
4416 Lee Road
Cleveland, OH 44128

MASONITE CORPORATION   **A,B,I,R**
29 N. Wacker Drive
Chicago, IL 60606

MARLITE   **J**
Dover, OH 44622

BIRD & SON   **A,X**
Washington Street
East Walpole, MA 02032

HUDSON HOME PUBLICATIONS   **V**
289 S. San Antonio Road
Los Altos, CA 94022

THE MAJESTIC COMPANY   **N**
Huntington, IN 46750

THE COLONIAL STAIR AND WOODWORK CO.   **U**
Jefferson, OH 43128

AMERICAN OLEAN   **K**
Lansdale, PA 19446

BRUCE FLOORING   **M**
P.O. Box 397
Memphis, TN 38101

ARMSTRONG CORK COMPANY   **L,M**
Lancaster, PA 17604

PLASTRONICS, INC.   **R**
586 Higgins Crowell Road
West Yarmouth, MA 02673

DACOR MANUFACTURING CO., INC.   **R**
Worcester, MA 01601

THE IRON SHOP   U
400 Reed Road
Broomall, PA 19008

CHRYSLER CORPORATION   T
Airtemp Division
Box 1205
Dayton, OH 45401

BORG-WARNER   T
Box 1592
York, PA 17405

WESTINGHOUSE ELECTRIC CO.   T
Box 2510
Staunton, VA 24401

WHIRLPOOL HEATING & COOLING   T
647 Thompson Lane
Nashville, TN 37204

GENERAL ELECTRIC CO.   T
Appliance Park
Louisville, KY 40225

GARDEN WAY LABORATORIES   S
Charlotte, VT 05445

ENERGY SYSTEMS, INC.   S
634 Crest Drive
El Cajon, CA 92021

SOL-THERM CORP.   S
7 West 14th Street
New York, NY 10011

UNIVERSAL SOLAR ENERGY CO.   S
1802 Madrid Avenue
Lake Worth, FL 33641

CAPITOL SOLAR HEATING, INC.   S
476 NW 25th Street
Miami, FL 33127

AMERICAN HELIOTHERMAL   S
3515 Tamarac
Denver, CO 80237

Continued next page

ENERGEX CORP.   **S**
5115 S. Industrial Road
Las Vegas, NV 89118

HEATILATOR FIREPLACE   **N**
4086 W. Saunders
Mt. Pleasant, IA 52641

RIDGE DOORS   **Y**
New Road
Monmouth Junction, NJ 08852

AZROCK FLOOR PRODUCTS   **M**
Box 531
San Antonio, TX 78292

TECO   **AA**
5530 Wisconsin Ave.
Washington, DC 20015

RECOMMENDED FURTHER READING

These books are available from Popular Science Book Club, 44 Hillside Ave., Manhasset, NY 11030.

● *Do-It-Yourself Plumbing* by Max Alth. Includes major section on planning and doing the plumbing for a new house or an addition. Also excellent on repairs and maintenance of existing plumbing. Well illustrated. 301 pages.

● *Basic House Wiring* by Monte Burch. Includes the gamut of wiring how-to from installing fixtures to wiring a new house, all in accord with codes. Well illustrated. 228 pages.

● *Home Energy How-To* by A. J. Hand. First part covers energy conservation — including insulation, caulking, weatherstripping, tuning-up conventional heating and cooling systems, using energy-efficient architecture. Second part explains how to produce energy from solar collectors, wind and water generators, biofuel plants, and woodburners. Well illustrated. 258 pages.

INDEX